ACS SYMPOSIUM SERIES **642**

Chemical Separations with Liquid Membranes

Richard A. Bartsch, EDITOR
Texas Tech University

J. Douglas Way, EDITOR
Colorado School of Mines

Developed from a symposium sponsored
by the Division of Industrial and Engineering Chemistry, Inc.

American Chemical Society, Washington, DC

Library of Congress Cataloging-in-Publication Data

Chemical separations with liquid membranes / Richard A. Bartsch, editor, J. Douglas Way, editor.

    p.   cm.—(ACS symposium series, ISSN 0097–6156; 642)

    "Developed from a symposium sponsored by the Division of Industrial and Engineering Chemistry, Inc. at the 209th National Meeting of the American Chemical Society, Anaheim, California, April 2–6, 1995."

    Includes bibliographical references and indexes.

    ISBN 0–8412–3447–7

    1. Membrane separation—Congresses. 2. Liquid membranes—Congresses.

    I. Bartsch, Richard A. II. Way, J. Douglas. III. American Chemical Society. Division of Industrial and Engineering Chemistry. IV. American Chemical Society. Meeting (209th: 1995: Anaheim, Calif.) V. Series.

TP159.M4C48   1996
660'.28424—dc20
                                     96–27197
                                          CIP

This book is printed on acid-free, recycled paper.

Advisory Board

ACS Symposium Series

TP159
M4C48
1996
CHEM

Foreword

The ACS Symposium Series was first published in 1974 to provide a mechanism for publishing symposia quickly in book form. The purpose of this series is to publish comprehensive books developed from symposia, which are usually "snapshots in time" of the current research being done on a topic, plus some review material on the topic. For this reason, it is necessary that the papers be published as quickly as possible.

Before a symposium-based book is put under contract, the proposed table of contents is reviewed for appropriateness to the topic and for comprehensiveness of the collection. Some papers are excluded at this point, and others are added to round out the scope of the volume. In addition, a draft of each paper is peer-reviewed prior to final acceptance or rejection. This anonymous review process is supervised by the organizer(s) of the symposium, who become the editor(s) of the book. The authors then revise their papers according to the recommendations of both the reviewers and the editors, prepare camera-ready copy, and submit the final papers to the editors, who check that all necessary revisions have been made.

As a rule, only original research papers and original review papers are included in the volumes. Verbatim reproductions of previously published papers are not accepted.

ACS BOOKS DEPARTMENT

Contents

CARRIER DESIGN, SYNTHESIS, AND EVALUATION

APPLICATIONS IN CHEMICAL SEPARATIONS

INDEXES

Preface

MEMBRANES that isolate distinct solutions and permit selective passage of chemical species are fundamental features of life processes. Considerable success has been achieved in the development of artificial membranes for the separation of a wide variety of ionic and molecular species. With sales of membrane separation systems surpassing $1 billion annually, membrane science is an extremely important topic for current and future technology.

The liquid membrane system involves an immiscible liquid that serves as a semipermeable barrier between two liquid or gas phases. The efficiency and selectivity of transport across the immiscible liquid may be markedly enhanced by the presence of a carrier in the membrane phase that reacts rapidly and reversibly with the desired chemical species. Liquid membrane systems have been and are being studied extensively by researchers in such fields as analytical, inorganic, and organic chemistry, chemical engineering, biotechnology, and biomedical engineering. Research and development activities within these disciplines involve diverse applications of liquid membrane technology, such as gas separations, recovery of valued or toxic metals, removal of organic compounds, development of sensing devices, and recovery of fermentation products. Transport of ionic species through synthetic liquid membranes also provides useful models for biological systems.

At national meetings of the American Chemical Society and the American Institute of Chemical Engineers in 1995, symposia on recent developments in liquid membrane technology were dedicated to Norman N. Li, who in 1968 invented the emulsion liquid membrane system. This book is based on papers that were presented in these two symposia and is also dedicated to Norman N. Li.

The 27 chapters of this book describe separations of metal ions, anionic species, organic molecules, and gas mixtures that involve liquid membrane processes. Scientists and engineers in academic, governmental, and industrial laboratories in eight countries contributed to this volume. An overview chapter provides an introduction to the various liquid membrane configurations, transport mechanisms, and experimental techniques. A tribute chapter follows, summarizing the many contributions of Norman N. Li in the field of membrane science. The remainder of the book is divided into sections on theory and mechanism (6 chapters), carrier design, synthesis, and evaluation (6 chapters), and applications in chemical separations (13 chapters). The applications section covers a very broad range of separation processes, including separation of carbon dioxide from nitrogen; unsaturated from saturated hydrocarbons in both the gaseous and liquid states; individual transition and heavy metal cations, such as mercury and lead, including the field testing of a system for copper recovery from mine solutions; selenium from contaminated waters; radiotoxic species such as cesium, uranium, and plutonium from nuclear wastes; and sugars from aqueous solutions.

These descriptions of liquid membrane technology provide state-of-the-art information for both the novice and the practitioner. Chemists who work in metal-ion complexation and separation, carrier synthesis, and nuclear chemistry will want a copy of this book. Chemical engineers involved in gas processing, metals recovery, and nuclear-waste processing will also find this book to be a valuable resource.

Acknowledgments

We acknowledge financial assistance from the Separation Science and Technology Subdivision of the ACS Division of Industrial and Engineering Chemistry, Inc.; the Donors of The Petroleum Research Fund, administered by the American Chemical Society; and Eichrom Industries, Inc., for financial support of the Symposium on Liquid Membrane Separations, held at the 209th ACS National Meeting in Anaheim, California, April 2–7, 1995. We also acknowledge the members of ACS Books Department for their efforts in assembling this volume. Finally, we thank the authors for preparing their manuscripts and the referees for reviewing each chapter.

RICHARD A. BARTSCH
Department of Chemistry and Biochemistry
Texas Tech University
Lubbock, TX 79409–1061

J. DOUGLAS WAY
Chemical Engineering and Petroleum Refining Department
Colorado School of Mines
1500 Illinois Street
Golden, CO 80401–1887

May 30, 1996

Chapter 1

Chemical Separations with Liquid Membranes: An Overview

Richard A. Bartsch[1] and J. Douglas Way[2]

[1]Department of Chemistry and Biochemistry, Texas Tech University,
Lubbock, TX 79409–1061
[2]Chemical Engineering and Petroleum Refining Department,
Colorado School of Mines, 1500 Illinois Street,
Golden, CO 80401–1887

Liquid membrane technology is introduced and its position within the broader field of separation science is delineated. The different types of liquid membrane configurations are described. Advantages of using a carrier species in the liquid membrane to enhance both the efficiency and selectivity of facilitated transport are summarized and two types of transport mechanisms are compared. Following a description of the general organization for the volume into sections on theory and mechanism, carrier design, synthesis, and evaluation, and applications in chemical separations, the contents of each chapter are briefly summarized.

A membrane is simply a barrier between two phases. If one component of a mixture moves through the membrane faster than another mixture component, a separation can be accomplished. Polymeric membranes are used commercially for many applications including gas separations, water purification, particle filtration, and macromolecule separations (1-4). There are several important aspects to this definition. First, a membrane is defined based on its function, not the material used to fabricate the membrane. Secondly, a membrane separation is a rate process and the separation occurs due to a chemical potential gradient, not by equilibrium between phases.

Liquids that are immiscible with the source (feed) and receiving (product) phases can also be used as membrane materials (5,6). Different solutes will have different solubilities and diffusion coefficients in a liquid. The product of the diffusivity and the solubility is known as the permeability coefficient, which is proportional to the solute flux. Differences in the permeability coefficient produce a separation between solutes at constant driving force. Because the diffusion coefficients in liquids are typically orders of magnitude higher than in polymers, a larger flux can be obtained with liquid membranes.

0097–6156/96/0642–0001$15.00/0

Facilitated Transport

As stated above, the use of a liquid phase can enhance the solute flux due to the higher diffusion coefficients in liquids than in solids. Further enhancement can be accomplished by adding a nonvolatile complexing agent to the liquid membrane (7,8). This carrier molecule can selectively and reversibly react with the solute. This reversible reaction provides a means of enhancing the solute flux and improving the selectivity at the same time. There are two basic mechanisms for this enhanced transport.

Figure 1 illustrates the concept of coupled transport where the reversible reaction is an ion-exchange reaction, and the solute flux is linked (coupled) to the flux of another ion (9). The carrier (HC) is an ion-exchange reagent. The ion-exchange reaction normally occurs at a liquid-liquid interface since metal ions are not soluble in the organic membrane phase. Coupled transport is analogous to performing solvent extraction in a thin liquid film. The majority of liquid membranes for metal ion separation involve a coupled transport mechanism.

As shown in Figure 2, facilitated transport is concerned with the reversible reaction between the solute and the carrier (C). It is not coupled to other components. This reaction normally takes place throughout the liquid membrane phase.

Four points demonstrate the benefits of using carriers in liquid membranes.

- High fluxes are possible.

- Very selective separations are possible. The selective nature of the carrier provides much better separations than those obtained solely from differences in the permeability coefficient, the product of the diffusivity and solubility.

- Ions can be concentrated since the coupled transport mechanism can pump one ion against its concentration gradient due to a gradient in the coupled ion.

- Expensive complexing agents can be used. Only small amounts of carrier are required because of the small solvent inventory in the membrane.

Liquid Membrane Configurations

Liquid membranes are encountered in three basic configurations which are depicted in Figure 3. In a bulk liquid membrane (BLM), a relatively thick layer of immiscible liquid is used to separate the source and receiving phases. To provide a thinner liquid membrane, the liquid can be impregnated in the pore structure of a microporous solid, such as a polymeric filter. This configuration is known as an immobilized or supported liquid membrane (ILM or SLM). In the third configuration which possesses the largest interfacial areas, the receiving phase is emulsified in an immiscible liquid membrane (9). The liquid surfactant membrane or emulsion liquid membrane (ELM) is then dispersed in the feed solution and mass transfer takes place from the feed phase to the internal receiving phase.

Bulk Liquid Membranes. Figure 4 shows four different cells which have been utilized in BLM transport experiments (11-13). The upper two are U-tube cells (12,13) and the lower two are so-called "tube-within-a-shell" cells (12,13). The apparatus for conducting bulk liquid membrane transport experiments has the advantage of simplicity. However due to the thickness of the membrane, the amount of species transported is very low. Therefore, bulk liquid membrane transport systems are utilized in studies of transport mechanisms and assessing the influence of carrier structure upon transport efficiency and selectivity, but have no potential for practical application.

Supported Liquid Membranes. A SLM can be fabricated in at least three different geometries. Planar or flat sheet SLMs are very useful for laboratory research and development purposes, but the surface area to volume ratio of flat sheets is too low for industrial applications. Spiral wound and hollow fiber geometries can

Coupled Transport

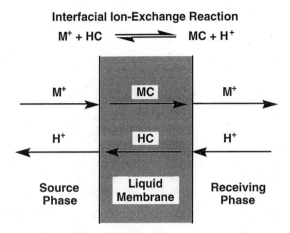

Figure 1. Schematic Diagram of the Coupled Transport Mechanism for Metal Ion Transport Through a Liquid Membrane. (M^+ is the metal ion, HC is the ion-exchange reagent (carrier), and H^+ is a hydrogen ion.)

Facilitated Transport

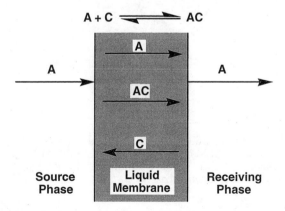

Figure 2. Schematic Diagram of the Facilitated Transport Mechanism for a Neutral Species Through a Liquid Membrane. (A is the solute, C is the complexing agent (carrier), and AC is the solute-carrier complex.)

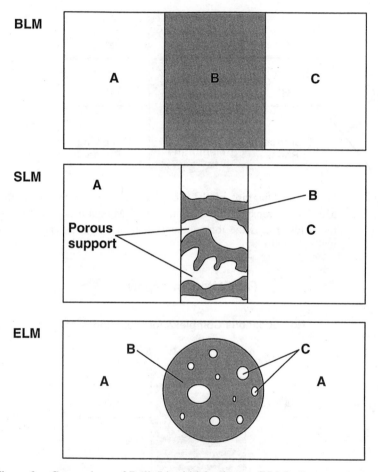

Figure 3.   Comparison of Bulk Liquid Membrane (BLM), Supported Liquid
            Membrane (SLM), and Emulsion Liquid Membrane (ELM)
            Systems. (A is the source (feed) phase, B is the liquid membrane,
            and C is the receiving phase.)

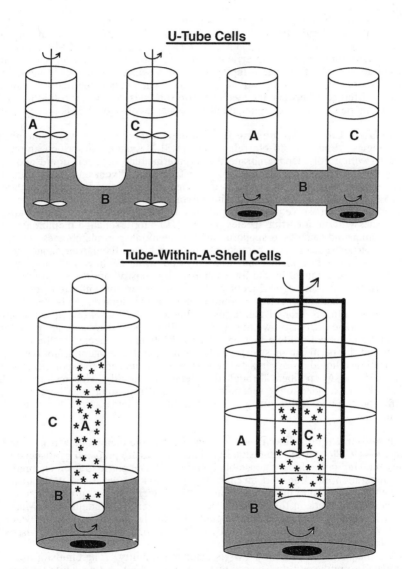

Figure 4.   Four Types of Cells Used to Study Transport Across Bulk Liquid
Membranes.

be used to increase the surface area of liquid membrane modules. Surface area to volume ratios for hollow fiber modules can approach 10^4 m^2/m^3 and 10^3 m^2/m^3 for spiral wound modules (14). Way et al. (15) discuss criteria for selecting supports for SLM systems.

For SLMs, there are several important stability issues including the integrity of the immobilized liquid film, the mechanical stability of the porous support, and chemical stability of the carrier. Danesi et al. (16) have correlated SLM lifetime with the following parameters: interfacial tension, carrier interaction with water, concentration of the feed and stripping solution, viscosity of the liquid membrane, and the water solubility in the liquid membrane. They concluded that low osmotic pressure differences between the feed and strip solutions, low water solubility in the liquid membrane, and low carrier solubility in water favor stable SLMs.

Related Liquid Membrane Technologies. There are several membrane technologies which are closely related to the SLM systems described above. A solvent swollen polymer membrane that is intermediate between liquid and solid phases can be made by swelling a polymer film in a solvent and introducing the carrier species by diffusion, or by ion exchange in the case of ionomer membranes (17,18). If the solvent used to swell the polymer film is a good physical solvent for the gas of interest, solvent swollen polymer films can be used as gas separation membranes without a carrier species present (19). Ion-exchange membranes have several unique advantages as supports for facilitated transport membranes. Once the charged carrier species is exchanged into the membrane, the carrier cannot be lost unless it is replaced by another ion. Secondly, the carrier loading in an ion-exchange membrane is determined by the ion-exchange site density, not the solubility of the carrier in the solvent. Consequently, the local carrier concentration obtained in heterogeneous materials, such as perfluorosulfonic acid ionomers, can be very high.

Another closely related membrane technology is the hollow fiber-contained liquid membrane (20). Non-wetting, porous, hollow fiber membranes are potted into a specially designed module that has half of the fibers connected to the feed gas stream and the permeate or sweep gas flows through the remaining hollow fibers. The membrane liquid resides in the shell side of the module and transport occurs from feed fibers to sweep fibers through the membrane liquid. This technology has recently been extended to the separation of organics from solution (Chapter 16) and has the potential to solve many of the stability problems associated with SLM systems.

Emulsion Liquid Membranes. The emulsion liquid membrane was invented by Li in 1968 (9, Chapter 2). Emulsion liquid membranes are prepared by dispersing an inner receiving phase in an immiscible liquid membrane phase to form an emulsion. The preparation and transport mechanisms of ELMs are described in detail in the Chapter 15. The liquid membrane phase can be either aqueous or organic although the majority of work in the literature describes water-in-oil emulsions. Marr and Kopp (21) created a set of quantitative guidelines for the formation of stable water-in-oil ELMs based on the surfactant HLB, surfactant concentration, organic viscosity, and volume ratios of the various phases.

The major problem with ELMs is emulsion stability. The emulsion must be formulated to withstand the shear which is generated by mixing while the desired species is being transported into the inner receiving phase, and the emulsion must be easily broken to recover the concentrated internal phase and to recycle and reformulate the emulsion. These two factors must be carefully balanced. Also, large osmotic pressure differences are encountered when there is a high metal ion concentration in the internal phase of the emulsion. This can lead to water transport from the dilute feed solution to the concentrated internal phase in a process known as osmotic swell (22). Osmotic swell can cause the emulsion globules to rupture, and the separation is lost.

Overview

Following this overview chapter and the dedication chapter entitled "A Tribute to Dr. Norman N. Li", the volume is divided into three sections: Theory and Mechanism; Carrier Design, Synthesis, and Evaluation; and, Applications in Chemical Separations. The first of these sections contains six chapters which describe theoretical and mechanistic aspects of the different liquid membrane types and configurations. In the next section, the design, synthesis, and evaluation of carrier species are described in six chapters. The applications sections begins with a survey chapter on "Recent Advances in Emulsion Liquid Membranes". This is followed by an overview chapter on hollow fiber-contained liquid membranes, which are utilized in both gas-phase and liquid-phase separations. The next three chapters report chemical separations of gas-phase species. Liquid-phase separations of various chemical species, which include cations, anions, and organic molecules, are described in the final eight chapters. It should be noted that a chapter assigned to one section may also contain relevant information for another section.

Theory and Mechanism. In Chapter 3, Reinhoudt and coworkers review recent mechanistic aspects of carrier-assisted transport through supported liquid membranes. Carriers for selective transport of neutral molecules, anions, cations, or zwitterionic species have been developed. Transport is described in terms of partitioning, complexation, and diffusion. Most of the mechanistic studies were focused on diffusion-limited transport, in which diffusion of the solute-carrier complex through the membrane phase is the rate-limiting step for transport. However, for some new carriers, the rate-limiting step was found to be decomplexation at the membrane phase-receiving phase interface.

The general types of liquid membranes are reviewed by Peterson and Lamb in Chapter 4, and factors which influence the effectiveness of a membrane separation system are summarized. These factors include the complexation/decomplexation kinetics, membrane thickness, complex diffusivity, anion type, solvent type, and the use of ionic additives. A novel membrane type, the polymeric inclusion membrane, is introduced.

Results from detailed studies of alkali metal cation transport across supported liquid membranes by fatty acid carriers are revealed by Kocherginsky in Chapter 5. An analytical expression for the transmembrane transport in this multi-step process is presented and good agreement between theory and the experimental data is obtained.

In Chapter 6, characteristic features of emulsion liquid membrane systems are examined by Yurtov and Koroleva. The effects of surfactant and carrier concentrations and external and internal phase compositions upon the properties of the extracting emulsions are discussed. Several mathematical models for the rheological curves are considered, and regions of applicability for the models are evaluated. An influence of nanodispersion formation on mass transfer through the interface and on the properties of extracting emulsions for cholesterol is demonstrated.

Draxler *et al.* describe the preparation and splitting of emulsions for emulsion liquid membrane systems in Chapter 7. An apparatus for the preparation of emulsions and various devices for the electrostatic splitting of emulsions are improved by investigation of the flow patterns using computational fluid field dynamics software.

For emulsion liquid membrane systems, a model which contains six differential and algebraic equations is developed by Huang *et al.* in Chapter 8. The model takes into account five steps in the transport process. For comparison with experiment, the arsenic concentration in the external phase versus time is measured for the removal of arsenic from water in an ELM system. Excellent correlation of the experimental data with the theoretical predictions is obtained.

Carrier Design, Synthesis, and Evaluation. In this section, new carriers for metal ions, anionic species, and neutral organic molecules are described. In addition to the six chapters contained in this section, the reader is also directed to Chapter 3 in which Reinhoudt and coworkers introduce a variety of novel carriers.

This section begins with Chapter 9 in which Peterson and Lamb discuss the various structural features of macrocyclic multidentate ligands (*e.g.* crown ethers) that influence the efficiency and selectivity of liquid membrane transport processes in which such molecules are utilized as metal-ion carriers.

In Chapter 10, Tsukube describes the use of computer chemistry in the design of new carriers with high specificity for lithium and silver ions. Also lipophilic lanthanide tris(ß-diketonates) are shown to be a new class of carriers which form 1:1 complexes with anionic species and with amino acid derivatives.

A variety of new crown and lariat ether compounds are evaluated as carriers in alkali metal cation and silver ion transport across bulk liquid membranes by Bartsch *et al.* (Chapter 11). For crown ethers substituted with one or more linear alkyl groups, the effects of varying the number and attachment sites for the lipophilic groups and replacement of one or more ring oxygen atoms with sulfur or nitrogen atoms are assessed. For lariat ethers with dibenzo-16-crown-5 rings, the influence of amide- and thioamide-containing side arms is investigated.

Hiratani and Kasuga describe in Chapter 12 the synthesis and evaluation of new acyclic amide carriers for heavy metal ion transport. Among the different types of acyclic amides, malonamide and glutaramide derivatives are found to exhibit high selectivity for copper(II) transport and different methionine derivatives for mercury(II) and silver(I) transport.

The transport of cobalt(II), copper(II), nickel(II), and zinc(II) from aqueous sulfate solutions by novel di(*p*-alkylphenyl)phosphoric acid carriers in bulk and emulsion liquid membrane transport processes is reported by Walkowiak and Gega in Chapter 13. To probe the mechanism of the liquid membrane transport processes, interfacial tension measurements are conducted. A multistage emulsion liquid membrane system for separation of the transition metal cation mixtures is developed.

In Chapter 14, Smith utilizes boronic acid carriers for the separation of hydrophilic sugars from aqueous solutions in bulk and supported liquid membrane systems. Boronic acids also function as carriers for the transport of sugars through lipid bilayers.

Applications in Chemical Separations. The scale of applications contained in this section vary widely from laboratory-scale testing to field testing to actual commercial installations.

In Chapter 15, Ho and Li briefly review recent advances in the theory for emulsion liquid membranes and applications in more detail. Commercial applications include the removal of zinc, phenol, and cyanide from wastewaters. Potential applications in wastewater treatment, biochemical processing, rare earth metal extraction, radioactive material removal, and nickel recovery are described.

The basic technique of hollow fiber-contained liquid membrane permeation-based separations of gas mixtures and of componenets from liquid solutions is reviewed by Sirkar in Chapter 16. Existing applications in organic acid and antibiotic recovery from fermentation broths, facilitated countertransport and co-transport of heavy metal ions, facilitated carbon dioxide-nitrogen (methane) separation, and facilitated sulfur dioxide-nitrogen oxides separation from flue gas are surveyed. New applications include the separation of organic mixtures (including isomers), highly selective pervaporation of organics from water, and a modified technique for gas separation that can treat feed at 175-200 psig.

Teramoto *et al.* describe in Chapter 17 the separation of carbon dioxide from methane with aqueous solutions of amines, such as monoethanolamine, diethanolamine, and ethylenediamine hydrochloride, in a supported liquid membrane system. The experimentally observed permeation rates for carbon dioxide can be simulated by a facilitated transport theory.

In Chapter 18, Matsuyama and Teramoto report the preparation of new types of cation-exchange membranes by grafting acrylic acid and methacrylic acid to substrates, such as microporous polyethylene, polytetrafluoroethylene, and poly[1-(trimethylsilyl)-1-propyne], by use of a plasma graft polymerization technique. Various monoprotonated amines are immobilized by electrostatic forces in the ion-exchange membranes and used as carriers for carbon dioxide. With these membrane systems carbon dioxide/nitrogen selectivities of greater than 4700 are obtained with high carbon dioxide flux.

For facilitated transport of ethylene through silver ion-containing perfluorosulfonic acid ion-exchange membranes, high ethylene/ethane separation factors are obtained by Way and coworkers in Chapter 19. Ethylene of greater than 99 percent purity is obtained from a 50:50 mixture of ethylene and ethane at ambient temperature with feed and permeate pressures of one atmosphere.

In Chapter 20, Koval *et al.* briefly review reports of the use of silver ion-exchanged membranes for separations of unsaturated hydrocarbons. Interesting properties of silver ion-exchanged Nafion are summarized and equations which define common transport performance parameters are presented. New performance data for liquid-phase separations of close boiling alkene/alkane mixtures and a variety of separations involving aromatic compounds are presented.

Selective separation of lead(II) and cadmium(II) chloride complexes by both liquid membranes containing liquid anion-exchangers as mobile carriers and solid polymeric membranes with anion-exchange sites as fixed carriers is described by Hayashita in Chapter 21. A novel polymeric plasticizer membrane, which is composed of cellulose triacetate polymer as a membrane support, *o*-nitrophenyl octyl ether as a membrane plasticizer, and trioctylmethylammonium chloride as an anion-exchange carrier, provides enhanced permeation selectivity and efficiency.

As described in Chapter 22, Wiencek *et al.* are developing improved emulsion liquid membrane systems. The possibility of employing microemulsions as liquid membranes to separate metals (especially mercury) from contaminated water is explored.

In Chapter 23, Nilsen *et al.* describe the field testing of an emulsion liquid membrane system for copper recovery from mine solutions. The small, pilot plant-scale, continuous circuit for the recovery of copper from mine waste waters and low-grade leach solutions was field tested at a copper mine. Formulation of the emulsion membranes was optimized to provide emulsions with good stability during extraction, but which could be easily broken in an electrical coalescer under mild conditions. Typical results from the tests were >90 percent copper recovery, while maintaining the membrane swelling in the range of 4-8 percent. Cost evaluations indicate the potential for cost-effective recovery of copper from such solutions.

An emulsion liquid membrane process is reported in Chapter 24 by Gleason *et al.* which rapidly reduces aqueous phase selenium concentrations to low levels within 15 minutes of contact time, while enhancing the concentration of selenium in the internal phase by more than a factor of 40. The emulsion formulation is very stable, and swelling does not significantly dilute the selenium in the internal phase. The presence of other anionic species, such as sulfate, is not a significant interference for removal of the selenium anions.

In Chapter 25, Misra and Gill survey the applications of supported liquid membranes in separations of transition metal, lanthanides, and actinides from aqueous solutions. Choices of membrane material and solvent which improve the membrane stability in a SLM system are discussed. A few pilot-scale studies of SLM processes are described which show the potential for large-scale utilization in the future.

Vicens and coworkers (Chapter 26) report the use of novel calix-*bis*-crown ether compounds as carriers in a supported liquid membrane system for the removal of cesium from nuclear waste water. Decontamination factors of greater than 20 are obtained in the treatment of synthetic acidic radioactive wastes. Very good stability (over 50 days) and high decontamination yields are achieved.

Carrier-facilitated transport of actinides across bulk, supported, and emulsion liquid membranes, as well as plasticized membranes and recently developed emulsion-free liquid membranes, are reviewed by Shukla *et al.* in Chapter 27. The discussion includes the effects of important experimental variables upon the solute flux for the various types of liquid membranes. Applications of liquid membranes in the recovery and removal of radiotoxic actinides from the nitric acid wastes generated during reprocessing of spent nuclear fuel by the PUREX process and wastes produced by other radiochemical operations are surveyed.

Summary

In this chapter, the reader has been introduced to liquid membrane technology. The various types of liquid membrane configurations are surveyed and the advantages and disadvantages of each type are described. The arrangement of chapters within this ACS Symposium Series volume is delineated and capsule summaries are provided for each chapter. These chapters encompass the entire breadth of liquid membrane technology and are intended to further the understanding and applications of this important method for performing chemical separations

Literature Cited.

1. Mulder, M. *Basic Principles of Membrane Technology*; Kluwer Academic Publishers: Norwell, MA, 1992.
2. *Membrane Handbook*; Ho, W.S.W.; Sirkar, K.K., Eds.; Chapman & Hall: New York, NY; 1992
3. *Membrane Science and Technology*; Osada, Y.; Makagawa, T., Eds; Marcel Dekker: New York, NY, 1992.
4. *Membrane Separation Technology*; Noble, R.D.; Stern, S.A., Eds; Elsevier: New York, NY, 1995.
5. *Liquid Membranes: Theory and Applications*, ACS Symposium Series 347; Noble, R.D.; Way, J.D., Eds.; American Chemical Society: Washington, DC, 1987.
6. *Liquid Membranes: Chemical Applications*; Araki, T.; Tsukube, H., Eds.; CRC Press: Boca Raton, FL, 1990.
7. Scholander, P.R. *Science* **1960**, *131*, 585.
8. Ward, J.J. III; Robb, W.L. *Science* **1967**, *156*, 1481.
9. Cussler, E. L. *AIChE J.* **1971,** *17*, 1300.
10. Li, N. N.; U. S. Patent 3,410,794, 1968.
11. Lamb, J.D.; Izatt, R.M.; Garrick, D.G.; Bradshaw, J.S.; Christensen, J.J. *J. Membr. Sci.* **1981**, *9*, 83.
12. Bartsch, R.A. In *Liquid Membranes: Theory and Applications*, ACS Sympsoium Series 347; Noble, R.D.; Way, J.D., Eds; American Chemical Society: Washington, DC, 1987; pp 86-97.
13. Tsukube, H. In *Liquid Membranes: Chemical Applications*; Araki, T.; Tsukube, H., Eds.; CRC Press: Boca Raton, Fl, 1990; pp 27-50.
14. Kesting, R. E. *Synthetic Polymeric Membranes: A Structural Perspective*. Wiley Interscience, New York, NY, 1985.
15. Way, J.D.; Noble, R.D.; Bateman, B.R. In *Materials Science of Synthetic Membranes*, ACS Symposium Series No. 269, Lloyd, D.L., Ed.; American Chemical Society: Washington, DC, 1984; pp 119-128.
16. Danesi, P.R.; Reichley-Yinger, L.; Rickert, P. G. *J. Membr. Sci.* **1987**, *31*, 117.
17. LeBlanc, O.H.; Ward, W. J.; Matson, S. L.; Kimura, S. G. *J. Membr. Sci.* **1980**, *6*, 339.
18. Way, J.D.; Noble, R. D.; Reed, D. L.; Ginley, G. M. *AIChE J.* **1987**, *33*, 480.
19. Matson, S. L.; Lee, E. K. L.; Friesen, D. T.; Kelly, D. J.; U. S. Patent 4,737,166, 1988.
20. Majumdar, S.; Sirkar, K. K.; Sengupta, A. In *Membrane Handbook*; Ho, W.S.W.; Sirkar, K.K., Eds.; Chapman & Hall: New York, NY, 1992; pp. 764-XXX.
21. Marr, R.; Kopp, A. *Int. Chem. Eng.* **1982**, *22*, 44.
22. Draxler, J.; Marr, R. *Chem. Eng. Process.* **1986**, *20*, 319

Chapter 2

A Tribute to Norman N. Li

J. Douglas Way[1] and W. S. Winston Ho[2]

**[1]Chemical Engineering and Petroleum Refining Department,
Colorado School of Mines, 1500 Illinois Street,
Golden, CO 80401–1887
[2]Corporate Research, Exxon Research and Engineering Company,
Route 22 East, Clinton Township, Annandale, NJ 08801**

Researchers in the fields of facilitated transport and liquid membranes owe much to Dr. Norman N. Li for his pioneering work on facilitated transport and the invention of liquid surfactant (emulsion) membrane technology when he was with Exxon Research and Engineering Company. A member of the National Academy of Engineering, Dr. Li became Chairman and President of NL Chemical Technology, Inc. after he recently retired as Director of Research and Technology at Allied Signal. Although his 30 year career has been spent entirely in industry, Dr. Li has an impressive record of publication. Dr. Li has also made important contributions to the profession through his participation in AIChE, ACS, and the North American Membrane Society. This chapter will highlight some of Dr. Li's accomplishments, briefly describe the liquid membrane technology he invented, and discuss commercial applications of the emulsion liquid membrane (ELM) and related technologies.

In 1968 Dr. Norman N. Li invented a new separation technology known as the liquid surfactant or emulsion liquid membrane (*1*). This technique combined the conventional separation unit operations of extraction and stripping into a single process capable of extremely rapid separations and high selectivity. The emulsion liquid membrane (ELM) process technology is capable of separating an extremely wide variety of solutes from organics to metal ions for a very diverse set of applications including wastewater treatment, biotechnology, and hydrometallurgy. This chapter will briefly describe ELM technology and Dr. Li's contributions to the profession through his work in professional societies. Extensive, recent reviews of all aspects of ELM technology are available including theory, design, applications and economics (*2-6*). For the theory and applications, see also the chapter in this volume entitled "Recent Advances in Emulsion Liquid Membranes" by W. S. Ho and N. N. Li.

ELM Technology.

As shown in Figure 1 in the chapter entitled "Recent Advances in Emulsion Liquid Membranes," ELMs are prepared by dispersing an inner receiving phase in an immiscible liquid membrane phase to form an emulsion. The emulsion is stabilized by surfactants with appropriate hydrophilic-lipophilic balance (HLB) number dissolved in the liquid membrane phase (7). The HLB is a parameter which is the percentage of hydrophilic functional groups in the surfactant molecule divided by five. The liquid membrane phase can be either aqueous or organic although the majority of work in the literature describes water-in-oil emulsions in which the liquid membrane phase is organic.

In use, the ELM is dispersed in a continuous phase in a stirred tank reactor and the liquid membrane phase separates two miscible phases. Under agitation, the ELM phase separates into spherical globules of emulsion which have typical diameters of 100 μm to 1 mm. Each globule contains many droplets of encapsulated inner or receiving phase with a typical size of 1 to 3 μm in diameter. The formation of many globules of emulsion produces large surface area/volume ratios of 1000 to 10^6 m^2/m^3 for very rapid mass transfer (6, 8). Due to this dispersed emulsion configuration, ELMs or liquid surfactant membranes are commonly referred to as double emulsions.

The transport of a solute from the continuous phase to the inner receiving phase can occur by a variety of mechanisms (2). The two primary mechanisms known as Type 1 and 2 facilitation are shown in Figure 2 in the chapter "Recent Advances in Emulsion Liquid Membranes." * The constituents of the ELM for extraction of a solute must be chosen in such a way that once the solute diffuses into the inner receiving phase it cannot diffuse back out into the continuous phase. The Type 1 mechanism is designed for the separation of nonionic solutes such as phenol (9-10). The phenol dissolves in the organic liquid membrane phase and diffuses to an internal aqueous droplet. A trapping reaction in the internal aqueous phase, such as the reaction of phenol with NaOH, creates an ionic species, sodium phenolate, which is insoluble in the organic liquid membrane phase. The trapping reaction also maintains the highest possible concentration driving force for diffusion.

Ionic solutes, such as metal ions, are not soluble in the organic liquid membrane phase. Consequently, the Type 2 mechanism involves the addition of an organic extractant, or chelating agent, to the organic liquid membrane. Figure 2 in "Recent Advances in Emulsion Liquid Membranes" illustrates the concept of coupled transport where an ion-exchange reaction takes place at the organic liquid membrane/aqueous phase interfaces and the solute flux is linked (coupled) to the flux of another ion, usually H^+ (11). Coupled transport is analogous to performing solvent extraction in a thin liquid film. The majority of liquid membranes for metal ion separation involve a coupled transport mechanism.

Commercial Applications for ELMs.

Draxler and Marr (12) discussed the application of ELM technology to remove zinc as zinc sulfate from low concentration wastewater from a textile plant in Lenzing,

* NOTE: Please see Figure 2 in Chapter 15, page 209.

Austria. The ELM process was chosen over solvent extraction and ion exchange due to the low zinc concentration and the presence of interfering Ca^{2+} ions. Furthermore, due to the very slow stripping kinetics, the extractant used was not suitable for solvent extraction; but the slow kinetics were overcome by the extremely large interal droplet surface area of the ELM (10^6 m^2/m^3). A commercial plant for Zn removal was constructed at the same location and began operation at the end of 1986. The plant used proprietary countercurrent extraction columns and electrostatic coalescers to break the emulsion prior to Zn recovery and recycle of the organic phase. The Zn concentration in the aqueous waste stream was reduced from 200 mg/L to 0.3 mg/L in less than 20 minutes average residence time in the extraction column. Concentrations in the inner droplet phase of the ELM of up to 50 g/L Zn were obtained. Proprietary design countercurrent stirred extraction columns (10 m high, 1.6 m diameter) similar to Oldshue-Rushton columns were used. The plant throughput was 75 to 100 m^3/hr.

Zhang and coworkers described a commercial application for ELMs in China for phenol removal from wastewater (*13-14*). At the Nanchung Plastics Factory in Guangzhou, China, the phenol concentration in a 250 L/h wastewater stream was reduced from 1000 mg/L to 0.5 mg/L. ELM technology was also employed at the Huang-hua Mountain Gold Plant near Tian-jin China (*14-15*). Cyanide was removed from waste liquors generated in a hydrometallurgy process. The cyanide concentration in the waste stream was reduced from 130 mg/L to 0.5 mg/L.

Finally, ELM technology was the basis for a commercial well control fluid developed by Exxon. A water-in-oil emulsion containing clay particles in the organic liquid membrane phase has a low viscosity allowing it to be pumped. Under the high shear conditions developed at the drilling bit nozzles which rupture the liquid membranes, the clay particles cause the internal water droplets to thicken the emulsion to a viscous paste. This viscous fluid prevents well blowout during drilling and can be used to seal loss zones in an oil or gas reservoir.

Potential Applications for ELMs.

There are a variety of potential applications for ELM technology that have yet to be exploited commercially. These include wastewater treatment, removal and recovery of metal ions, biochemical processing, a fracturing fluid for oil and gas production. Marr and Draxler (1992a) describe the removal of heavy metals including Zn, Cd, Cu, Pb, and Hg from wasterwaters and the recovery of Ni from spent electroplating solutions and rinse waters. ELMs have been investigated for the removal of acetic acid, nitrophenols, and ammonia from wastewater (*16-18*). More recently, Huang et al. (*19*) and Gleason et al. (*20*) have applied ELM technology to the removal of arsenic from wastewater. Solvent extraction (e. g. the TRUEX Process) (*21*) has been used very successfully in the processing of radioactive metals such as U, Pu, and Am. ELM technology could be a complementary process to solvent extraction (SX) in the separation of radionuclides, especially in the cases where the reaction kinetics are not fast enough for SX processing. Another potential application in metal recovery is the extraction of rare earth metals from dilute solutions in hydrometallurgical processing (*22-23*). Hatton and coworkers (*24-25*) have shown that ELM technology can be used to recover aminoacids and other fermentation products in biochemical processes. Recently, Ho and Li (*26*) have reported on an emulsion-based fracturing fluid to increase oil and gas production via core-annular flow.

Contributions to the Profession.

Researchers in the fields of facilitated transport and liquid membranes owe much to Dr. Norman N. Li for his pioneering work on facilitated transport and his invention of liquid surfactant (emulsion) membrane technology while at Exxon Research and Engineering Company. Dr. Li became Chairman and President of NL Chemical Technology, Inc. after he recently retired as Director of Research and Technology at Allied Signal. Although his 30 year career has been spent entirely in industry, Dr. Li' s publication record is impressive, including 44 U. S. patents, 100 publications and 13 edited books. His contributions to the profession by very active participation in professional societies are equally impressive. Dr. Li is a former Director of AIChE and a Fellow of AIChE. He was awarded the AIChE Alpha Chi Sigma Award in chemical engineering research in 1988 and the Ernest Thiele Award by the AIChE Chicago section. The American Chemical Society recognized Dr. Li's contributions by awarding him the ACS Award in Separation Science and Technology in 1988. He served as the Chairman of the Industrial and Engineering Chemistry Division of the American Chemical Society. He organized and chaired the Gordon Research Conference on Separation and Purification in 1973, the first Gordon Research Conference on Membranes in 1975, and the first and second Engineering Foundation Conferences on Separation Technologies in Switzerland in 1984 and in Germany in 1987. Dr. Li was chairman of the 1990 International Congress on Membranes and Membrane Processes and was president of the North American Membrane Society from 1990 to 1993. Dr. Li was recognized for both his technical and professional contributions by induction into the National Academy of Engineering.

References

1. Li, N. N. U. S. Patent 3,410,794, 1968.
2. Ho, W. S. W.; Li, N. N. In *Membrane Handbook*; Ho, W. S. W.; Sirkar, K. K., Eds.; Chapman & Hall: New York, NY, 1992, pp. 597-610.
3. Ho, W. S. W.; Li, N. N. In *Membrane Handbook*; Ho, W. S. W.; Sirkar, K. K., Eds.; Chapman & Hall: New York, NY, 1992, pp. 611-655.
4. Gu, Z.; Ho, W. S. W.; Li, N. N. In *Membrane Handbook*; Ho, W. S. W.; Sirkar, K. K., Eds.; Chapman & Hall: New York, NY, 1992, pp. 656-700.
5. Marr, R. J.; Draxler, J. In *Membrane Handbook*; Ho, W. S. W.; Sirkar, K. K., Eds.; Chapman & Hall: New York, NY, 1992, pp. 701-717.
6. Marr, R. J.; Draxler, J. In *Membrane Handbook*; Ho, W. S. W.; Sirkar, K. K., Eds.; Chapman & Hall: New York, NY, 1992, pp. 718-724.
7. Adamson, A. W. *The Physical Chemistry of Surfaces.* Wiley: New York, NY, 1977.
8. Marr, R.; Kopp, A. *Int. Chem. Eng.* **1982**, *22*, 44-60.
9. Matulevicius, E. S.; Li, N. N. *Sep. Purif. Methods* **1975**, *4*, 73-96.
10. Li, N. N. *J. Membr. Sci.* **1978**, *3*, 265.
11. Cussler, E. L. *AIChE J.* **1971**, *17*, 1300-1303.
12. Draxler, J.; Marr, R. *Chem. Eng. Process.* **1986**, *20*, 319-329.
13. Zhang, X.-J.; Liu, J.-H.; Fan, Q.-J.; Lian, Q.-T.; Zhang, X.-T.; Lu, T.-S. In *Separation Technology*; Li, N. N..; Strathmann, H. Eds.; United Engineering Trustees: New York, NY, 1988, pp. 190-203.

14. Jin, M.; Zhang, Y. In Proc. International Congress on Membranes and Membrane Processes, Chicago, IL, August 20-24, 1990, Vol. I; pp. 676-678.
15. Jin, M.; Wen, T.; Lin, L.; Liu, F.; Liu, L.; Zhang, Y.; Zhang, C.; Deng, P.; Song, Z. *Mo Kexue Yu Jishu* **1994**, *14*, 16.
16. Yan, N.-X.; Huang, S.-A.; Shi, Y.-J. *Sep. Sci. Technol.* **1987**, *22*, 801-818.
17. Gadekar, P. T.; Mukkolath, A. V.; Tiwari, K.K. *Sep. Sci. Technol.* **1992**, *27*, 427.
18. Lee, C. J.; Chan, C. C. *Ind. Eng. Chem. Res.* **1990** *29*, 96-100.
19. Huang, C. R.; Zhou, D. W.; Ho, W. S.; Li, N. N. Paper No. 89b Presented at the AIChE National Meeting, Houston, TX, March 19-23, 1995.
20. Gleason, K. J.; Yu, J.; Bunge, A. L.; Wright, J. D.; In *Chemical Separations with Liquid Membranes*; Bartsch, R. A.; Way, J. D. Eds.; American Chemical Society: Washington, DC, 1996.
21. Horwitz, E. P.; Kalina, D. G.; Diamond, H.; Vandegrift, G. F.; Schulz, W. W *Solvent Extr. Ion Exch.* **1985**, *3*, 75-109.
22. Tang, J.; Wai, C. M.; *J. Membr. Sci.* **1989**, *46*, 349.
23. Goto, M.; Kakoi, T.; Yoshii, N.; Kondo, K.; Nakashio, F. *Ind. Eng. Chem. Res.*, **1993**, *32*, 1681.
24. Thien, M. P.; Hatton, T. A. *Sep Sci. Technol.* **1988**, *23*, 819.
25. Itoh, H.; Thien, M. P.; Hatton, T. A.; Wang, D. I. C. *J. Membr. Sci.* **1990**, *51*, 309.
26. Ho, W. S. W.; Li, N. N. *AIChE J.* **1994**, *40*, 1961-1968.

Theory and Mechanisms

Chapter 3

Mechanistic Studies of Carrier-Mediated Transport Through Supported Liquid Membranes

Lysander A. J. Chrisstoffels, Feike de Jong, and David N. Reinhoudt

Department of Organic Chemistry, University of Twente, P.O. Box 217, 7500 AE Enschede, Netherlands

Carrier-assisted transport through supported liquid membranes is one of the important applications of supramolecular chemistry. Carriers for the selective transport of neutral molecules, anions, cations or zwitter ionic species have been developed. The transport can be described by subsequent partitioning, complexation and diffusion. Mechanistic studies are mainly focussed on diffusion limited transport, in which diffusion of the complex through the membrane phase is the rate-limiting step of the transport. Recently, kinetic aspects in membrane transport have been elucidated with new carriers for which the rate of decomplexation determines the rate of transport. In this chapter, recent mechanistic aspects for carrier-assisted transport through supported liquid membranes are reviewed.

Membrane separation is a relatively new and fast growing field in supramolecular chemistry. It is not only an important process in biological systems, but may become a large-scale industrial activity. For industrial applications, many synthetic membranes have been developed. Important conventional membrane technologies are microfiltration, ultrafiltration, electro- and hemodialysis, reverse osmosis and gas separations (*1*). The main advantages are the high separation factors that can be achieved under mild conditions and the low energy requirements.

In living matter, biological membranes provide a barrier between an intracellular and extracellular aqueous environment. The membranes are bilayers of amphiphilic phospholipids and glycolipids which support the membrane proteins that can induce selective transport through the cell membrane (*2*). These proteins complex certain species and carry these through the membrane to the other compartment where they are released. The first antibiotic recognized to facilitate selective transport is valinomycine (*3*). It binds K^+ a thousand times more strongly than Na^+, and it induces the selective uptake of K^+ in mitochondria.

0097–6156/96/0642–0018$19.75/0

An artificial membrane set-up that mimics carrier-facilitated transport is the liquid membrane. It consists of two aqueous phases that are separated by a water-immiscible organic phase to which receptors (carriers) are added to facilitate selective transport.

Natural receptors, such as valinomycin or beauverin, can be used as carriers. Also synthetic receptors have been developed. The first generation of synthetic carriers were the crown ether macrocycles, which selectively bind alkali metal ions (*4*). Since their discovery in 1967, many other types of macrocycles have been synthesized and used for the selective recognition of neutral, charged, or zwitter-ionic species. Many have been used as carriers in liquid membranes (*5*).

Liquid membrane separations can be achieved by different configurations (Figure 1). The main types will be discussed briefly. Bulk liquid membranes (BLM) have often been used to investigate novel carriers and transport mechanisms, because of their relative simplicity (Figure 1a). They consist of an aqueous source and receiving phase separated by a water-immiscible organic solution in a U-tube. Both aqueous phases and the membrane phase are stirred in order to achieve homogeneous solute concentrations and to minimize the thickness of the unstirred boundary layers. Disadvantages of the BLM are the relative instability and the large amounts of organic solvent and carrier needed.

In an emulsion liquid membrane (ELM), a dispersion of water containing oil droplets is stirred in a bulk aqueous source phase (Figure 1b) (*6*). Separation occurs very rapidly, because the organic layer is very thin while the surface area is large. Furthermore, the source phase volume is at least ten times larger than that of the receiving phase. A disadvantage is the instability of the emulsions. This is due to pH changes, ionic strength and physical forces. Furthermore, it is difficult to separate the oil droplets from the source phase and to break up the oil droplets to recover the receiving phase.

A supported liquid membrane consists of an organic carrier solution immobilized in a porous hydrophobic polymeric support. Two frequently used configurations are shown in Figure 1. The flat sheet supported liquid membrane is clamped between two cells that are filled with either an aqueous source or a receiving phase (Figure 1c). Because of its simplicity, the small amounts of solvent and carrier required, and its well defined diffusion layer (membrane thickness), this is a suitable system for mechanistic studies. The hollow fiber supported liquid membrane which is shown in Figure 1d has a cylindrical geometry. It consists of a few hundred hollow fibres in which the organic carrier solution is immobilized (*7*). The hollow fiber configuration is the most suitable for industrial application due to its large exchanging surface and long-term stability.

Carrier-Assisted Membrane Transport of Salts

Assisted transport through a liquid membrane takes place via the following consecutive steps: partitioning of a species from the aqueous source phase into the membrane phase, complexation by the carrier inside the membrane phase, and transport of the complex through the membrane. At the receiving side of the membrane, decomplexation and partitioning of the species into the aqueous receiving phase takes place. The performance of a liquid membrane is strongly related to the characteristics of the carrier, hence it is important to describe its properties with the

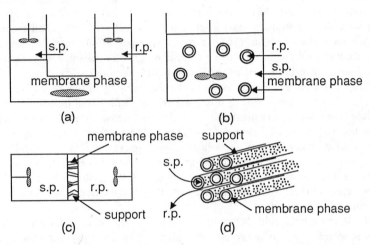

Figure 1. Schematic Representation of Different Liquid Membrane Configurations: (a) Bulk Liquid Membrane, (b) emulsion liquid membrane, (c) flat-sheet supported liquid membrane, (d) hollow fiber supported liquid membrane.

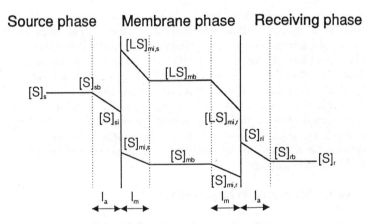

Figure 2. Concentration Profile for the Transport of Substrate S Assisted by Ligand L Through a BLM. (Adapted from ref. 5).

physical parameters D_m and K_{ex}. D_m is the apparant diffusion coefficient of the complex, which is a measure of the diffusion velocity of the complex through the membrane. The extraction constant K_{ex} is a measure of the ability of a carrier to extract a certain species from the aqueous phase into the membrane phase. Both parameters are described more fundamentally in the next section.

To determine D_m and K_{ex}, various models have been applied in mechanistic studies of cation-assisted transport. Accurate mechanistic studies are carried out with a SLM, because of its known diffusional layer and negligible external transport resistances. A detailed description of its configuration is given in the section on Transport Through a Supported Liquid Membrane. Models for assisted transport are differentiated by their rate-limiting step for transport, which is dependent on the rate of decomplexation at the receiving phase-membrane interface. Two types of transport mechanisms are distinguished. First, diffusion-limited transport, in which the diffusion process of the complex through the membrane phase is the rate-limiting step (see the section on Diffusion-Limited Transport of Salts). Second is kinetically limited transport in which the rate of decomplexation at the receiving side of the membrane phase is the rate-limiting step (see the section on Reaction-Rate Limited Transport of Monovalent Ions). To distinguish both mechanisms, a dimensionless number α is introduced as is described in the section on Kinetically Limited Transport: Determination of α.

Optimization of the flux is investigated through the variation of external parameters which affect transport, such as the membrane polarity, viscosity, anion used and operating temperature (see the section on Optimization of the Flux: External Influences on Transport). The influence on the transport kinetics has been investigated as well. Until recently, the transport of salts has only been facilitated through the use of a cation carrier. The cation is complexed by the carrier while the anion accompanies the complex as a free ion, hence a lipophilic anion is used to accompany the complex.

Another situation arises when the transport of the anion is assisted. In a later section, the transport of salts using an anion carrier, a mixture of a cation and anion carrier, or a ditopic carrier is discussed. Finally, the transport of neutral molecules will be described in the section on Diffusion-Limited Transport of Neutral Molecules. This is illustrated with our recent investigations on the transport of urea.

Transport Through a Bulk Liquid Membrane. All theoretical models concerning carrier-assisted transport through SLMs are based on the theoretical work for carrier-assisted transport through BLM systems reported by Reusch and Cussler (*8*). They described the transport of different alkali salt mediated by dibenzo-18-crown-6 through a BLM.

For the simplest case, *i.e.* the transport of a species (S) assisted by a ligand (L), the concentration profile may be visualized as shown in Figure 2. A BLM contains three bulk phases, *viz.* an aqueous source phase, a bulk liquid membrane, and an aqueous receiving phase. In these phases, transport takes place by convection and there are no concentration gradients. There are unstirred aqueous and organic layers at the source and receiving phase interfaces with the bulk liquid membrane. Processes which play a role in these interfacial regions are partitioning, complexation, and diffusion. The function of a carrier can be illustrated by comparison of the "blank" transport with carrier-assisted transport. For the diffusional processes at the interface, Fick's first law (Equation 1) is used to describe the flux through a stagnant layer with thickness l.

$$J_s = \frac{D_S}{l}\Delta c_S \tag{1}$$

The fluxes through the aqueous interfaces at the source phase (J_{as}) and the receiving phase (J_{ar}) are described according to Equation 2.

$$J_{ar} = \frac{D_a}{l_a}([S]_{ri} - [S]_{rb}) \quad ; \quad J_{as} = \frac{D_a}{l_a}([S]_{sb} - [S]_{si}) \tag{2}$$

In case of "blank" transport, only partitioning at both interfaces takes place according to Equation 3.

$$S_{si} \rightleftarrows S_{mi,s} \quad ; \quad S_{ri} \rightleftarrows S_{mi,r} \quad ; \quad K_p = \frac{[S]_{mi,s}}{[S]_{si}} = \frac{[S]_{mi,r}}{[S]_{ri}} \tag{3}$$

The fluxes of species "S" through the membrane interface layers at the source side (J_{ms}) and the receiving side (J_{mr}) are written according to Equation 4.

$$J_{mr} = \frac{D_m}{l_m}([S]_{mi} - [S]_{mb}) \quad ; \quad J_{ms} = \frac{D_m}{l_m}([S]_{mb} - [S]_{mi}) \tag{4}$$

In the steady state situation, all fluxes are equal and can be summarized to obtain a relation which is expressed in terms of a resistance = (chemical potential/flux).

$$\frac{([S_s] - [S_r])}{J} = \frac{2l_a}{D_a} + \frac{2l_m}{D_m K_p} \tag{5}$$

The first term of Equation 5 refers to the resistance due to presence of an external interface layer. The internal resistance is described by the second term ($2l_m/D_m K_p$). In the presence of a carrier, both diffusion of uncomplexed and complexed species through the membrane phase takes place. The latter involves the extraction of a species S from the aqueous phase to form a complex "LS" in the membrane phase. In this process both the stability of the complex (K_a) and the partition (K_p) play a role.

$$S_{ai} + L_{mi} \rightleftarrows LS_{mi} \quad ; \quad K_{ex} = \frac{[LS]_{mi}}{[S]_{ai}[L]_{mi}} \tag{6}$$

The extraction constant K_{ex} is defined as $K_{ex} = K_a K_p$. The association constant K_a is defined as the equilibrium constant for the complexation of species S by ligand L in the membrane phase. Combination of the extraction Equation with the mass-balance for the carrier gives Equation 7, which shows that the flux of species S increases upon the addition of ligand L_0 via the term $(K_{ex}L_0)/((1+K_{ex}[S]_s)(1+K_{ex}[S]_r))$. If no carrier is present, Equation 7 reduces to Equation 5. The rate-limiting step of the process is the diffusion of the complex through the organic layer.

$$\frac{([S]_s - [S]_r)}{J} = \frac{2l_a}{D_a} + \frac{2l_m}{D_m} \frac{1}{(K_p + \frac{K_{ex}L_0}{(1 + K_{ex}[S]_s)(1 + K_{ex}[S]_r)})} \tag{7}$$

The thicknesses of both the interfacial layers (l_a and l_m) are unknown. Therefore a quantitative description of the transport is difficult. This is the main disadvantage of the BLM for mechanistic studies.

Transport Through a Supported Liquid Membrane. Mechanistic studies of carrier-assisted membrane transport have been carried out in our group using the supported liquid membrane (SLM) configuration. In comparison with the configuration of a BLM, the SLM has a few differences which favor its use for mechanistic studies. Usually we have conducted transport experiments through SLMs in a permeation cell as shown in Figure 3. The source phase (SP) and the receiving phase (RP) consist of an identical cylindrical compartment with a half-cell volume of 50 mL and an effective membrane area of about 13 cm². Both compartments are double-walled and thermostated at T = 25 °C. A flat-bladed turbine positioned in the center is driven by a magnet outside the compartment at a stirring rate of 750 rpm. Compared to BLM transport, stirring of the aqueous phases can take place more vigorously, due to the use of a solid support. For mechanistic studies this implies that the influence of an external transport resistance can be neglected (*9*). Baffles, which are set radially along the wall of the compartments provide a large top-to-bottom circulation, creating well-mixed solutions without significant concentration gradients.

Hydrophobic polymers as polyethylene, polypropylene, or polysulfone are often used as supports. For fast transport-rates, the exchanging surface between the aqueous phases and the membrane phase should be large. Therefore thin films with high porosity are used (*10*). The dimensions of the support are known, which is an advantage of the SLM in mechanistic studies. The stagnant diffusional layer of the complex is defined as the length of the polymeric support ($2l_m = d_m$). Commonly used commercial supports are Celgard and Accurel. The membrane thickness of Celgard 2500 is 25 μm with a porosity $\Theta = 0.45$ and a tortuosity $\tau = 2.35$. Accurel has a thickness of 100 μm with a porosity $\Theta = 0.64$ and a tortuosity $\tau = 2.1$.

Solvent characteristics that influence the stability of the membrane are surface tension and water solubility (*12*). Processes that destabilize liquid membranes are the loss of membrane material due to osmotic (*11*) or hydrostatic (*12*) pressure differences over the membrane, formation of emulsions, or formation of inverted micelles. A frequently used organic membrane solvent is *o*-nitrophenyl octyl ether (NPOE). It has a low viscosity, high lipophilicity and a high polarity. Other solvents used are phenylcyclohexane, decanol, and diphenylether.

In order to measure the salt transport, the receiving phase contains a conductivity cell. A typical experiment lasts 24 hours in which the conductivity is monitored with a time-interval of 60 seconds. The activities are calculated according to the Debye-Hückel theory (13). Fluxes are determined as the first derivative of the activity vs. time according to Equation 8. When competitive transport experiments are carried out, the receiving phase is analysed by atomic absorption spectrometry.

$$J(t) = \frac{V_r}{A_m} \frac{da_r(t)}{dt} \tag{8}$$

where V_r = volume of receiving phase; A_m = membrane surface.

Membrane instability due to leaching of the carrier into the aqueous phases has been studied in our laboratory. This phenomenon is strongly related to the lipophilicity of the carrier. Enhancement of the lipophilicity has been achieved by attachment of carriers to a polymeric backbone (14) or covalent linkage of aliphatic chains to carriers (15, 16). Recently, both the stability of the membrane phase and the solubility of the carrier have been improved by attachment of one or more solvent molecules to the carrier (17). Long-term membrane stability was demonstrated for a period of 5 weeks, while carrier concentrations up to 50 wt% could be used.

Diffusion-Limited Transport of Salts. In the case of fast extraction equilibria for salts at the interfaces, the rate of transport is determined by diffusion of the complex through the membrane and the flux for steady state transport through an SLM is given by Equation 9.

$$J = \frac{D_m}{d_m} ([LS]_{ms} - [LS]_{mr}) \tag{9}$$

where J = flux through the membrane; D_m = apparent membrane diffusion coefficient; d_m = membrane thickness; $[LS]_{mr}$ = internal complex concentration at the receiving side of the membrane; $[LS]_{ms}$ = internal complex concentration at the source side of the membrane.

Two limiting situations can be distinguished based on the presence of complex at the receiving side of the membrane. In the case that $[LS]_{mr}$ is zero, the situation of initial transport is studied and Equation 9 reduces to the so-called "initial flux equation" $J_0 = (D_m/d_m)[LS]_{ms}$. In practice, this assumption is valid when the measured flux remains constant with time. Using this assumption, many transport studies have been carried out in which the flux is related to the initial salt concentration in the aqueous source phase, D_m and K_{ex}. The initial flux model for the transport of ion-pairs is described in the next section. The initial flux model for assisted transport of free-ions is described in the section after that.

When the flux decreases with time, the concentration of complex $[LS]_{mr}$ is no longer zero. This situation arises when the salt in the aqueous receiving phase is significantly extracted. Hence, the concentration of complex at the receiving side of the membrane deviates from zero. This situation is described later for the diffusion limited transport of salts as free-ions.

Co-Transport of Monovalent Ions as Ion-Pairs. Ion-pair formation is a process which occurs when solvents are of low polarity (dielectric constant ε is low). Consequently, transport through non-polar solvents occurs via ion-pairs, whereas in polar solvents ions are solvent separated and transport takes place as free ions. Hence, the polarity of the solvent determines the extraction equilibrium.

In BLM transport, the most common solvents are chlorinated hydrocarbons. These have low dielectric constants and, as a consequence, the transport is described by ion-pairs. Izatt *et al.* have extensively studied ion-pair transport through BLMs (*18*). In SLM transport, cyclohexyl phenyl ether is an important non-polar membrane solvent (*19*). The extraction equilibrium at the interface in a non-polar membrane

$$M_a^+ + L_m + A_a^- \rightleftharpoons MLA_m \; ; \quad K_{ex} = \frac{[MLA]_m}{[M^+]_a [A^-]_a [L]_m} \tag{10}$$

solvent is written as Equation 10.
Reusch and Cussler were the first to describe crown ether-mediated transport of alkali metal cations through supported liquid membranes as ion pairs and derived for the initial flux Equation 11.

$$J = \frac{D_m K_{ex} L_0}{d_m} \left[\frac{[M^+]_s^2 - [M^+]_r^2}{(1 + K_{ex}[M^+]_s^2)\,(1 + K_{ex}[M^+]_r^2)} \right] \tag{11}$$

where L_0 = initial carrier concentration in the membrane; $[M^+]_s$ = initial salt activity in the source phase; $[M^+]_r$ = initial salt activity in the receiving phase.

Izatt *et al.* also presented transport studies in which ion-pair formation in the aqueous phase was taken into account (*20*). The extraction equilibrium was adjusted to Equation 12.

$$MA_a + L_m \rightleftharpoons MLA_m \; ; \quad K_{ex} = \frac{[MLA]_m}{[MA]_a [L]_m} \tag{12}$$

Equation 13 is the resulting initial flux equation.

$$J_0 = \frac{D_m K_{ex} L_0}{d_m} \left[\frac{[MA]_s}{(1 + K_{ex}[MA]_s^2)} \right] \tag{13}$$

In fact, this equation is similar to the flux equation for neutral molecules (see section on Diffusion-Limited Transport of Neutral Molecules).

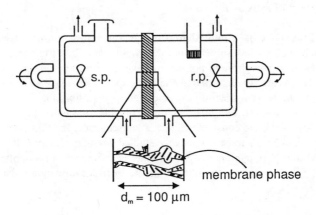

Figure 3. Representation of the Transport Cell and a Flat Sheet Supported Liquid Membrane.

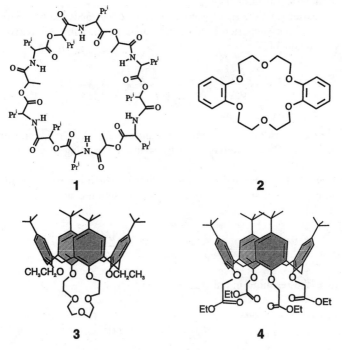

Chart 1. Structures of Carriers 1-4.

Co-transport of Monovalent Ions as Free Ions. Frequently used polar solvents in supported liquid membranes are o-nitrophenyl octyl ether (NPOE, $\varepsilon_r=24$), o-nitrodiphenyl ether (NDPE, $\varepsilon_r=30$) and 2-fluorodiphenyl ether (FNDPE, $\varepsilon_r=50$). When the complex concentration in the membrane phase is low, the complex and anion are solvent separated and Equilibrium 14 is obtained.

$$M_a^+ + L_m + A_a^- \rightleftharpoons ML_m^+ + A_m^- \; ; \qquad K_{ex} = \frac{[M^+]_a[A^-]_a[L]_m}{[ML^+]_m[A^+]_m} \qquad (14)$$

Besides the mass-balance, an additional assumption concerning the electroneutrality in the membrane phase is included ($[A^-]_m = [ML^+]_m$). The flux (Equation 9) is rewritten to obtain Equation 15.

$$J = \frac{D_m}{2d_m}[- A_s + \sqrt{A_s^2 + 4A_sL_0} + A_r - \sqrt{A_r^2 + 4A_rL_0}\,] \qquad (15)$$

$$A_s = K_{ex}[M^+]_s^2 \quad A_r = K_{ex}[M^+]_r^2$$

where D_m = apparent membrane diffusion coefficient; L_0= initial carrier concentration.

In case of initial transport ($\lim t \to 0$; $A_r=0$) an expression for the initial flux is obtained (Equation 16).

$$J_0 = \frac{D_{ML^+}}{2d_m}[- K_{ex}a_s^2 + \sqrt{K_{ex}a_s^2 + 4L_0K_{ex}a_s^2}\,] \qquad (16)$$

where a_s= activity of salt in the source phase.

The model has been verified experimentally for the transport of alkali metal cations (*21*) mediated by valinomycin (**1**), dibenzo-18-crown-6 (**2**) and calix[4]crown-5 derivative **3** and calix[4]arene tetraester derivative **4** (Chart 1). D_m and K_{ex} for the transport of guanidinium ion (*22, 23*) mediated by calix[6]arene derivatives have been obtained as well.

The transport parameters of carrier **1-4** are summarized in Table 1. Transport parameters D_m and K_{ex} (Table 1) are determined by describing the initial fluxes obtained from transport experiments as a function of the initial salt activity with the initial flux model (Equation 16) according to the best fit of the model to the experimental data.

The results presented in Table 1 indicate that diffusion coefficients show little variation ($\sim 1 \times 10^{-11}$ m^2 s^{-1}), while the extraction constants are strongly dependent on the carrier.

Table 1. Transport Parameters Based on Diffusion-Limited Free-Ion Transport Through Supported Liquid Membranes (Adapted from ref. 14)

Carrier	Salt	$10^{12} D_m$ $[m^2.s^{-1}]$	K_{ex} $[l.mol^{-1}]$
1	$KClO_4$	9.4	2300
2	$KClO_4$	19	1.3
3	$KClO_4$	12.5	2.6
4	$NaClO_4$	12.6	10.6

The characteristics of membrane transport are illustrated in Figure 4. The curve represents the calculated relation between the initial flux and the initial activity of the salt in the source phase. Based on the shape of the curve, two conclusions can be drawn. The steepness of the initial slope is a measure for the extraction coefficient, K_{ex}. The height of the plateau-value is a measure of the diffusion coefficient, D_m. When the flux reaches the plateau value, the carrier at the source phase interface is saturated. Hence, the diffusion coefficient is calculated from the corresponding flux (J_{max}) according to $J_{max} = (D_m/d) L_0$.

Time Dependency of the Flux. In the above discussions, only initial transport is examined. The flux equation is a simplified version, because conditions are chosen in such a way that the salt concentration in the receiving phase can be neglected. When time-dependency of transport is examined, the concentration of salt in the receiving phase can no longer be neglected and the influence on transport has to be taken into account.

Lehn *et al.* addressed the time dependency in BLM counter-transport. They performed numerical simulations to describe the competitive transport of two species as a function of time (*24*). Ibáñez *et al.* investigated time-dependent transport through a BLM system and used a model to obtain the membrane permeability. An analytical equation was derived which describes the concentration of salt in the aqueous receiving phase as a function of time (*25*). Recently, measurements of time-dependent, single-ion transport using carriers with a high K_{ex} ($K_{ex} > 1000$ l mol^{-1}) indicated that fluxes decrease even in an early stage of transport ($a_r < 10\%$ a_s). Using the law of conservation for the receiving phase (Equation 17), time dependency was introduced into the model equation for cation-assisted transport (Equation 15).

$$V_r \frac{da_r}{dt} = A_m \times J_M. \tag{17}$$

The flux equation J_m (Equation 15) is rewritten as a function of $a_r(t)$ by using the mass-balance for the total amount of salt: $a_s(t) = a_{s,0} - V_r/V_s a_r(t)$. Subsequent introduction in Equation 17 leads to a differential equation which can be integrated by numerical integration to obtain the salt concentration in the receiving phase as a function of time (*26*).

The influence of the transport characteristics D_m and K_{ex} has been investigated by simulation of the transport as a function time as a function of D_m and K_{ex}. From Equation 15, it is evident that the transport rate $J_m(t)$ linearly increases with the diffusion coefficient. The influence of K_{ex} on the transport rate is more difficult to predict. In Figure 5, the flux is depicted as a function of time. Figure 6 shows the resulting time-dependent activity in the receiving phase. The decrease of flux with time is explained by complex formation at the receiving side of the membrane. When a carrier has a high affinity for a salt (*i.e.*, K_{ex} is high), even at low salt activities in the receiving phase, complexation at the receiving phase interface takes place which lowers the transport rate, because the driving force is reduced.

We have used the time-dependent model to develop a system (Cursim) which determines the transport parameters by curve-fitting of a single-ion transport experiment $a_{r,exp}$ as a function of time (24 hrs). From an initial starting point (D^i_m, K^i_{ex}) the parameters are varied in such a way that the model describes the experiment, *i.e.* the deviation between the experimental data and the model is minimized. The deviation is expressed in terms of the least squares σ^2 value. Having obtained the best-fit parameters D^*_m and K^*_{ex}, the optimum is examined by a search for other "best fits" around the best-fit parameter set (D^*_m, K^*_{ex}). This is represented by a 3-dimensional plot of the transport parameters and the inverse least squares value ($1/\sigma^2$). In the case of a precise estimation, a single and sharp peak in D^*_m and K^*_{ex} can be observed.

To illustrate the validity of this approach with respect to the initial flux determination as described in the section on Co-transport of Monovalent Ions as Free Ions, both initial fluxes have been determined and time-dependent measurements have been carried out for $KClO_4$ transport with the 1,3 diethoxy calix[4]crown-5 (**5**) and the 1,3-bis isopropoxy calix[4]crown-5 (**6**) in the 1,3 alternate conformation (Chart 2). The results of both methods are presented in Table 2. The parameters obtained by the time-dependent approach and initial flux approach agree well. The 3-dimensional surface plots, which illustrate that the optimum is indeed global, are depicted in Figure 7.

Table 2. **The Transport Parameters D_m and K_{ex} Obtained from the Initial Flux Approach and Cursim in the Determination of the Transport Parameters for Carriers 5 and 6**

Carrier	$10^{11} D_m{}^a$ [m^2 s^{-1}]	$10^{-3} K_{ex}{}^a$ [l mol^{-1}]	$10^{11} D_m{}^b$ [m^2 s^{-1}]	$10^{-3} K_{ex}{}^b$ [l mol^{-1}]
5	1.1	28	1.0	41
6	1.2	40	1.4	36

[a]Determined by the initial flux approach.
[b]Determined by the Cursim system.

The main advantage of Cursim is that it only requires one experiment, whereas the initial flux approach requires at least six measurements at different salt concentrations.

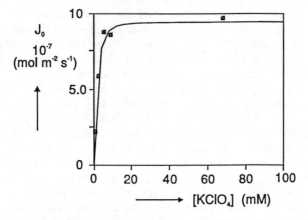

Figure 4. Transport of $KClO_4$ by Valinomycin [1] at 298 °K. Flux as a Function of the $KClO_4$ Activity. (Adapted from ref. 21).

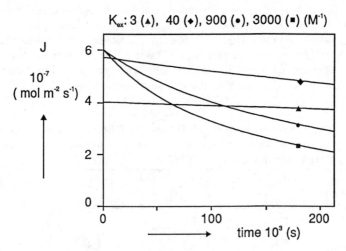

Figure 5. Simulation of the Flux Through an SLM as a Function of Time and Extraction Constant. (Adapted from ref. 25).

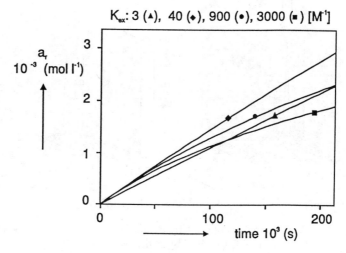

Figure 6. Simulation of the Activity in the Receiving Phase a_r as a Function of Time and Extraction constant. (Adapted from ref. 25).

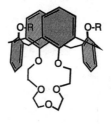

R
5 CH_2CH_3
6 $CH_2(CH_3)_2$

Chart 2. Structures of Calix[4]crown-5 Carriers 5 and 6.

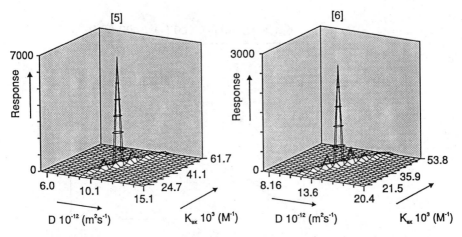

Figure 7. 3-Dimensional Representations of the Surface Area Around the Optimum Determined by Curve-Fitting of the Down-Hill Transport of $KClO_4$ ($a_{s,0}$=0.1 [M]) by Carriers **5** and **6**. (Adapted from ref. 25).

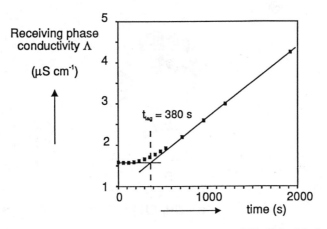

Figure 8. Lag-Time Experiment for the Transport of $NaClO_4$ Mediated by Carrier **3**. (Adapted rom ref. 9).

Independent Determination of Diffusion Coefficients. Independent methods for the determination of diffusion coefficients serve as a reference for the determination of transport parameters. Different methods can be applied to determine diffusion coefficients independently from transport experiments, *e.g.* determination of the lag-time (*27*), pulsed-field gradient NMR (*28*), and permeability measurements (*9*).

The diffusional process through an SLM is affected by the porosity and tortuosity of the polymeric support. Direct comparison of fluxes J and the corresponding diffusion coefficients D_m when using different supports is not possible and D_m has to be corrected for the membrane characteristics in order to obtain the bulk diffusion coefficient D_b (*29*).

$$D_b = D_m \frac{\theta}{\varepsilon} \qquad (18)$$

The bulk diffusion coefficient D_b is described by the Stokes-Einstein equation (Equation 19) and the Wilke-Chang relation (Equation 20) (*30*).

$$D_b = \frac{kT}{6\pi\eta r} \qquad (19)$$

$$D_b = 7.4 \cdot 10^{-8} [\frac{(M)^{0.5}T}{\eta V^{0.6}}] \qquad (20)$$

The bulk diffusion coefficient D_b obtained from membrane transport can differ from the Stokes-Einstein equation and Wilke-Chang relation. Both equations are valid under free-isotropic conditions, whereas the diffusion through the membrane can be reduced by an additional resistance of the pores (*31*). Significant reduction of the diffusion process takes place when pores are less then 10 times larger than the diffusing species.

A lag-time is defined as the time required for the complex to diffuse across the membrane from the source phase to the receiving phase, assuming dilute conditions.

$$t_{lag} = \frac{d_m^2}{6D_{lag}} \qquad (21)$$

Lag-times can be obtained from lag-time experiments of which an example is shown in Figure 8. The resulting diffusion coefficient D_{lag} has to be corrected for the tortuosity τ, to obtain the bulk diffusion coefficient ($D_b = D_{lag}\tau$).

Determination of the diffusion coefficient by permeability experiments has been investigated in our group recently (*9*). In these experiments, a liquid membrane is clammed between a source and receiving phase, which both contain a membrane solvent. At time t=0, a carrier which is substituted with a chromophoric group is added to the source phase ($c_s(0)$). The carrier diffuses through the membrane and the increase of concentration $c_r(t)$ in the receiving phase is monitored by UV/Vis spectroscopy as a function of time. The transport through the pores of the membrane is assumed to be rate-limiting and Equation 22 is derived.

$$\ln \frac{c_s(0) - 2c_r(t)}{c_s(0)} = -2 \frac{A_m D_m}{V_r d_m} t \tag{22}$$

The bulk diffusion coefficient is obtained by correction of D_m for the tortuosity and porosity of the support ($D_b = D_m \tau/\Theta$).

Pulse Field Gradient NMR spectroscopy can be applied to investigate self-diffusion of molecules in solution, through membranes, and through zeolites. PFG NMR is a direct method to measure the mean square distance $<r^2(t)>$ which is traveled by a tracer during a time period Δt. Under the conditions of free-isotropic diffusion in three dimensions (dilute solutions), the replacement is related to the self-diffusion coefficient D_{sd} by Equation 23.

$$<r^2(t)> = 6D_{sd}\Delta t \tag{23}$$

With PFG NMR the molecular displacements by self-diffusion can be measured to give the microscopic diffusion coefficient D_{sd} in the range of 10^{-6}-10^{-14} m^2s^{-1} with an optimum between 10^{-9}-10^{-13} m^2s^{-1}. Geringer and Sterk have determined the diffusion constant of 18-crown-6 and its Cs^+ complex in D_2O (32). The diffusion constant of ovalbumine inside porous particles, such as those used for gel filtration, has been investigated as a function of the protein concentration by Gibbs et al. (33). Our group has investigated the self-diffusion of NPOE inside the pores of various membranes (34). Through comparison of D_{sd} in the bulk liquid and in the pores of the membrane, it was found that the diffusion coefficient in Accurel was reduced by a factor of about 2.5, and in the case of Celgard by a factor of about 10. The radial and axial diffusion coefficients in the pores were measured by changing the orientation of the magnetic field compared to the pores of the membrane. For Celgard the radial self-diffusion coefficient was about 4 times lower than the axial self-diffusion coefficient, while for Accurel no significant difference was found. These results indicate that not only the porosity and tortuosity of the support have a large influence on the diffusion process, but the morphology of the support as well.

Reaction-Rate Limited Transport of Monovalent Ions. From ^1H NMR studies it is known that decomplexation rates can be very slow and, as a consequence, complexes can be kinetically stable (35,36). Until recently the role of slow rates of cation release in SLM transport was unclear. Lehn et al. (24) and Fyles (37) theoretically raised the question of the influence of slow rates of alkali metal cation release on transport through a BLM. Experimentally, this phenomenon has only been observed by Yoshida et al. (38). They showed that cation transport through a BLM mediated by polynactin was limited by the rate of cation release from the membrane. In 1994, Echegoyen stated that in SLM transport the rate of cation release from the membrane could never be rate limiting due to the large diffusional layer in the membrane (*39*). However, in cation transport experiments with different calix crown ether derivatives, we have proven recently that the rate of decomplexation can be indeed rate-determining in the transport through SLMs (*9, 40*).

Mechanism of Kinetically Limited Transport. The flux of $KClO_4$ for the two very similar calix[4]crown-5 derivatives **6** and **7** (Chart 3) was examined as a function of the initial $KClO_4$ concentration. As was discussed earlier, the diffusion coefficient D_m can be obtained from the maximum flux, assuming that transport is diffusion limited. Figure 9 shows that the maximum flux and resulting the diffusion coefficient of carrier **7** (3.7×10^{-12} m^2 s^{-1}) is almost three times lower than that for carrier **6** (11.6×10^{-12} m^2 s^{-1}). Because both carriers are of comparable size, one would expect the diffusion coefficients of both carriers to be almost equal. Also from transport experiments using other calix[4]arene based ligands (Table 1) an average diffusion coefficient of about 10^{-11} $m^2.s^{-1}$ was found.

From a lag-time experiment, the diffusion coefficient D_m of carrier **7** could be obtained independently (Equation 21). D_m is 9.4×10^{-12} m^2 s^{-1} which deviates from the diffusion coefficient D_m as obtained by applying the diffusion limited model ($D_m = 3.7 \times 10^{-12}$ m^2 s^{-1}). Apparently, the diffusion limited model can not be applied to determine the transport parameters of carrier **7**. Hence, kinetics must influence the transport mechanism that lowers the transport rate as an additional resistance.

Kinetically limited transport; determination of α. In this section the influence of decomplexation kinetics in membrane transport will be discussed. An overview will be given of the different methods to distinguish diffusion-limited transport from kinetically limited transport. To illustrate these techniques, they will be applied to a diffusion limited (carrier **7**) and a kinetically limited (carrier **6**) transport system. A general model, that also includes the rate of cation release, was reported by Reichwein-Buitenhuis *et al.* (*9*). The flux is described by Equation 24.

$$J = \frac{D_m}{2d_m} \{ \frac{-A + \sqrt{A^2 + 4AL_0 \frac{(1+2\alpha)}{(1+\alpha)}}}{1+2\alpha} \} \tag{24}$$

where α = dimensionless number $D_m/kd_m\theta$, k rate of release at the membrane-water interface, θ = porosity of the membrane, $A = K_{ex}a_s^2$.

The basic transport parameters in the model given above are D_m, K_{ex}, and α. The parameter α is a dimensionless number which relates diffusion limited transport to kinetically limited transport. When the transport is purely limited by diffusion ($\alpha \to 0$), D_m and K_{ex} can be obtained from measurements of the flux as a function of a_s (*21*). In principle, all parameters can be derived by measuring the flux

$$J_{max} = (\frac{D_m}{d_m}L_0)(\frac{1}{1+\alpha}) \tag{25}$$

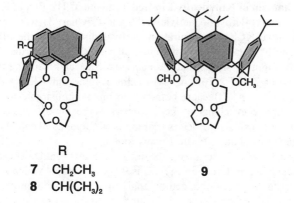

R

7 CH$_2$CH$_3$ **9**
8 CH(CH$_3$)$_2$

Chart 3. Structures of Calix[4]crown-5 Carriers **7-9**.

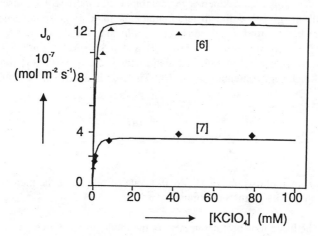

Figure 9. Transport of KClO$_4$ by di-isopropoxycalix[4]crown-5 in the 1,3-Alternate (Carrier **6**) and Partial Cone (Carrier **7**) Conformations with Variation of the Initial Salt Activity $a_{s,0}$.

while varying a_s and d_m. Equation 24 can be simplified in the case when the carrier at the source phase interface of the membrane is fully loaded by cations (large value for A). This situation is achieved when the flux reaches its maximum value (J_{max}). The first term in Equation 25 describes the diffusion-limited flux, while the second term $1/(1+\alpha)$ is a correction factor for slow kinetics of release. Because α is defined as the ratio $D_m/kd_m\theta$, the values of D_m and k have to be determined independently. We have developed two methods to determine the transport regime, *viz.* *(i)* independent determination of diffusion constants from lag-time measurements and *(ii)* direct determination of k and D_m from flux measurements as a function of the membrane thickness. As a result α is obtained. If α is lower than 1, the transport is mainly limited by diffusion of the complex; while the transport is primarily controlled by the cation release in the case that α is larger than 1.

From lag-time experiments, the diffusion coefficient D_{lag} obtained as shown in the section on Independent Determination of Diffusion Coefficients. D_{lag} is related to the membrane diffusion coefficient after correction for the porosity of the support ($D_m = D_{lag}\ \theta$). By combining this relationship with Equation 25, α is obtained from D_{lag} and J_{max} (Equation 26).

$$\alpha = \frac{D_{lag}\theta L_0}{d_m J_{max}} - 1 \tag{26}$$

Separation of the diffusional and kinetic terms is achieved by expressing the flux as the ratio of driving force and flux (Equation 27).

$$\frac{L_0}{J_{max}} = \frac{d_m}{D_m} + \frac{1}{\theta k} \tag{27}$$

Consequently, a plot of L_0/J_{max} vs. d_m should give a straight line with a slope $1/D_m$ and an intercept of $1/\theta k$. The value of α can be determined directly by calculation of the ratio of the kinetic and diffusional resistances via $\alpha = (1/\theta k)/(d_m/D_m)$. The values for the rate of cation release (k) and the diffusion coefficient D_m are determined by measuring the flux at different membrane thicknesses and consequently α is obtained by application of Equation 31.

The results shown in Table 3 and 4 both indicate that the transport of $KClO_4$ by carrier **7** is mainly limited by a slow decomplexation rate ($\alpha>1$), while carrier **6** is mainly diffusion limited ($\alpha < 1$).

Table 3.   Calculated α Values from Lag Times and Maximum Fluxes for Carrier 6 and 7 at 298 °K

Carrier	t_{lag} [s]	$10^{11}\ D_{lag}$ [$m^2\ s^{-1}$]	$10^{11}\ D_m$ [$m^2\ s^{-1}$]	$10^7\ J_{max}$ [mol $m^{-2}\ s^{-1}$]	α
6	260	2.6	1.6	10.6	0.56
7	452	1.5	9.5	3.2	2.0

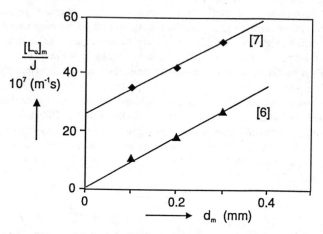

Figure 10. Influence of the Membrane Thickness (n*100 μm) on L_0/J for Carrier 6 and 7. $[Carrier]_m = 0.01$ [M], $[KClO_4] = 0.1$ [M] , T = 298 ^{o}K.

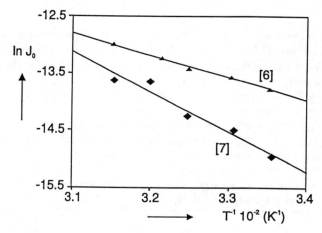

Figure 11. Eyring plots for $KClO_4$ Transport by Carrier 6 and 7 Across a NPOE-Accurel Membrane.

**Table 4. D_m, K and α-Values Calculated from L_0/J vs. d_m
for Carriers 6 and 7 at T=298 °K (Figure 10)**

Carrier	$10^{11}\ D_m$ $[m^2\ s^{-1}]$	$10^7\ k$ $[m.s^{-1}]$	α
6	1.1	61	<0.1
7	0.9	0.8	1.8

Activation energy, E_a, for transport. Apart from the calculation of α as discussed above, determination of the apparent activation energy of transport gives information about the rate-limiting step in the transport process. Lazarova *et. al.* varied the temperature in order to determine the activation energy E_a from an Eyring plot (*41*). They indicated that E_a values below ca. 20 kJ mol^{-1} are generally accepted as indicative of pure diffusion limited transport. Generally, activation energies are above 40 kJ mol^{-1} when chemical reactions do play a role in the transport (*42*).

To obtain a clearer indication of the activation energy for diffusion-limited transport, the activation energy for the self-diffusion coefficient of NPOE can be measured. The activation energy for self-diffusion of a solvent often correlates well with the activation energy for diffusion of a solute species, since on a molecular level diffusion of a solute can be considered as a process in which either a solute or solvent molecule jumps from solvent cavity to cavity. Since the activation energy for self-diffusion varies with the solvent used, it is important to determine the activation energy E_a for the self-diffusion of NPOE first. The temperature dependency of the viscosity of organic solvents η has an Arrhenius-type behaviour.

$$\eta(T) = \eta_0 \exp(E_a/RT) \qquad (28)$$

An activation energy for self-diffusion or viscous flow may be calculated from Equation 28 by measuring the kinematic viscosity as a function of temperature. The activation energy for the self-diffusion through NPOE is 24 kJ mol^{-1}.

The influence of the temperature on the transport rate is related to the Arrhenius equation: $J = J_0\exp(-E_a/RT)$. In order to obtain activation energies for the transport of KClO$_4$ by carriers **6** and **7**, the maximum flux J_{max} as a function of operating temperature was measured. Figure 11 shows an Eyring plot of ln J_{max} as a function of 1/T. Values of the apparent activation energies are obtained from the slopes of the curves: $E_a= 32 \pm 2$ kJ mol^{-1} and $E_a= 59 \pm 7$ kJ mol^{-1}. Consequently, it can be concluded that the transport of KClO$_4$ by carrier **7** is diffusion limited and that of carrier **6** is determined by the slow kinetics of release.

Optimization of the Flux; External Influences on Transport. Apart from the properties of the carrier transporting a cation selectively through the membrane phase, external factors can influence the transport. Properties of the membrane phase that influence transport are solvent polarity and viscosity. The operating temperature has also a large influence on the transport, as well as the anion that accompanies the complex through the membrane. In this section, the external factors which influence transport are summarized.

Parameters that Affect Diffusion-Limited Transport. Brown *et al.* varied the length of the alkyl chain of alkyl *o*-nitrophenyl ether membrane solvents and examined its influence on the proton-coupled transport of alkali metal cations by a crown ether derivative. By comparison of solvent characteristics, such as the dielectric constant, viscosity, and surface tension, they concluded that hexyl *o*-nitrophenyl ether is a better membrane solvent than NPOE (*43*). The effect of the solvent on the transport of NaClO$_4$ by carrier **4** has been studied by Visser *et al.* (*44*). Transport parameters D_m and K_{ex} were determined. A series of octyl phenyl ethers containing an electron withdrawing group (NO$_2$, CN) of various positions on the phenyl ring were used. Data are presented in Table 5.

Table 5. Influence of the Membrane Solvent on the Transport of NaClO$_4$ Mediated by Carrier 4 at T = 298 K (Adapted from ref 43)

Membrane solvent[a]	ε_r	η [mPa s]	$10^8 J$[b] [mol m^{-2} s^{-1}]	$10^{12} D_m$[c] [m^2 s^{-1}]	K_{ex}[c] [l mol^{-1}]
o-NPOE	23.1	11.5	121	12.6	10.6
m-NPOE	16.3	13.6	88.6	9.47	0.35
p-NPOE	16.7	19.0	64.2	6.52	1.68
o-CPOE	23.4	14.9	99.6	9.94	8.38
p-CPOE	14.6	15.8	80.3	8.47	0.46
o-NDPE	28.1	18.3	72.5	7.50	40.0
o-CDPE	29.1	22.9	63.1	6.61	24.4
FNDPE	50	25.9	55.8	5.64	220

[a] NPOE = nitrophenyl octyl ether; CPOE = cyanophenyl octyl ether; NDPE = nitrodiphenyl ether; CDPE = cyanodiphenyl ether; FNDPE = *o*-fluoro *o*'-nitrodiphenyl ether.
[b] Flux measured at [NaClO$_4$]$_{s,0}$= 1.0 M.
[c] Transport parameters obtained by curve-fitting of the initial flux model.

Solvent characteristics that influence the diffusion and extraction are found to be viscosity (η) and polarity (ε). For spherical solutes, the diffusion coefficient depends on the solvent according to the Stokes-Einstein relation (Equation 19). From this, it follows that the diffusion coefficient linearly increases with T/η. Hence, the permeability increases linearly with the reciprocal viscosity of the membrane solvent (*12*). Dozol *et al.* plotted the permeability P vs. $\eta/MW^{0.5}$ and obtained a linear fit

(*45*). Visser *et al.* related the diffusion coefficient D_m to the solvent viscosity η (Figure 12). The solvent effect on K_{ex} is a combination of the influence on salt partition (K_p) and association (K_a). Both processes are influenced by the polarity of the solvent. An empirical relation, the Kirkwood function, describes the relation between polarity (dielectric constant ε_r) and extraction coefficient well. From Figure 13 it follows that a more polar membrane solvent promotes extraction. In general, polar solvents favor salt partition, but the tendency towards complexation diminishes. Since the overall effect of solvent polarity on the extraction is positive, the polarity appears to affect the partition coefficient to a higher degree.

At higher temperatures, both the diffusion and decomplexation processes are accelerated, and transport rates increase. The transport parameters D_m and K_{ex} were determined at elevated temperatures. The diffusion coefficient D_m increases, while the extraction coefficient K_{ex} decreases with increasing temperature. D_m correlates well with T/η as described by the Stokes-Einstein equation and Wilke-Chang relation.

The anion effect on BLM transport of metal salts was examined by Lamb *et al.* (*46*). From determination of the transport rates for different alkali metal salts by dibenzo-18-crown-6, it became clear that the rate of transport was strongly dependent on the co-transported anion. The relative fluxes of different potassium salts were related to the hydratation energy. Cations which are accompanied by a hydrophilic anion ($-\Delta G_{g-w}$ is large) show low fluxes, due to a decrease in the partitioning. A linear relation was obtained by plotting ln J vs $-\Delta G_{g-w}$ of the anion. For transport through an SLM, a similar relationship was found (*47*).

Parameters that Affect Kinetically Limited Transport. Both the diffusional process and cation release are influenced by the membrane solvent, support and anion in kinetically limited transport (*40, 44*). Using the membrane solvents listed in Table 6, the diffusion coefficient D_m and the decomplexation rate constant k were determined for the transport of $KClO_4$ through NPOE/Accurel by 1,3-dimethoxycalix[4]crown-5 (carrier **9**). The value for α is 1.9 and the activation energy E_a is 61 kJ mol^{-1}. The transport is therefore considered as a kinetically limited process (*9*).

From comparison of the transport parameters in *o*-NPOE and *p*-NPOE, it follows that the diffusion coefficient in *p*-NPOE is twice as high as in *o*-NPOE. Since the viscosity of *p*-NPOE is almost twice as high as that of *o*-NPOE, this is in accordance with the Stokes-Einstein equation.

From the α-values, it is clear that the transport regime strongly depends on the solvent. The transport varies from purely diffusion limited to mainly kinetically limited. Both transport parameters D_m and k have been correlated to the viscosity η and polarity ε_r (Kirkwood function) of the solvent, respectively.

The diffusion coefficient correlates linearly with the viscosity η as was already shown in the case of a diffusion limited transport system (Figure 14). When using the empirical Kirkwood function, a linear relationship between k and ε_r is obtained. Figure 15 shows that the use of a more polar solvent slows down the rate of decomplexation. Hence, the regime of the transport can be influenced by the solvent polarity.

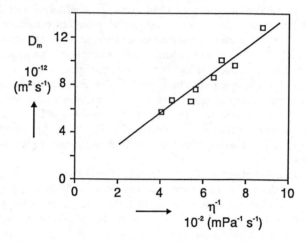

Figure 12. Relationship Between the Diffusion Coefficient and Solvent Viscosity η for the Transport of $NaClO_4$ by Carrier **3** (Adapted from ref. 43).

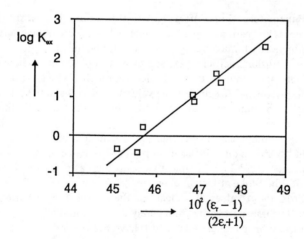

Figure 13. Relationship Between the Extraction Constant and Solvent Dielectric Constant ε_r for the Transport of $NaClO_4$ by Carrier **3** (Adapted from ref 43).

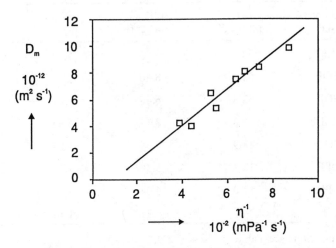

Figure 14. Relationship Between the Diffusion Constant and Solvent Viscosity for Transport of $KClO_4$ by Carrier **9** (Adapted from ref. 43).

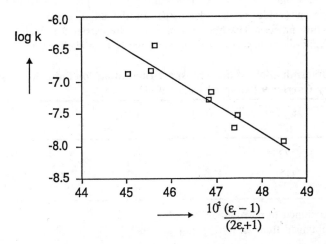

Figure 15. Relationship Between the Decomplexation Rate Constant and the Solvent Polarity for the Transport of $KClO_4$ by Carrier **9** (Adapted from ref. 43).

Table 6.   Diffusion and Decomplexation Rate Constants for the Transport of
            KClO$_4$ by Carrier 9 as Influenced by Variation of the Membrane
            Solvent (Adapted from ref 43)

Membrane solvent[a]	$10^8 \, J_{max}$ [mol m^{-2}s^{-1}]	$10^{12} \, D_m{}^b$ [m^2 s^{-1}]	$10^8 \, k^b$ [m s^{-1}]	α
o-NPOE	34.1	9.85	5.24	1.9
m-NPOE	52.7	8.40	13.8	0.6
p-NPOE	52.8	6.48	33.1	0.2
o-CPOE	35.3	8.12	6.44	1.3
p-CPOE	48.9	7.54	12.8	0.6
o-NDPE	13.4	5.38	1.82	3.0
o-CDPE	17.0	4.04	2.94	1.4
FNDPE	9.1	4.31	1.14	3.8

[a] For abbreviations see Table 5.
[b] Parameters obtained from Equation 27.

      Although it is assumed that transport takes place via solvent separated
ions, it is seen (Table 7) that the nature of the anion affects the rate of the
decomplexation process. The transport of KSCN by carrier 9 is mainly limited by
diffusion, while the transport of KClO$_4$ is primarily kinetically controlled.

Table 7. Determination of the Transport Limitation for Potassium Transport
          by Carrier 9 (Adapted from ref. 43)

Salt	t_{lag} [s]	$10^{11} \, D_{lag}$ [m^2 s^{-1}]	$10^7 \, J_{max}$ [mol m^{-2} s^{-1}]	α^a	$E_a{}^b$ [kJ mol^{-1}]
KClO$_4$	395	1.7	3.3	2.3	63
KSCN	415	1.6	6.3	0.6	41

[a] α-values obtained from Equation 26.
[b] E_a-value obtained from an Eyring plot.

No clear explanation was given for the influence of the co-transported counter ion on
the kinetics. It can be due to ion-pair formation, influence of size and shape of the
anion, or independent partition of the anion.
      The effect of the support on the transport kinetics has been examined by
normalizing the decomplexation rate constant for the porosity of the membrane: k_n =
$k*\theta$ (40, 44). Consequently, α can be calculated from the rate of decomplexation k_n
and the diffusion coefficient D_m and compared with other supports. From the

calculation of α for the transport of $KClO_4$ by carrier **9** through a NPOE/Celgard-2500 ($\alpha = 4.9$) and a NPOE/Accurel membrane ($\alpha = 1.9$), it follows that the support may influence the transport regime (*40, 44*).

Transport Selectivity. For separation purposes, the selectivity of transport by macrocyclic carriers through supported liquid membranes has been investigated extensively. The transport selectivity of a carrier towards two different cations M_1^+ and M_2^+ is obtained by competition experiments in which both M_1^+ and M_2^+ were present in the receiving phase.

The selectivity not only depends on the relative fluxes, but is related to the relative concentration in the aqueous source phase as well. Therefore, the selectivity is often expressed as ratio of permeabilities $S = P_{M1}/P_{M2}$ (*48*).

In 1973, Cussler related the "selectivity" of dibenzo-18-crown-6 for potassium over other alkali and alkaline earth metal cations to the correlation between the radius of the unsolvated cation and the radius of the internal cavity of the crown ether (*8*). He obtained the highest flux for potassium, which has the best fit with the cavity of the carrier. However, his observations were based on the results of single-ion transport experiments.

In 1975, Lehn *et. al.* reported the selective transport of alkali metal ions by cryptands through BLMs by competitive experiments (*49*). The cation concentration inside the membrane did increase when the association constant was larger. However, the resulting flux in competitive experiments did not correlate with the association constant and reversed selectivities were found. Strong binding of the substrate has a negative effect on the transport selectivity. Therefore, they concluded that an optimal association constant for transport lies in the range of log K_a(methanol) ≈ 5. Behr *et al.* reported numerical simulations to investigate selective transport theoretically (*24*).

Lamb *et al.* carried out competitive experiments of monovalent and divalent cations mediated by various crown ethers and cryptands through chloroform BLMs. Relative fluxes from binary mixtures were explained in terms of ring substituents and ring size of the crown ether carriers (*50,51*). Saito studied the transport of alkali and alkaline earth metal cations as their ClO_4^- salts through SLMs by tripentyl phosphate and observed a selective transport of Li^+ in the presence of Na^+, K^+, and Mg^+ (*52*).

When the transport is limited by diffusion, the initial transport selectivity is predicted from the results of single-ion transport experiments (*21*). The flux ratio can be expressed by Equation 29.

$$\frac{J_{M_1^+}}{J_{M_2^+}} = \frac{D_{m,LM_1^+}}{D_{m,LM_2^+}} \cdot \frac{K_{ex,M_1^+}}{K_{ex,M_2^+}} \cdot \frac{a_{M_1^+}}{a_{M_2^+}} \tag{29}$$

Assuming that the sizes of the complexes are nearly the same, the ratio of the diffusion coefficients is 1. The transport ratio and hence the selectivity is dependent on the ratio of the extraction constants. A high extraction selectivity ($K_{ex,M1}/K_{ex,M2}$) does not always imply a high transport selectivity. With high extraction constants, the maximum flux is reached at low salt concentrations and the carriers at the source phase side are nearly 100% complexed.

Since transport rates change with time due to the presence of salt in the receiving phase, selectivity may also change in time. In order to investigate the time dependency of selectivity, transport of potassium in the presence of sodium has been investigated using different calix[4]crown-5 carriers and valinomycin. The amount of Na^+ and K^+ in the receiving phase was measured by atomic absorption spectroscopy after 2 and 24 hours of transport. The results are presented in Table 8.

Table 8. **Selective Transport of Potassium in the Presence of Sodium. Determination of the Selectivity after 2 and 24 hours of transport[a]**

Carrier	$t_{tsp.}$ [hours]	% $NaClO_4$ transported	% $KClO_4$ transported	Selectivity
1	2	8.9×10^{-3}	91.5	10300
1	24	0.293	99	338
5	2	4.72×10^{-3}	94.8	20100
5	24	0.232	98.8	426
6	2	4.2×10^{-3}	93.8	22300
6	24	0.252	95.2	378

[a] Source phase: 10^{-4} M $KClO_4/10^{-1}$ M $NaClO_4$, 298 °K.

From the 2-hour transport data, it is clear that both carriers 5 and 6 are extremely selective for potassium in the presence of sodium. However, the selectivity drops dramatically with time. This is rationalized by the fact that after 2 hours virtually all of the $KClO_4$ has been removed from the source phase and only $NaClO_4$ remains to be transported. Additional proof for this explanation was obtained from the conductivity in the receiving phase as a function of time. After approximately 2 hours a sharp levelling of in the conductivity increase was observed, indicating that $NaClO_4$ was being transported instead of $KClO_4$ (Figure 16).

When kinetics play a role, transport selectivity can not simply be determined by the ratio of the extraction constants. The operating temperature and membrane thickness influence the transport selectivity. Selectivity experiments ($S_{K/Na}$) have been reported for 1,3-diethoxycalix[4]crown-5 in the partial cone conformation (carrier 7). It is a potassium-selective carrier for which the $KClO_4$ transport is kinetically limited ($\alpha_K = 3.6$) (9). With a membrane thickness of 100 μm the potassium selectivity over sodium is 160. The selectivity $S_{K/Na}$ increases when the temperature (at T = 43°C, $S_{K/Na} = 420$) or membrane thickness (with $d_m = 300$ μm, $S_{K/Na} = 300$) is increased. Under these conditions, transport of K^+ is determined more by the diffusional resistance in the membrane and to a smaller extent by the complexation kinetics. Apparently, the selectivity drops when a kinetic resistance influences the rate of transport.

Uphill Transport. Cussler *et al.* have shown that common-ion pumping can be used to transport a salt against its concentration gradient (*53*). When a secondary salt with the same anion and a low extraction constant is present in large excess, the anion gradient acts as a "pump" for the primary salt. The ideal carrier for uphill transport of M_1^+ by common-ion pumping has a high affinity for M_1^+, but shows *no* significant transport for M_2^+.

Nijenhuis *et al*. have derived a model for competitive transport of two cations and a common ion in the case of 1:1 complexation and free ions in the membrane phase (*21*). When the extraction constant for the secondary salt is negligible, the flux for the primary cation is given by Equation 30.

$$J_{M_1} = \frac{D_m(a_{M_1} + a_{M_2})K_{ex,1}a_{M_1}}{d_m} \left(-1 + \sqrt{1 + \frac{4L_0}{a_{M_1}K_{ex,1}(a_{M_1} + a_{M_2})}}\right) \tag{30}$$

Using the common-ion principle, uphill transport of $KClO_4$ by valinomycin was achieved by adding an excess of $LiClO_4$ or HNO_3 to the aqueous source phase. The common ion effect as described above can be useful for the removal of small amounts of heavy metal ions or radioactive salts from waste water streams.

Uphill transport of silver nitrate was reported by Sirlin *et al*. (*54*). Plutonium(IV) nitrate was transported completely from a HNO_3 aqueous solution through an SLM containing dicyclohexyl-18-crown-6 in toluene. At the receiving phase side, a stripping agent was added (0.5 M Na_2CO_3). Plutonium selectivity was observed in the presence of Cs^{137}, Ru^{106}, and Sb^{125} (*55*). Cesium was separated completely (≥99.8%) (Figure 17) from a radioactive waste stream within 25 hours (*56*) using different calix[4]crown-6 carriers (Chart 4). An overall transport selectivity for Cs^+ over Na^+ of about 400 was obtained.

In nature uphill transport is often achieved by proton gradients. In liquid membrane transport studies, counter transport of protons is generated by applying a pH gradient and by using proton-ionizable carriers (azacrown ethers) in the membrane phase.

Carrier-Assisted Transport of Anions. Assisted transport of anions has been investigated less thoroughly than cation transport. The measured transport rates are lower and selectivity mostly follows the Hoffmeister series (*57*): $ClO_4^- > I^- > SCN^- > NO_3^- > Br^- > Cl^- >> CO_2^-, H_2PO_4^-, SO_4^{2-}$.

Assisted anion transport through a BLM was first reported by Behr and Lehn, who investigated the transport of amino acids at high pH with a quaternary ammonium salt as carrier (*58*). Selectivity was based only on the difference in partition coefficients of the amino acids. Since then, many charged carriers have been utilized for the transport of anions via a counter-transport mechanism.

Cussler *et al*. showed that alkyl ammonium cations are suitable for the transport of Cl^- when an opposing OH^- gradient was applied (*59*). Neplenbroek *et al*. transported NO_3^- with alkyl ammonium cations by an opposing Cl^- gradient (*60*). Phosphates were transported through a BLM containing an tetraphenylporphyrin-oxomolybdenum(V) complex, when coupled to a counter flow of halide ions (*61*). Selective counter transport of CN^- was achieved by Sumi *et al*. They reported the

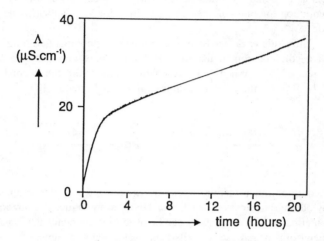

Figure 16. Time-Dependent Na$^+$/K$^+$-Transport by Carrier **5**; Conductivity in the Receiving Phase as a Function of Time. Source phase: 10^{-4} M KClO$_4$/10^{-1} M NaClO$_4$, 298 $^\circ$K.

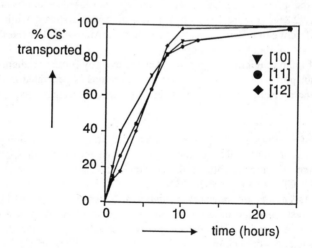

Figure 17. Transport Rates of CsNO$_3$ as a Function of Time for Carriers **10**, **11**, **12** [L$_0$] = 5 mM. The source phase was a mixture of 1 mM CsNO$_3$, 4 M NaNO$_3$, and 1 M HNO$_3$ (Adapted from ref. 55).

counter-transport of CN^- against OH^- upon the addition of a tetraphenylporphyrinato-manganese(III) complex through a BLM and SLM. Selective transport was achieved in the presence of various anions, Cl^-, F^-, NO_3^- (*62*).

Co-transport by charged carriers can be achieved by introducing a pH-gradient into the system. The carrier is protonated at the source phase side of the membrane, which is then charged and therefore capable of complexing an anion. At the receiving phase side deprotonation and decomplexation takes place to regain a neutral carrier. Using this concept, Sessler *et al.* enhanced the transport of nucleotide monophosphates (AMP and GMP) by a sapphyrin carrier through a BLM (*63*). Araki *et al.* obtained selective SCN^- transport through a BLM containing a 6,6'-bis(4-hexylbenzoylamino)-2,2'-bipyridine complex of Cu(II). The metal center serves as a binding site for SCN^-. Via reverse dissociation/association of the amide proton(s), a neutral or charged carrier can be obtained. Hence, the uptake and release of SCN^- can be regulated (*64*).

Selective co-transport by neutral carriers was reported recently by our group (*65*). The transport of phosphates through an SLM was achieved by using uranyl salophene carrier **13** which is functionalized with two additional amide groups (Chart 5). The transport rate of KH_2PO_4 was measured. In the membrane phase the anion is complexed, whereas the cation diffuses through the membrane as a free ion. By the addition of the K^+-selective carrier **9**, both the complexed anion and cation are transported through the membrane as complexes and the transport is further enhanced. By measuring the transport rate while varying the ratio of cation and anion receptor from 100% anion receptor to 100% cation receptor, a bell-shaped curve was obtained with a maximum at a 1:1 ratio. Selectivity ratios of phosphate over chloride of 140 were achieved.

Covalent binding of an anion and a cation receptor to obtain a ditopic receptor and subsequent application in an SLM has been reported by Rudkevich *et al.* (*66*). They synthesized a receptor (carrier **14**) which is capable of binding a cation and an anion simultaneously. The crown ether part binds Cs^+, while the salophene moiety can complex Cl^-. For carrier **14** $CsNO_3$-fluxes ($J = 0.89$ x 10^{-7} mol m^{-2} s^{-1}) and CsCl-fluxes ($J = 1.2$ x 10^{-7} mol m^{-2} s^{-1}) were measured. In the case of cation assisted transport, $CsNO_3$ is expected to give a higher flux than CsCl, due to the higher lipophilicity of the NO_3^- anion. This proves that both the anion and the cation are involved in the complexation of CsCl, and that selective transport of a hydrophilic salt over a lipophilic salt can be obtained with a ditopic receptor.

Diffusion-Limited Transport of Neutral Molecules. Contrary to the transport of charged molecules, only a few receptors that assist the transport of neutral compounds have been reported. Pirkle *et al.* reported the enantioselective transport of different *N*-(3,5-dinitrobenzoyl)amino acid derivatives through supported liquid membranes containing (*S*)-*N*-(1-naphthyl)leucine octadecyl ester carriers in dodecane.   T h e transport of neutral amines assisted by a 4-octadecyloxybenzaldehyde carrier was investigated by Yoshikawa *et al.* (*67,68*). The transport process is based on the reversible formation of a Schiff base between the amino group and the aldehyde. The assisted transport rate of *D*-glucosamide is about two times the blank transport and the transport is described using Michaelis-Menten kinetics, which in principle describes catalysis in product formation when using an enzyme able to complex the substrate involved in the reaction.

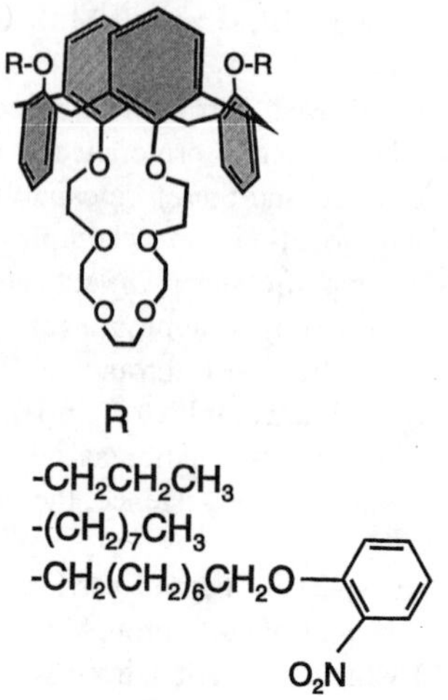

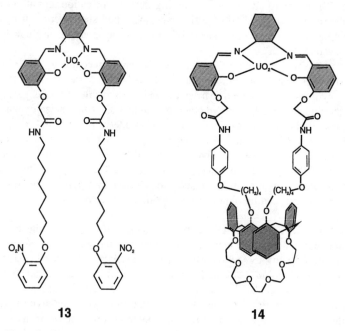

R

10   -CH₂CH₂CH₃
11   -(CH₂)₇CH₃
12   -CH₂(CH₂)₆CH₂O—

Chart 4. Structures of Calix[4]crown-6 Carriers **10-12**.

13   **14**

Chart 5. Structures of Carriers **13** and **14** for Transport of Anions.

Nijenhuis *et al.* were the first to report assisted transport of neutral molecules by macrocyclic receptors (*69*). They investigated the transport of urea by neutral metallosalophenes through an SLM (Chart 6). A model for diffusion limited transport of neutral molecules has successfully been applied to describe the transport of urea (*70*). For neutral molecules, the extraction equilibrium (1:1 host-guest complex) at the interface is expressed by Equation 31.

$$U_a + L_m \rightleftharpoons UL_m \; ; \qquad K_{ex} = \frac{[UL]_m}{[U]_w[L]_m} \tag{31}$$

The extraction coefficient K_{ex} is the product of the partition coefficient and the association coefficient $K_{ex} = K_p K_a$. Partitioning of the carrier to the aqueous phases was neglected, because of the high lipophilicity of the carrier. The total carrier concentration is written as Equation 32.

$$[L]_{m,0} = [L]_m + [LU]_m \tag{32}$$

Assuming that all equilibria are fast, the rate-limiting step of transport is the diffusion of the complex through the membrane. The flux is described by Fick's first law. When examining initial transport, the complex concentration at the receiving phase side of

$$J_{ass,0} = \frac{D_m}{d_m} [LU]_{m,0} \tag{33}$$

the membrane is again neglected.
Equation 33 can be rearranged using the mass-balance (Equation 32) and the extraction coefficient K_{ex} (Equation 31) to obtain Equation 34.

$$J_{ass,0} = \frac{D_m}{d_m} \frac{K_{ex}[U]_{w,0}[L]_{m,0}}{1 + K_{ex}[U]_{w,0}} \tag{34}$$

The blank flux of urea can not be neglected ($J_{blank} = 4.4 \times 10^{-8}$ mol m^{-2} s^{-1}, and the assisted flux was obtained by substraction of the blank flux from the measured flux. The model has been applied for the transport of urea using two types of carriers, *e.g.* lipophilic metallomacrocycles **15** and **16** (*69*), and polyaza (cleft-type) receptors **17**, **18**, and **19** (*71*). The transport data are presented in Table 9. Transport experiments were carried out by varying the urea concentration to obtain D_m and K_{ex}. In Figure 18, the assisted flux is plotted vs. the initial urea concentration for carrier **15**. Diffusion constants and extraction constants for macrocycles **15** and **16** are also given in Table 9.

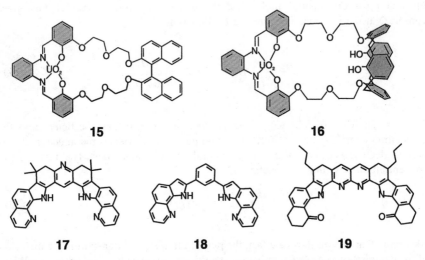

Chart 6. Structures of Carriers **15-19** for Transport of Neutral Molecules.

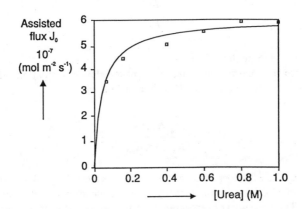

Figure 18. Influence of the Urea Concentration in the Source Phase on the Assisted (Initial Minus Blank) Flux Through a Supported Liquid Membrane Using Carrier **15** ($[CA]_{m,0}$ = 6 mM). The line is drawn according to the model and the symbols are measure values (Adapted from ref. 69).

Table 9. Total Urea Fluxes Carrier-Assisted Urea Transport Through Supported Liquid Membranes, D_m and K_{ex} (Adapted from ref. 69, 70)

Carrier[a]	$J_{initial}$[b] [mol m^{-2}s^{-1}]			$10^{12}D_m$ [m^2 s^{-1}]	K_{ex} [l mol^{-1}]
15	61	64	64	9.7	20
16	50	50	50	6.1	22
17	25	23	24	-	-
18	36	36	36	-	-
19	16	14	14	-	-

[a] $[U]_{w,0}$ = 1.0 M; $[CA]_{m,0}$ = 6 mM.
[b] Initial fluxes after 0, 24, and 48 hrs[c].
[c] Second replacement of the water phase after 48 h.

The flux strongly increases as the urea concentration is enhanced and reaches a plateau value (J_{max}) at higher urea concentrations. The diffusion coefficient and the extraction constant were obtained by curve-fitting of the flux as a function of the urea concentration. With values for D_m and K_{ex}, the percentage of carrier complexed at the source interface can be calculated. For carrier **15**, 67% is complexed using a 0.1 M urea source phase solution and 92 % with a 0.5 M urea solution. This clearly indicates that the maximum flux can be explained by complete loading of the carrier.

Conclusions

Several models have been developed for the transport of different types of species through supported liquid membranes. The transport is described by subsequent partitioning, complexation and diffusion. Mechanistic studies are mainly focussed on diffusion limited transport of cations in which diffusion of the complex through the membrane phase is the rate-limiting step of the transport. Recently, kinetic aspects in membrane transport have been elucidated for new carriers with which the rate of decomplexation determines the rate of transport.

New topics in the field are the assisted transport of anions and transport assisted by ditopic receptors through SLMs. In the future, new studies have to be undertaken to elucidate the corresponding transport mechanisms.

Acknowledgement

The authors like to thank all co-workers and colleagues, the names of whom are given in the literature cited. Financial support of the Netherlands Foundations for Chemical Research (SON) and from the Netherlands Technology Foundation (STW) is gratefully acknowledged.

Literature cited

1. Lonsdale, H .K. *J. Membr. Sci.* **1982**, *10*, 81.
2. Singer, J.; Nicolson, G. L. *Science* **1972**, *175*, 723.
3. Moore, C.; Pressmann, B. C. *Biochem. Biophys. Res. Comm.* **1964**, *15*, 562.
4. Pedersen, C. J. *J. Am. Chem. Soc.* **1967**, *89*, 7017.
5. de Jong, F.; Visser, H. C. In *Supramolecular chemistry*; Vol 10, in press.
6. Li, N. N. *U.S. Patent 3 410 794*, **1968**.
7. de Haan, A. B.; Bartels, P. V.; de Graauw, J. *J. Membr. Sci.* **1989**, *45*, 281.
8. Reusch, C. F.; Cussler, E. L. *AIChE J.* **1973**, *19*, 736.
9. Reichwein-Buitenhuis, E. G.; Visser, H. C.; de Jong, F.; Reinhoudt, D. N. *J. Am. Chem. Soc.* **1995**, *117*, 3913.
10. van Straaten-Nijenhuis, W. F., de Jong, F., Reinhoudt, D. N. *Recl. Trav. Chim. Pays-Bas* **1993**, *112*, 317.
11. Fabiani, C.; Merigiola, M.; Scibona, G.; Castagnola, A. M. *J. Membr. Sci.* **1987**, *30*, 97.
12. Deblay, P.; Delepine, S.; Minier, M.; Renon, H. *Sep. Sci. Technol.* **1991**, *35*, 123.
13. Kolthoff, I. M.; Elving, P. J. In *Treatise on Analytical Chemistry*; Part 1, Vol. 1; Eds.; Interscience Publishers, Inc.: New York, NY 1959.
14. Wienk, M. M.; Stolwijk, T. B.; Sudhölter, E. J. R.; Reinhoudt, D.N. *J. Am. Chem. Soc.* **1990**, *112*, 797.
15. Stolwijk, T. B.; Sudhölter, E. J. R.; Reinhoudt, D. N. *J. Am. Chem. Soc.* **1987**, *109*, 7042..
16. Nijenhuis, W. F.; Walhof, J. J. B.; Sudhölter, E. J. R.; Reinhoudt, D. N. *Recl. Trav. Chim. Pays-Bas* **1991**, *110*, 265.
17. Visser, H. C.; Vink, R.; Snellink-Ruel, B. H. M.; Kokhuis, S. B. M.; Harkema, S.; de Jong, F.; Reinhoudt, D. N. *Recl. Trav. Chim. Pays-Bas* **1995**, *114*, 285.
18. Izatt, R. M.; Clark, G. A.; Bradshaw, J. S.; Lamb, J. D.; Christensen, J. J. *Sep. Purif. Methods* **1986**, *15*, 21.
19. Lamb, J. D.; Bruening, R. L.; Izatt, R. M.; Hirashima, Y.; Tse, P.-K.; Christensen, J.J. *J. Membr. Sci.* **1988**, *37*, 13.
20. Izatt, R. M.; Bruening, R. L.; Bruening, M. L.; LindH, G. C.; Christensen, J. J. *Anal Chem.*, **1989**, *61*, 1140.
21. Nijenhuis, W. F.; Buitenhuis, E. G.; de Jong, F.; Sudhölter, E. J. R.; Reinhoudt, D. N. *J. Am. Chem. Soc.* **1991**, *113*, 7963.
22. Stolwijk, T. B.; Sudhölter, E. J. R.; Reinhoudt, D. N.; *J. Am. Chem. Soc.* **1989**, *111*, 6321.
23. Casnati, A.; Minari, P.; Pochini, A.; Ungaro, R.; Nijenhuis, W. F.; de Jong, F.; Reinhoudt, D. N. *Isr. J. Chem.* **1992**, *32*, 79.
24. Behr, J.-P.; Kirch, M.; Lehn, J.-M. *J. Am. Chem. Soc.* **1985**, *107*, 241.

25. Ibáñez, J.; Victoria, L.; Hernández, A. *Sep. Sci. Technol.* **1990**, *25*, 739.
26. Chrisstoffels L. A. J.; Struijk, W.; de Jong, F.; Reinhoudt, D. N. *J. Chem. Soc., Perkin 2.* accepted for publication.
27. Bromberg, L; Levin, G.; Kedem, O. *J. Membr. Sci.* **1992**., *71*, 41.
28. Zhu, X. X.; MacDonald, P. M. *Macromolecules* **1992**, *25*, 4345.
29. Kiani, A.; Bhave, R. R.; Sirkar, K. K. *J. Membr. Sci.* **1986**, *26*, 79.
30. Wilke, C. R.; Chang, P. *AIChE J.* **1955**, *1*, 264.
31. Anderson, J. L.; Quinn, J. A. *Biophys. J.* **1974**, *14*, 130.
32. Geringer, M.; Sterk, H. *Magn. Res. Chem.* **1989**, *27*, 1148.
33. Gibbs, S. J.; Lightfoot, E. N.; Root, T. W. *J. Phys. Chem.* **1992**, *96*, 7458.
34. van den Berg, R.; Schulze, D.; Bolt-Westerhoff, J. A.; de Jong, F.; Reinhoudt, D. N.; Velinova, D.; Buitenhuis, L. *J. Phys. Chem.* **1995**, *99*, 7760.
35. Ghidini, E.; Ugozzoli, F.; Ungaro, R.; Harkema, S.; Abu El-Fadl, A.; Reinhoudt, D. N. *J. Am. Chem. Soc.* **1990**, *112*, 6979.
36. Reinhoudt, D. N.; Dijkstra, P. J.; in 't Veld, P. J. A.; Bugge, K. E.; Harkema, S.; Ungaro, R.; Ghidini, E. *J. Am. Chem. Soc.* **1987**, *109*, 4761.
37. Fyles, T. M. *J. Membr. Sci.* **1985**, *24*, 229.
38. Yoshida, S.; Hayano, S. *J. Membr. Sci.* **1982**, *11*, 157.
39. Li, Y.; Gokel, G.; Hernández, J.; Echegoyen, L. *J. Am. Chem. Soc.* **1994**, 3087.
40. Casnati, A.; Pochini, A.; Ungaro, R.; Bocchi, C.; Ugozzoli, F.; Egberink, R. J. M.; Struijk, W.; Lugtenberg, R. J. M.; de Jong, F.; Reinhoudt, D. N. *Chem. Eur. J.* accepted for publication.
41. Lazarova, Z.; Boyadzhiev, L. *J. Membr. Sci.* **1993**, *78*, 239.
42. O'Hare P. A.; Bohrer, M. P. *J. Membr. Sci.*, **1989**, *44*, 273.
43. Brown, P. R.; Hallman, J. L.; Whaley, L. W.; Desai, D. H.; Pugia, M. J.; Bartsch, R. A. *J. Membr. Sci.* **1991**, *56*, 195.
44. Visser, H. C.; de Jong, F.; Reinhoudt, D. N. *J. Membr. Sci.* **1995**, accepted for publication.
45. Dozol, J. F.; Casas, J.; Sastre, A. M. *Sep. Sci. Technol.* **1993**, *28*, 2007.
46. Lamb, J. D.; Christensen, J. J.; Izatt, S. R.; Bedke, K.; Astin, M. S.; Izatt, R. M. *J. Am. Chem. Soc.*, **1980**, 3399.
47. Visser, H. C., *Thesis*, University of Twente, Enschede, The Netherlands, **1994**.
48. Deblay, P; Minier, M.; Renon, H. *Biotech. Bioeng.* **1990**, *35*, 123.
49. Kirch, M.; Lehn, J.-M. *Angew. Chem.* **1975**, *15*, 542.
50. Lamb, J. D.; Izatt, R. M.; Garrick, D. G.; Bradshaw, J. S.; Christensen, J. J. *J. Membr. Sci.* **1981**, *9*, 83.
51. Lamb, J. D.; Brown, P. R; Christensen, J. J.; Bradshaw, J. S.; Garrick, D. G.; Izatt, R. M. *J. Membr. Sci.* **1983**, *13*, 89.
52. Saito, T. *Sep. Sci. Technol.* **1993**, *28*, 1629.

53.   Caracciolo, F.; Cussler, E. L.; Evans, D. F. *AIChE J.* **1975**, *21*, 160.
54.   Sirlin, C.; Burgard, M.; Leroy, M. J. F. *J. Membr. Sci.* **1990**, *54*, 299.
55.   Shukla, J. P.; Kumar, A.; Singh, R. K. *Sep. Sci. Technol.*, **1992**, *27*, 477.
56.   Casnati, A.; Pochini, A.; Ungaro, R.; Ugozzoli, F.; Arnaud, F.; Fanni,
      S.; Schwing, M.-J.; Egberink, R. J. M.; de Jong, F.; Reinhoudt, D. N.
      J. Am. Chem. Soc. **1995**, *117*, 2767.
57.   Wegmann, D.; Weiss, H.; Amman, D.; Morf, W. E.; Pretsch, E.;
      Sugahara, K.; Simon, W. *Mikrochim. Acta.* **1984**, *3*, 1.
58.   Behr, J.-P.; Lehn, J.-M. *J. Am. Chem. Soc.* **1973**, *95*, 6108.
59.   Molnar, W. J.; Wang, C. P.; Fennel-Evans, D.; Cussler, E. L. *J.
      Membr. Sci.* **1978**, *4*, 129.
60.   Neplenbroek, A. M.; Bargeman, D.; Smolders, C. A. *J. Membr. Sci.*
      1992, *16*, 107.
61.   a)Kukufuta, E.; Nobusawa, M. *Chem. Lett.* **1988**, 425; b) Kokufuta,
      E.; Nobusawa, M. *J. Membr. Sci.* **1990**, *48*, 141.
62.   Sumi, K.; Kimura, M.; Kokufuta, E.; Nakamura, I. *J. Membr. Sci.*
      1994, *86*, 155.
63.   Furuta, H.; Cyr, M. J.; Sessler, J. L. *J. Am. Chem. Soc.* **1991**, *113*,
      6677.
64.   Araki, K.; Lee, S. K.; Otsuki, J.; Seno, M. *Chem. Lett.* **1993**, 493.
65.   Visser, H. C.; Rudkevich, D. M.; Verboom, W.; de Jong, F.;
      Reinhoudt, D. N. *J. Am. Chem. Soc.* **1994**, *116*, 11554.
66.   Rudkevich, D. M.; Mercer-Chalmers, J. D.; Verboom, W.; Ungaro,
      R.; de Jong, F.; Reinhoudt, D. N. *J. Am. Chem. Soc.* **1995**, *117*,
      6124.
67.   Yoshikawa, M.; Mori, Y.; Tanigaki, M.; Eguchi, W. *Bull. Chem. Soc.
      Jpn.* **1990**, 63, 304.
68.   Yoshikawa, M.; Kishida, M.; Tanigaki, M.; Eguchi, W. *J. Membr.
      Sci.* **1989**, *47*, 53.
69.   Nijenhuis, W. F.; van Doorn, A. R.; Reichwein, A. M.; de Jong, F.;
      Reinhoudt, D. N. *J. Am. Chem. Soc.* **1991**, *113*, 3607.
70.   van Straaten-Nijenhuis, W. F.; van Doorn, A. R.; Reichwein A.M.;,
      de Jong, F.; Reinhoudt, D. N. *J. Org. Chem.* **1993**, *58*, 2265.
71.   van Straaten-Nijenhuis, W. F.; de Jong, F.; Reinhoudt, D. N.;
      Thummel, R. P.; Bell, T. W.; Liu, J. *J. Membr. Sci.* **1993**, *82*, 277.

Chapter 4

Rational Design of Liquid Membrane Separation Systems

Randall T. Peterson and John D. Lamb

Department of Chemistry and Biochemistry, Benson Science Building, Brigham Young University, Provo, UT 84602

Liquid membrane separation systems possess great potential for performing cation separations. Many factors influence the effectiveness of a membrane separation system including complexation/ decomplexation kinetics, membrane thickness, complex diffusivity, anion type, solvent type, and the use of ionic additives. The role that each of these factors plays in determining cation selectivity and flux is discussed. In an effort to arrive at a more rational approach to liquid membrane design, the effect of varying each of these parameters is established both empirically and with theoretical models. Finally, several general liquid membrane types are reviewed, and a novel membrane type, the polymeric inclusion membrane, is discussed.

The emergence of life was marked by the development of membranes which isolate distinct solutions and permit selective passage of chemical species. In recent decades, man has made significant strides in mimicking nature's membranes for his own benefit. With sales of membrane separation systems topping $1 billion annually (*1*), membrane science has taken a place among the most important topics of research in the world.

When Charles Pedersen first synthesized crown ethers in 1967 (*2*), their tremendous potential for incorporation into membranes was quickly noted. Since then, hundreds of papers and many reviews have outlined novel applications of macrocycles to membrane separation systems (*3-22*). Because of the ease with which macrocycle separation characteristics can be adjusted, both by structural changes in the ligand and changes in separation system parameters, the possibilities for application seem enormous. As the wealth of data concerning the complex science of macrocycle-facilitated membrane separation has increased, so has our ability to define governing principles and create models which predict results for untried systems. The result is the emergence of a rational approach to membrane system design encompassing both design of the ligand itself and design of the other parameters that influence membrane performance. This chapter attempts to further the advance toward an understanding

0097–6156/96/0642–0057$15.00/0
© 1996 American Chemical Society

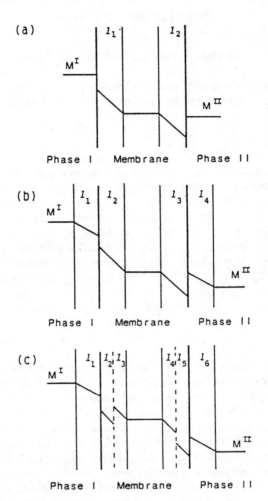

Figure 1. Diagrammatic Representation of Nernst Films that Form at Membrane Interfaces. (a) One-, (b) two-, and (c) three-film models are represented. (Reproduced with permission from ref. 31. Copyright 1990 CRC Press.)

of the science which allows us to optimize membrane separations. We will examine the factors that influence cation transport rates and other factors influencing membrane effectiveness. We will also survey well established membrane separation systems and describe novel systems which are just beginning to emerge. In a later chapter, we will take a closer look at the ligand structural features that influence the strength of cation binding by macrocycles, such as donor atom number and type, macrocycle and chelate ring size, and ligand preorganization.

Transport Rates

Most membrane separation systems involve stirring or continuous flow of the source and receiving solutions to minimize the time for diffusion of dissolved species toward and away from the membrane (23-26). Nevertheless, even the most vigorously stirred systems possess thin films at the aqueous/organic interface that are essentially stagnant (22-30). (Figure 1) These films, often referred to as Nernst films, vary from 10^{-2} to 10^{-4} cm thick (30) and can be crossed only by diffusion processes. Diffusion across Nernst films is only part of the process that must occur during membrane transport. Assuming that both the source and receiving phases are rapidly mixed, virtually all membrane transport processes can be broken down into five parts: diffusion through the thin aqueous source film, a chemical complexation reaction at the source-membrane interface, diffusion through the membrane, a chemical decomplexation reaction at the membrane-receiving interface, and diffusion through the aqueous receiving phase film. These five processes can be categorized as diffusion steps or chemical reaction steps.

Studies (29, 31) indicate that for many systems the combined diffusion steps are rate determining and that the chemical reaction aspect is relatively insignificant. However, Juang and Lo (28) report that in a supported liquid membrane containing bis(2-ethylhexyl)phosphoric acid under conditions of high pH and $[VO^{2+}]$ in the feed and low pH and $[VO^{2+}]$ in the stripping phase, the interfacial chemical reaction governs the transport rate. Juang and Lo cite two other sources which report similar results but also acknowledge the sparseness of confirming reports. Certainly in the majority of cases, diffusion is the rate-determining factor in membrane transport. We will discuss both the influence of diffusion and the influence of reaction rates at the membrane interfaces.

Diffusion. The barriers to transport imposed by the need for diffusion across Nernst films can be minimized by decreasing film thickness or by increasing the mobility of the diffusible species. Film thickness, up to a certain limiting value (10^{-3} to 10^{4} cm) is inversely related to mechanical energy supplied (e.g. by stirring) (30). Viscosity and density of the liquids used as well as equipment geometry also affect film thickness (32), but interfacial films apparently cannot be completely eliminated.

Decreased film thickness is not the only way to increase diffusion. Structural features of macrocyclic carriers can themselves alter diffusion rates. Several models have been devised to predict the role of various membrane features on membrane flux (32-37). The simplest relationship is the Stokes-Einstein relationship (33) in which the diffusion coefficient, D, is given by:

$$D = kT/6\pi\eta r \tag{1}$$

where k is the mass transfer coefficient based on concentration, T is temperature, η is the solvent viscosity, and r is the molecular radius. This relationship is accurate for neutral molecules. For ionic species, models are required which are much more complex, taking into account such factors as ionic charge, ionic strength, the presence of electric fields and others. In either case, the range of values for D is quite narrow, usually 10^{-5} to 10^{-6} cm^2/s. Thus, manipulation of carrier structural features offers minimal benefit for increasing transport rates. Macrocyclic structure has a great deal of influence over membrane selectivity but little influence over membrane flux.

Complexation/Decomplexation Kinetics. Though most researchers report that diffusion predominates over complexation/decomplexation kinetics in determining transport rates, reaction kinetics also play a role (*37*). The rate at which macrocycles bind to cations is determined principally by two factors-- the charge density of the cation and the degree of preorganization inherent in the macrocycle.

Gas-phase studies of macrocycle-cation interaction demonstrate that smaller cations display greater complexation rates with 18-crown-6 than larger cations (*38-39*). Apparently, this trend is not merely a consequence of their superior rates of collision over heavier cations. The decrease in complexation rates with increasing cationic size is greater than the decrease in collision rate. It has been postulated (*39*) that charge density plays a critical role. The smaller cations, with greater charge densities, are involved in more pronounced ion-dipole interactions with the ligand and can therefore more rapidly rearrange the ligand to a conformation favorable to binding. This clearly results in increased rates of complexation.

Preorganization also plays a role in complexation/decomplexation kinetics. Other factors being equal, a highly preorganized macrocycle will often bind cations more slowly than a macrocycle lacking preorganization because its rigid structure prevents it from undergoing the rearrangement necessary to bring the cation into the macrocycle's binding site. Thus, while preorganization may thermodynamically favor cation binding, it may kinetically disfavor function as a membrane carrier.

Other Factors Influencing Transport

Solvent Type. Nearly every macrocycle-mediated membrane system devised to date involves the solvation of macrocycles in an organic solvent. Because transport of any substance requires that it pass through this organic solvent, transport rates and selectivities depend heavily upon the properties of this solvent. In an earlier review (*27*) we divided the free energy of transport, ΔG_T, into four components in a thermodynamic cycle as follows:

$$\Delta G_T = \Delta G_1 + \Delta G_2 + \Delta G_3 + \Delta G_4 \tag{2}$$

where ΔG_1 is the free energy of desolvation of the cation, ΔG_2 is the free energy of desolvation of the ligand, ΔG_3 is the free energy for the gas phase interaction between cation and ligand, and ΔG_4 is the free energy of solvation of the macrocycle/cation complex. Under this method of describing membrane transport, three of the four components are intimately related to the nature of the organic solvent. Izatt, *et al. (40)* outline four ways in which a solvent exerts its influence on membrane transport. Solvent characteristics influence the thickness of the Nernst films at the membrane interfaces, equilibrium constants for cation-macrocycle interaction in the membrane, partition coefficients, and the diffusivities of the species in the system. Not only is the influence of the solvent type rather large, it is quite complex.

Many groups have investigated the suitability of various solvents for use in membrane systems and have attempted to describe the relationship between solvent characteristics and transport properties *(40-44)*. Of all solvent properties, dielectric constant seems to be most predictable in its effect on transport *(40)*. For related solvents such as the halocarbons, log K usually decreases with increasing dielectric constants, resulting in reduced transport *(27)*. Figure 2 shows this trend for alkali metal binding by dicyclohexano-18-crown-6 in a number of alcohols. While this trend holds true for many simple systems, it breaks down under more complex conditions. Matsuura *(45)* among others *(46)* has established the importance of solvent donor number. Other factors in the equation are solvent molecule size, solvent viscosity, ligand solubility in the solvent, permanent and induced dipole moments, and heats of vaporization. Another group *(40)* uses a general rule that the solvent must be able to accommodate as much water as possible and still retain the ligand. Equations have been developed that consider many of these factors in predicting transport rates *(27)*, but their predictions still suffer from considerable inaccuracy. Moreover, the issue of solvent effect is a complex one and considerable room remains for further study of this subject.

Recently Dernini, *et al. (44)* and Szpakowska and Nagy *(42)* have made some significant advances in the incorporation of mixed solvents in membrane systems. Szpakowska's work has been primarily theoretical and lends credence to the existence of mixed solvation shells in binary solvent mixtures. Future research in this area promises to bring more economical use of expensive but efficient solvents. It also purports to open the door to simultaneous use of solvents with both general and specific solvent effects in order to optimize solvent characteristics. In a similar report, Dernini, *et al. (44)* report greatly improved transport by employing mixed solvents. Figure 3 shows the synergistic effect of binary mixtures of chloroform and nitrobenzene: maximum Na^+ transport occurs with an equimolar solution of the two solvents. Parthasarathy and Buffle *(35)*, in a similar study, report optimal transport of Cu^{2+} with an equimolar solution of phenylhexane and toluene. Certainly membrane systems will continue to improve as our understanding of the complex influences of solvent properties increases.

Anion Type. Typically, discussion of liquid membranes focuses on the transport of cations with little mention of the anionic species involved. However, in order to maintain electroneutrality, many membrane carrier systems require that an anion be cotransported along with the cation. Because the anion must also enter and cross the organic phase, it is bound to influence transport efficiency. In fact, for K^+ transport by

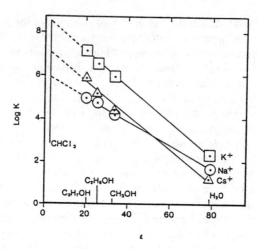

Figure 2. Effect of Solvent Dielectric Constant ∈ on Macrocycle-cation Complex Stability. A plot of log K versus ∈ is shown for three cations in several pure solvents. (Reproduced with permission from ref. 27. Copyright 1981 Wiley.)

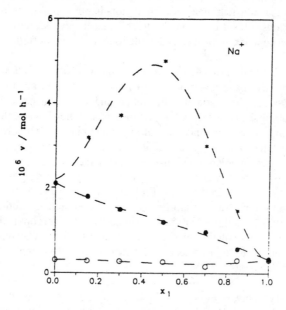

Figure 3. Synergistic Effect of Mixed Solvents on Na⁺ Transport Through a Bulk Liquid Membrane. The source phase is 1 M NaCl. The receiving phase is distilled water and dibenzo-18-crown-6 is the carrier. Three mixed solvent systems were tested: (*) chloroform(1)-nitrobenzene(2);(●) dichloroethane(1)-nitrobenzene(2); and (○) chloroform(1)-dichloroethane(2). (Reproduced with permission from ref. 44. Copyright 1992 American Chemical Society.)

18-crown-6 in a bulk liquid membrane, the anion effect accounts for transport efficiencies that differ by almost 10^8 *(47)*. Many studies of the anion effect on transport efficiency have been conducted *(47-53)*. The effects of anion hydration free energy, anion lipophilicity, and anion interaction with benzo groups on benzo-substituted crown ethers have all been cited *(43)*, although anion hydration free energy seems to be the major determinant of transport efficiency. Using dibenzo-18-crown-6, K^+ transport decreased in the order picrate > PF_6^- > ClO_4^- > IO_4^- > BF_4^- > I^- > SCN^- > NO_3^- > Br^- > BrO_3^- > Cl^- > OH^- > F^- > acetate > SO_4^- (Figure 4a). This order is almost identical to that for increasing anion hydration free energy *(52)* which is ClO_4^- < BF_4^- < I^- < NO_3^- < SCN^- < Br^- < Cl^- < BrO_3^- < OH^- < F^- < SO_4^- (Figure 4b). This simple example demonstrates the strong correlation between anion hydration and transport efficiency. The larger anions are more easily dehydrated and thus more readily enter the membrane to facilitate transport.

While nearly all investigations of anion effects have focused on transport efficiency, a few recent manuscripts suggest a correlation between anion type and *selectivity*. A communication from Olsher, *et al.* *(48)* reports that for extraction of alkali metals into chloroform by dicyclohexano-18-crown-6, selectivity for K^+ over both Rb^+ and Cs^+ decreases dramatically depending upon anion type in the order NO_3^- > SCN^- > ClO_4^- ≥ I^- > Br^-. K^+/Cs^+ selectivity decreases from 16.0 for nitrate to 3.5 for bromide. The authors were unable to tie this trend to any particular parameter, though they discounted the possibility that it is correlated to anion radius, hydration enthalpy, or anion softness.

Further research that links anion type to membrane selectivity was done by us using 18-crown-6 derivatives and a variety of membrane types *(47)*. We demonstrated that anions are capable not only of altering selectivities but indeed of reversing them. Typically, 18-crown-6 analogs show a strong preference for Hg^{2+} over Cd^{2+}, and this preference is reflected in larger transport rates for Hg^{2+}. However, by using SCN^- as a counter-anion, we were able to completely reverse the usual selectivity, resulting in highly selective transport of Cd^{2+} over Hg^{2+}. This result is due to the fact that the SCN ion forms coordination complexes with these cations -- Hg^{2+} and Cd^{2+} complex with SCN^- to different degrees. Hg^{2+} tends to be present as the ionic species $Hg(SCN)_4^{2-}$ while Cd^{2+} is more frequently present as the neutral species $Cd(SCN)_2$, which is more readily transported through the membrane than the Hg^{2+} charged complex. The result is a reversal of selectivity in the presence of SCN^-. A similar result is present when Br^- is used as the anion for altering selectivity between Cd^{2+} and Zn^{2+}. These results suggest that careful consideration of anion is crucial when designing a membrane system or when comparing results listed in the literature.

Ionic Additives. Cotransport of anions is the most obvious way to maintain electroneutrality, but alternative means have been explored. In recent years, many studies have been conducted which examine the use of anionic membrane additives for maintenance of electroneutrality *(35, 54-58)*. The anionic additives can be either stationary or mobile and are typically lipophilic carboxylic, phosphoric, or sulfonic acids. Neutral macrocyclic carriers coupled with anionic additives result in a synergistic transport of cations which exceeds that accomplished by each component individually.

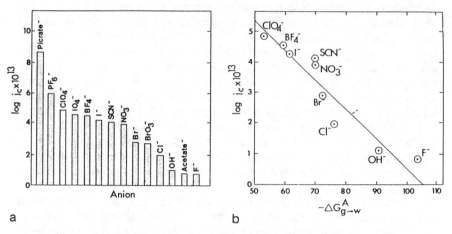

a   b

Figure 4. Correlation Between Anion Free Energy of Hydration and Transport Efficiency. a) Variation of K^+ transport across a chloroform-supported liquid membrane with different anions. Anion source phase concentration is adjusted to 2.0 mM. b) Variation of K^+ transport across a chloroform-supported liquid membrane with increasing anion hydration free energy. Literature values are used for thermodynamic data. (Reproduced with permission from ref. 52. Copyright 1980 American Chemical Society.)

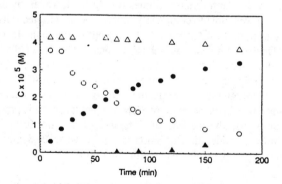

Figure 5. Transport of Cu^{2+} Through a Toluene Supported Liquid Membrane in the Presence and Absence of Laurate. The carrier was 0.1 M 1,10-didecyl-1,10-diaza-18-crown-6. The source phase was 5×10^{-5} M Cu^{2+} in 1×10^{-2} M N-morpholinoethanesulphonic acid-LiOH buffer (pH 6). The receiving phase was 5×10^{-4} M cyclohexanediaminetetraacetic acid. Circles represent transport with 0.1 M laurate in the membrane. Triangles indicate transport without laurate. (Reproduced with permission from ref. 35. Copyright 1994 Elsevier Science B.V.)

This synergism was demonstrated by Moyer, *et al. (58)*. They impregnated a resin composed of strong-acid poly(styrene-divinylbenzene) cation-exchange beads with tetrathia-14-crown-4 and observed a 10-100 fold enhancement of Cu^{2+} extraction. In such a system enhanced extraction is achieved by way of cation coordination by the macrocycle and cation exchange by the anionic group *(58)*. While proton-ionizable macrocycles can single-handedly accomplish both of these tasks, use of anionic additives is simpler, more economical, and affords greater flexibility to the separation system *(58)*.

Parthasarathy and Buffle *(35)* have systematically varied the chain length of a series of lipophilic carboxylic acids in a supported liquid membrane with 1,10-didecyldiaza-18-crown-6 as carrier. Chain lengths ranged from 10 to 18 carbons. Optimal Cu^{2+} transport was achieved with additives from 12 to 14 carbons in length, and lauric acid (n=12) yielded the best results due to its decreased tendency to form precipitates with Cu^{2+}. Others *(56-57)* also report success upon addition of tetraphenylborate derivatives.

The transport enhancement which results from incorporation of anionic additives can be striking. Figure 5 shows Cu^{2+} transport through a supported liquid membrane (SLM) in the presence and absence of an anionic additive (laurate). In this system, transport is enhanced by more than ten times by the addition of laurate.

The amount of anion additive that should be added has not been fully resolved, though it appears that the ratio of additive to macrocycle is the most critical measure. Schaller, *et al. (56)* report that a molar additive to ionophore ratio of 0.3 to 0.6 is optimal while 10 mole % is sufficient. Parthasarathy *(35)* recommends additive concentrations of 0.1 M in the membrane solvent. While the optimal amount of additive varies from system to system, the potential benefit of the additives is well established.

That anionic sites are present in PVC-based membranes is well accepted *(59-62)*. These anionic sites clearly influence transport, and since the quantity of these sites varies with material quality and age, anionic additives offer the additional advantage of overriding inherent anionic influences which could otherwise render data unreliable.

Crown Lipophilicity. Crown lipophilicity is necessary to maintain the carrier in the membrane, but its influence extends even further in that it is also related to cation flux through the membrane. Stolwijk, *et al. (63)* compared the flux values of guanidinium thiocyanate through a SLM using 18 macrocycles of varying lipophilicity. The results, shown in Figure 6, indicate a direct relationship between crown lipophilicity and flux. However, there are several outliers, and while a general trend certainly exists, lipophilicity alone does not determine flux. Better correlation is seen when flux is considered in light of both crown lipophilicity and log K for macrocycle-cation binding. Figure 7 groups the data points by approximate log K value, with A, B, and C corresponding to log K values of 3, 4, and 5, respectively. This means of comparison accounts for the apparent outliers and makes the correlation between lipophilicity and cation flux more obvious.

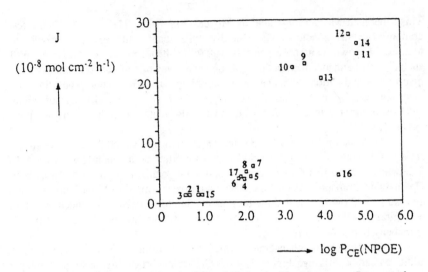

Figure 6. The Effect of Carrier Lipophilicity on Flux Across a Supported Liquid Membrane. The source phase was 10^{-1} M guanidinium thiocyanate and the carrier concentration was 10^{-2} M. The carrier lipophilicity was calculated using an equation given in reference 63. (Reproduced with permission from ref. 63. Copyright 1989 American Chemical Society.)

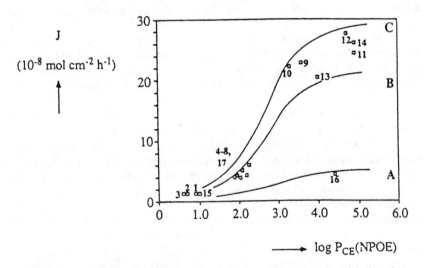

Figure 7. The Effect of Carrier Lipophilicity on Flux After Considering the Influence of Complex Stability. Conditions are as stated in Figure 11. Cations are grouped according to their log K values. A, B, and C represent approximate log K values of 3, 4, and 5 respectively. (Reproduced with permission from ref. 63. Copyright 1989 American Chemical Society.)

Membrane Types

A variety of membrane types exists, with each type giving rise to variations and specific alterations. The general membrane types are discussed below with their respective advantages and disadvantages.

Bulk Liquid Membranes. Bulk liquid membranes (BLMs) consist of a bulk, stirred organic phase that separates the aqueous and receiving phases. (Figure 8a) The exact configuration has several forms including the Shulman Bridge and the U-tube. Usually, the aqueous to organic ratio is 2:1, and the transport mechanism is very similar to the simple model of simultaneous extraction and back-extraction of solvent extraction.

BLMs are useful for screening macrocycles for carrier effectiveness due to their simplicity and because only small quantities of the macrocycle are necessary if the phase volumes are kept small. Furthermore, this system is more forgiving than others because of its low aqueous to organic ratio -- carrier and organic solvent loss are not generally issues. This makes BLMs good candidates for the screening of primary carrier structures before hydrophobic substituents have been added. However, BLMs are not a commercially viable option mainly because of their low transport rates. Large data standard deviations are also common *(64)*.

Emulsion Liquid Membranes. The emulsion liquid membrane (ELM) was developed by Li and Cussler in the late 1960's and early 1970's *(65-66)*. A water-in-oil emulsion is formed by an organic solvent and water, often containing acid. The emulsion can be stabilized by the addition of a surfactant. This emulsion is then stirred into an aqueous source solution, and transport occurs across the extremely thin membrane bubbles (Figure 8b). When extraction is complete the emulsion is collected and broken to obtain the concentrated target substance.

The large surface area and thinness of the ELM result in rapid transport, making this system viable for commercial application. The high source to receiving volume ratio also allows a high degree of concentration *(64)*. The major disadvantage of this type of system is the necessity of breaking the emulsion to recover its contents after transport is complete. This can be a difficult and messy process, greatly reducing the appeal of ELMs for practical application.

Supported Liquid Membranes. Supported liquid membranes (SLMs) consist of a hydrophobic, porous plastic sheet or hollow fiber (usually polypropylene or polysulfone). The pores are filled with an organic solvent in which the carrier is dissolved. This membrane separates the aqueous source and receiving phases (Figure 8c). The liquid-containing pores allow transport of the target species via the dissolved carrier, while the plastic sheet offers support for the liquid membrane. Pore sizes generally range from 0.02 to 1.0 μm.

This system offers the advantage of requiring only small quantities of macrocycle and solvent. It also is easy to model and yields small standard deviations. However, it possesses significant flaws *(64)*. The macrocyclic carrier and solvent are easily lost to the aqueous phases. As a result, both carrier and solvent must be

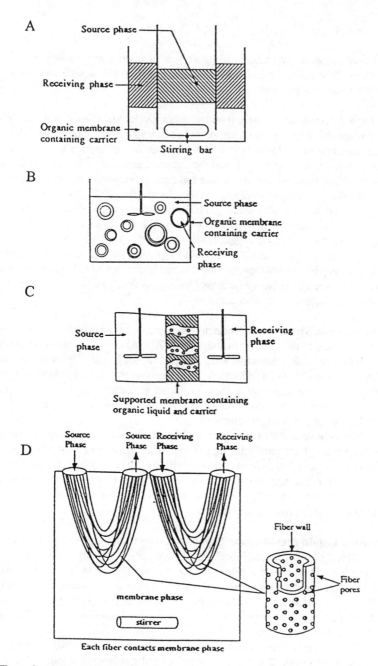

Figure 8. Frequently Used Membrane Types. A, B, C, and D are bulk liquid membrane, emulsion liquid membrane, supported liquid membrane, and dual module hollow fiber membrane configurations respectively. (Reproduced with permission from ref. 47. Copyright 1990 CRC Press.)

extremely hydrophobic. Furthermore, transport rates are small, and fouling of the pores can eliminate transport altogether. The short lifetime of these membranes has been the principle impediment to their practical application. SLMs can be implemented in both flat sheet and hollow fiber formats.

Dual Module Hollow Fiber Membrane. Dual module hollow fiber (DMHF) systems are similar to SLMs except that the membrane solvent is not confined to the pores of the plastic layer. Rather, the hollow fibers serve to carry the flowing source and receiving phases (Figure 8d). The fibers provide a large surface area which is always in contact with organic solvent containing carrier. This lessens the chance of losing membrane integrity due to solvent loss during transport. Through the lumen of one of the bundles flows the source phase and through the other bundle flows the receiving phase. Transport occurs as the target cation crosses the source fibers and enters a bulk organic phase that contains carrier. The cation is then transported across the receiving fibers where it enters the aqueous receiving phase.

Benefits of this system include its potential for practical application due to the enhanced stability. Disadvantages include those inherent in any SLM including the potential for fouling of membrane pores due to surface effects. Also, solvents and macrocycles must be rather hydrophobic for optimal membrane function *(64)*.

Polymeric Inclusion Membranes. Very recently a novel type of membrane system has been developed that combines the virtue of rapid transport with high selectivity and ease of setup and operation. At the same time, it exhibits excellent durability. We have called these membranes polymeric inclusion membranes (PIMs).

PIMs have been modelled after the β-diketone-containing membranes used by Sugiura *(67-69)* and are formed by the polymerization of cellulose triacetate (CTA) to form a thin film. Polymerization takes place in the presence of a macrocyclic carrier, and as the thin sheet forms, carrier molecules are trapped within the CTA matrix. The resultant membrane is then placed between aqueous source and receiving solutions and selectively mediates transport of a desired species from one phase to the other. While PIMs can effectively separate two aqueous phases, they are not dependent upon organic solvents to maintain phase separation and allow transport. Thus they are simpler to use than SLMs and do not suffer from loss of organic solvent nor as much leaching of carrier into the aqueous phases.

Three lines of evidence suggest that transport across PIMs is carrier mediated. First, CTA membranes which contain no macrocyclic carrier do not exhibit measurable cation transport. Second, transport selectivity is entirely consistent with macrocycle selectivity. In our experimental system, dicyclohexano-18-crown-6 (DC18C6) was included in the membranes. Binding constants for DC18C6 with the alkali metal cations decreases in the order $K^+ > Rb^+ > Na^+$. In a similar fashion, transport in our DC18C6-containing PIMs was highest for K^+ followed by Rb^+ and finally by Na^+ (Figure 9). The final line of evidence that supports a carrier-mediated mechanism is the second-order variation in transport rates with varying source solution concentration. The amount of KNO_3 in the source solution was varied from 0.0 mM to 500 mM. As shown by Figure 10, transport increased as a function of the square of the source phase

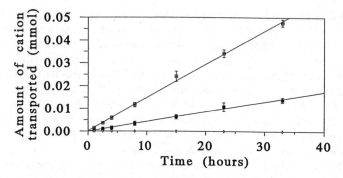

Figure 9. Selectivity Exhibited by Polymeric Inclusion Membranes with Dicyclohexano-18-crown-6 as Carrier. The source phase was 100 mM for each of the alkali metal nitrates. The receiving phase was deionized water. The receiving phase concentration is shown versus time. (Reproduced with permission from ref. 70. Copyright 1996 Elsevier Science B.V.)

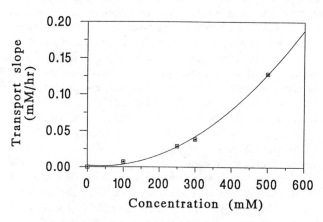

Figure 10.   Dependence of Cation Flux on Source Phase Cation Concentration. The membrane was polymerized in the presence of 10.0 mM DC18C6. The receiving phase was deionized water.

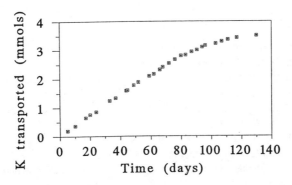

Figure 11. Long Term Stability of Polymeric Inclusion Membranes. The receiving phase K^+ concentration is plotted against time. The source phase contained 100 mM KNO_3; the receiving phase was deionized water. The membrane was polymerized in the presence of 10.0 mM DC18C6.

concentration. This relationship has been observed in all membrane systems which function by a carrier-mediated mechanism, and its presence in the PIM systems we have studied further supports the idea that PIMs function by a traditional facilitated-transport mechanism.

One great advantage of PIMs over many other membrane separation systems is the superior flux values they can produce. As in most membrane systems, flux is influenced greatly by membrane thickness. To date, we have achieved the polymerization of membranes as thin as 28 μm which maintain structural integrity. The flux value for K^+ transport in these membranes is 7.4×10^{-7} mol/s·m^2, which compares favorably to values of 5.3×10^{-8} mol/s·m^2 for a related SLM reported by Izatt, *et al.* (*64*).

Perhaps the greatest virtue of PIMs is their durability. SLMs quickly break down as carrier and organic solvent are lost to the aqueous phases. ELMs can be used only once. In contrast, PIMs are not dependent upon maintenance of organic solvent in the membrane and appear to hold the macrocyclic carrier with enough tenacity that leaching into the aqueous phases is not a concern. As a result, the longevity of PIMs is greatly increased. One PIM in our laboratory recently ran for more than 100 days with little variation in cation flux and no signs of structural weakening (Figure 11). After 100 days, cation flux began to taper off, reflecting the approach of equilibrium concentrations in both the source and receiving phases. When coupled with their high flux values, the durability of PIMs makes them amenable to practical application in ways that other membrane separation systems are not.

Acknowledgement. The authors express appreciation to the United States Department of Energy, Office of Basic Energy Sciences who funded this work through Grant No. DE-FG03-95-ER14506.

Literature Cited

1. *Membrane Separation Systems: Recent Developments and Future Directions*; Bakker, R. W.; Cussler, E. L.; Eykamp, W.; Koros, W. J.; Riley, R. L.; Strathmann, H., Eds.; Noyes Data Corporation: Park Ridge, NJ, 1991.
2. Pedersen, C. J. *J. Am. Chem. Soc.* **1967**, *89*, 7017.
3. Wang, C.; Huszthy, P.; Bradshaw, J. S.; Lamb, J. D.; Olenyuk, B.; Bearss, D.; Izatt, R. M. *Sep. Sci. Tech.*, in press.
4. Izatt, R. M.; Lamb, J. D.; Bruening, R. L.; Wang, C.; Edge, N.; Bradshaw, J. S. *Sep. Sci. Technol.* **1993**, *28*, 383-395.
5. Izatt, R. M.; Roper, D. K.; Bruening, R. L.; Lamb, J. D. *J. Membr. Sci.* **1989**, *45*, 73-84.
6. Nijenhuis, W. F.; van Doorn, A. R.; Reichwein, A. M.; de Jong, F.; Reinhoudt, D. N. *J. Am. Chem. Soc.* **1991**, *113*, 3607-3608.
7. Aranda, P.; Casal, B.; Fripiat, J. J.; Ruiz-Hitaky, E. *Langmuir* **1994**, *10*, 1207-1212.
8. Uddin, M. S.; Hidajat, K.; Woo, F. *J. Chem. Tech. Biotechnol.* **1993**, *58*, 123-128.
9. Izatt, R. M.; Bruening, R. L.; Geng, W.; Cho, M. H.; Christensen, J. J. *Anal. Chem.* **1987**, *59*, 2405-2409.
10. Hosseini, M. W.; Lehn, J.; Jones, K. C.; Plute, K. E.; Mertes, K. B.; Mertes, M. P. *J. Am. Chem. Soc.* **1989**, *111*, 6330-6335.
11. Sirlin, C.; Burgard, M.; Leroy, M. J. F.; Prevost, M. *J. Membr. Sci.* **1990**, *54*, 299-305.
12. Nigenhuis, W. F.; Buitenhuis, E. G.; de Jong, F.; Sudholter, E. J. R.; Reinhoudt, D. N. *J. Am. Chem. Soc.* **1991**, *113*, 7963-7968.
13. Chen, Z.; Gokel, G. W.; Echegoyen, L. *J. Org. Chem.* **1991**, *56*, 3369-3372.
14. Horwitz, E. P.; Dietz, M. L.; Fisher, D. E. *Anal. Chem.* **1991**, *63*, 522-525.
15. Horwitz, E. P.; Chiarizia, R.; Dietz, M. L. *Solv. Extr. Ion Exch.* **1992**, *10*, 313-336.
16. Attiyat, A. S.; Christian, G. D.; Cason, C. V.; Bartsch, R. A. *Electroanalysis* **1992**, *4*, 51-56.
17. Attiyat, A. S.; Christian, G. D.; McDonough, J. A.; Strzelbicka, B.; Goo, M.; Yu, Z.; Bartsch, R. A. *Anal. Lett.* **1993**, *26*, 1413-1424.
18. Buschmann, H. *Thermochim. Acta* **1986**, *102*, 179-184.
19. Gupta, O. D.; Kirchmeier, R. L.; Shreeve, J. M. *Inorg. Chem.* **1994**, *33*, 2161-2166.
20. Chen, L.; Thompson, L. K.; Bridson, J. N.; Xu, J.; Ni, S.; Guo, R. *Can. J. Chem.* **1993**, *71*, 1805.
21. Shukla, J. P.; Jeon, E.; Knudsen, B. E.; Pugia, M. J.; Bradshaw, J. S.; Bartsch, R. A. *Thermochim. Acta* **1988**, *130*, 103-113.
22. Sukhan, V. V.; Kronikovskii, O. I.; Nazarenko, A. Y. *Akad. Nauk SSSR* **1988**, *299*, 921.
23. *Membrane Processes in Separation and Purification;* Crespo, J. G.; Boddeker, K. W., Eds.; NATO ASI Series E: Applied Sciences 272; Kluwer Academic Publishers: Boston, MA, 1994.
24. *Membrane Handbook*; Ho, W. S. W.; Sirkar, K. K., Eds.; Van Nostrand Reinhold: New York, 1992.

25. *Basic Principles of Membrane Technology*; Mulder, M., Ed.; Kluwer Academic Publishers: Boston, MA, 1991.

26. *Liquid Membranes: Chemical Applications*; Araki, T.; Tsukube, H., Eds.; CRC Press: Boca Raton, FL, 1990.

27. Lamb, J. D.; Izatt, R. M.; Christensen, J. J. In *Progress in Macrocyclic Chemistry*, Vol. 2; Izatt, R. M.; Christensen, J. J., Eds.; Wiley: New York, 1981; pp 41-90.

28. Juang, R.; Lo, R. *Ind. Eng. Chem. Res.* **1994**, *33*, 1011-1016.

29. Juang, R. *Ind. Eng. Chem. Res.* **1993**, *32*, 911-916.

30. *Principles and Practices of Solvent Extraction;* Rydberg, J.; Musikas, C.; Choppin, G. R., Eds.; Marcel Dekker: New York, 1992.

31. Inoue, Y. In *Liquid Membranes: Chemical Applications;* Araki, T.; Tsukube, H., Eds; CRC Press: Boca Raton, FL, 1990, pp 77-102.

32. Izatt, R. M.; Bruening, R. L.; Bruening, M. L.; LindH, G. C.; Christensen, J. J. *Anal. Chem.* **1989**, *61*, 1140-1148.

33. Noble, R. D.; Koval, C. A.; Pellegrino, J. J. *Chem. Eng. Prog.* **1989**, March, 58-70.

34. Basaran, O. A.; Burban, P. M.; Auvil, S. R. *Ind. Eng. Chem. Res.* **1989**, *28*, 108-119.

35. Parthasarathy, N.; Buffle, J. *Anal. Chim. Acta* **1994**, *284*, 649-659.

36. Nahir, T. M.; Buck, R. P. *J. Phys. Chem.* **1993**, *97*, 12363-12372.

37. Li,Y.; Gokel, G.; Hernandez, J.; Echegoyen, L. *J. Am. Chem. Soc.* **1994**, *116*, 3087-3096.

38. Chu, I.; Zhang, H.; Dearden, D. V. *J. Am. Chem. Soc.* **1993**, *115*, 5736-5744.

39. Wong, P. S. H.; Antonio, B. J.; Dearden, D. V. *J. Am. Soc. Mass Spectrom.* **1994**, *5*, 632-637.

40. Izatt, R. M.; McBride, D. W.; Brown, P. R.; Lamb, J. D.; Christensen, J. J. *J. Membr. Sci.* **1986**, *28*, 69-76.

41. Izatt, R. M.; Bruening, R. L.; Bruening, M. L.; Lamb, J. D. *Isr. J. Chem.* **1990**, 30, 239-245.

42. Szpakowska, M.; Nagy, O. B. *J. Phys. Chem.* **1989**, *93*, 3851-3854.

43. Izatt, R. M.; Clark, G. A.; Bradshaw, J. S.; Lamb, J. D.; Christensen, J. J. *Sep. Purif. Meth.* **1986**, *15*, 21-72.

44. Dernini, S.; Palmas, S.; Polcaro, A. M.; Maronglu, B. *J. Chem. Eng. Data* **1992**, *37*, 281-284.

45. Matsuura, N.; Umemoto, K.; Takeda, Y.; Sasahi, A. *Bull. Chem. Soc. Jap.* **1976**, *49*, 1246.

46. Shchori, E.; Jagur-Grodzinski, J. *Isr. J. Chem.* **1973**, *11*, 243.

47. Izatt, R. M.; Bradshaw, J. S.; Lamb, J. D.; Bruening, R. L. In *Liquid Membranes: Chemical Applications*; Araki, T.; Tsukube, H., Eds; CRC Press: Boca Raton, FL, 1990, pp 123-140.

48. Olsher, U.; Hankins, M. G.; Kim, Y. D.; Bartsch, R. A. *J. Am. Chem. Soc.* **1993**, *115*, 3370-3371.

49. Christensen, J. J.; Lamb, J. D.; Izatt, S. R.; Starr, S. E.; Weed, G. C.; Astin, M. S.; Stitt, B. D.; Izatt, R. M. *J. Am. Chem. Soc.* **1978**, *100*, 3219-3220.

50. Okahara, M.; Nakatsuji, Y.; Sakamoto, M.; Watanabe, M. *J. Inclusion. Phenom.* **1992**, *12*, 199-211.

51. Christensen, J. J.; Christensen, S. P.; Biehl, M. P.; Lowe, S. A.; Lamb, J. D.; Izatt, R. M. *Sep. Sci. Technol.* **1983**, *18*, 363-373.

52. Lamb, J. D.; Christensen, J. J.; Izatt, S. R.; Bedke, K.; Astin, M. S.; Izatt, R. M. *J. Am. Chem. Soc.* **1980**, *102*, 3399-3403.

53. Nakatsuji, Y.; Inoue, T.; Wada, M.; Okahara, M. *J. Inclusion. Phenom.* **1991**, *10*, 379-386.

54. Ensor, D. D.; McDonald, G. R.; Pippin, C. G. *Anal. Chem.* **1986**, *58*, 1814-1816.

55. Eugster, R.; Spichiger, U. E.; Simon, W. *Anal. Chem.* **1993**, *65*, 689-695.

56. Schaller, U.; Bakker, E.; Spichiger, U. E.; Pretsch, E. *Anal. Chem.* **1994**, *66*, 391-398.

57. Bakker, E.; Malinowska, E.; Schiller, R. D.; Meyerhoff, M. E. *Talanta* **1994**, *41*, 881-890.

58. Moyer, B. A.; Case, G. N.; Alexandratos, S. D.; Kriger, A. A. *Anal. Chem.* **1993**, *65*, 3389-3395.

59. Thoma, A. P.; Vivianii-Nauer, A.; Aravantis, S.; Morf, W. E.; Simon, W. *Anal. Chem.* **1977**, *49*, 1567-1572.

60. Morf, W. E.; Simon, W.; *Helv. Chim. Acta* **1986**, *69*, 1120-1131.

61. Horvai, H.; Graf, E.; Toth, K.; Pungor, E.; Buck, R. P. *Anal. Chem.* **1986**, *58*, 2735-2740.

62. Van den Berg, A.; van der Wal, P. D.; Skowronska-Ptasinska, M.; Sudholter, E. J. R.; Reinhoudt, D. N.; Bergveld, P. *Anal. Chem.* **1987**, *59*, 2827-2829.

63. Stolwijk, T. B.; Sudholter, E. J. R.; Reinhoudt, D. N. *J. Am. Chem. Soc.* **1989**, 111, 6321-6329.

64. Izatt, R. M.; Lamb, J. D.; Bruening, R. L. *Sep. Sci. Tech.* **1988**, *23*, 1645-1658.

65. Li, N. N. U.S. Patent 3 410 794, 1968.

66. Hochhauser, A.; Cussler, E. L. *AIChE Symp. Ser.* **1975**, *71*, 136.

67. Sugiura, M.; Kikkawa, M. *J. Membr. Sci.* **1989**, *42*, 47-55.

68. Sugiura, M. *Sep. Sci. Technol.* **1990**, *25*, 1189-1199.

69. Sugiura, M. *Sep. Sci. Technol.* **1992**, *27*, 269-276.

70. Lamb, J.D.; Schow, A.J. *J. Membr. Sci.,* in press.

Chapter 5

Facilitated Transport of Alkali Metal Cations Through Supported Liquid Membranes with Fatty Acids

N. M. Kocherginsky[1]

N. N. Semenov Institute of Chemical Physics, Moscow, Russia

This review describes the transport properties of liquid membranes with fatty acids as mobile carriers of ions. It is demonstrated that fatty acids induce coupled counter transport of alkali metal cations and protons. This electrically neutral process has a stoichiometry 1:1. Competing with this process is the generation of an electrical transmembrane potential. A theoretical description of transmembrane ion transport from one aqueous solution into another through an organic liquid membrane is presented. The model assumes that transport is facilitated by a low molecular weight acid carrier which is distributed in both the aqueous and membrane phases. Ion-exchange reactions are taking place simultaneously with diffusion in the aqueous solutions and can be the rate limiting steps for the transport process. An analytical expression describing the rate of transmembrane transport in this multi step process is presented. The experimental rate of metal ion transport by fatty acids is analyzed in terms of the theory. All parameters that are necessary to describe the kinetics, including ion-exchange rate constants, diffusion coefficients in all phases, and the thickness of reaction layers in both aqueous solutions, are determined.

Ion transport through liquid organic membranes from one aqueous solution into another is a process which has a sequence of at least three steps, i.e. extraction from an aqueous sourse solution into the organic membrane, diffusion through the membrane due to a concentration gradient, and reextraction into a receiving aqueous solution. In the case of inorganic ion transport facilitated by a low molecular weight carrier, the total process is complicated by reversible chemical interactions of the hydrophilic ion and usually much more hydrophobic carrier. As a result a

[1]Current address: Department of Chemical Engineering, National University of Singapore, 10 Kent Ridge Crescent, Singapore 0511

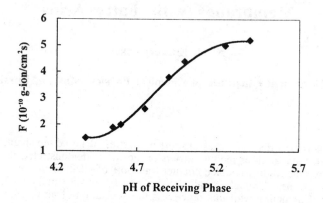

Figure 1. Rate of H⁺-transport Due to pH difference as a Function of the pH in the Receiving Solution. The filter was impregnated with oleic acid. (Adapted from ref. 4.).

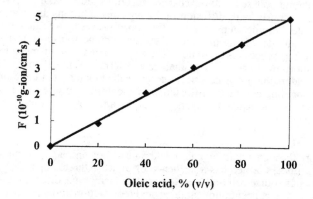

Figure 2. Maximum Rate of H⁺-transport as a Function of the Oleic Acid Content in the Membrane. The second component of the impregnating solution was isobutyl ester of lauric acid. (Adapted from ref. 4.).

detailed kinetic mechanism of membrane transport should be based on knowledge and correct description of all of these steps.

The purpose of this paper is:

a) to summarize the main facts of our previously published experiments describing alkali metal transport through supported liquid membranes (*1-5*);

b) to present a theoretical model of facilitated transport, taking into account the possibility of the transfer of a dissolved carrier from the liquid membrane into the aqueous phases and its diffusion with simultaneous ion exchange reactions in unstirred boundary layers of water (*6,7*);

c) to demonstrate the applicability of this theory for the description of fatty acid-facilitated transport.

Our interest in fatty acids was determined essentially by the facts that:

a) these molecules are the simplest acidic low molecular carriers and can be used as a model system;

b) they can be utilized for industrial separation of different metals (*8*); and

c) they play an important role in regulation of ion transport in biomembranes (*9*).

Materials and Methods.

The experiments were carried out with a liquid membrane immobilized in pores of nitrocellulose filters with a mean pore diameter of $2.5 * 10^{-4}$ cm and a thickness $l \sim 0.01$ cm. The filters were impregnated by simple immersion into liquid oleic acid or its mixture with isobutyl laurate. Then these impregnated filters were vertically clamped in a temperature-controlled Teflon cell. The cell was divided by this membrane into equal parts having 15 ml volumes. The depth T of each half-cell is 3.5 cm. If necessary the membrane thickness was changed by the use of stacks of impregnated filters. Aqueous solutions in both half-cells were preequilibrated with the liquid organic phase used for impregnation the filter. The solutions in both chambers were stirred with plastic stirrers connected to an electrical motor. Calculations of the rate of H^+-ion transport were based on pH changes with time and the buffer capacity of solutions. The rate of K^+ transport was measured with a K^+-selective electrode. The transmembrane electrical potential was measured with Ag/AgCl electrodes and agar bridges.

Experimental Part. Transport of Ions with Fatty Acid as the Carrier.

Electroneutral Transport. When the membrane was impregnated with oleic acid, it was possible to observe the decrease of pH in the receiving solution, due to H^+ transport from the more acidic source to the more alkaline receiving solution. The rate of this process reached a maximum after a relatively small lag period (~20 min) and H^+ transport was stimulated by increasing the KCl concentration in the receiving solution. Addition of 1M KCl to the source solution practically did not change the rate of transport.

If pH $_{source}$ was 2.5 and pH $_{receiving}$ was 5.3, the maximum value of the specific rate of H^+-transport F_{max} was $(5-6) \times 10^{-10}$ g-ion/cm^2 s in the presence of 1M KCl in receiving solution. The plot of the dependence of F on the pH of the receiving phase is shown in Figure 1. The dependence is essentially a titration curve with F equal to half of the maximal rate at pH 4.8, which is similar to the pK_a of a fatty acid carboxylic group in water. There was no transport of H^+ if the membrane was impregnated only with isobutyl laurate, and the rate of H^+ transport was proportional to the oleic acid content in impregnating liquid (Figure 2).

It was also possible to obtain the transport of H^+ if the source and receiving solutions had the same pH, but different concentrations of K^+. In these experiments the solution with the higher KCl concentration finally had a lower pH,

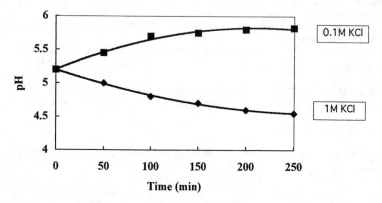

Figure 3. Kinetics for the pH Gradient Formation Due to a KCl
Concentration Difference. The filter was impregnated with oleic acid.
(Adapted from ref. 4.).

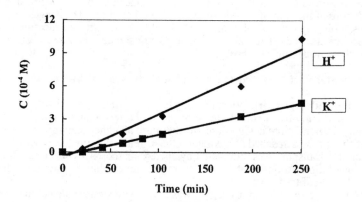

Figure 4. Kinetics of H^+ and K^+ Transport Through the Membrane
Impregnated with Oleic Acid. (Adapted from ref. 4.).

while the other solution became more alkaline (Figure 3). This means that oleic acid facilitates K^+-H^+ exchange and the gradient of one ion can be used as a source of energy for the active transport of the other ion against its concentration gradient.

Kinetics of K^+ transport from a 1 M source solution into the receiving solution containing 0.1 M NaCl is presented in Figure 4. Both solutions initially had the same pH, equal to 5.8. With time the concentration of H^+ in the source solution increased. It is interesting to note that the flux of K^+ was almost twice higher than that of H^+ in the opposite direction; whereas for a 1:1 stoichiometry, the fluxes have to be same. The difference can be explained by simultaneous transport of Na^+, which was added to the system and was transported by oleate in the same direction as H^+. $MgCl_2$ had a different effect in comparison to KCl. Initially, if added into more alkaline solution in the presence of KCl, it induced fast release of H^+. Then the rate of acidification decreased to a value which was lower than that without $MgCl_2$. It seems that Mg^{+2} is able to form a stable complex with oleate anion, thus decreasing the content of free and mobile carrier in the system. Stimulation of K^+-H^+ exchange was demonstrated also with saturated capric acid. In this case we used another type of liquid membrane, which was a bulk solution of the fatty acid in chloroform, separating two aqueous phases in U-shaped tube or similar cells (*1*). For this bulk liquid membrane system the rate of transport was an order of magnitude lower that that in the experiments with impregnated filters.

The rate of transport could be regulated by temperature. In experiments with impregnated filters, the rate of facilitated transport decreased to zero when temperature was below the phase transition temperature of the oleic or capric acid. Another way to change the rate of transport was to change the rate of stirring. Switching off the stirring decreased the **F** value by a factor of 3 due to a growth of the thickness of unstirred water layers near the membrane. An increase in the membrane thickness also reduces the transport rate. The value of 1/**F** was a linear function of the membrane thickness both with and without stirring (Figure 5). The slopes of the lines in Figure 5, which characterize the "resistivity" of the membrane without aqueous layers in the facilitated ion transport, were the same in both cases. As will be demonstrated below, the resistivity is determined by the diffusion coefficient of the carrier in the membrane. The lines do not pass through the origin, which indicates that the resistance to transport as determined by the membrane-water interphase and the unstirred water layers is also significant. The thickness of these layers and the corresponding segment on Y-axis increases 10-fold when the aqueous phases are not stirred. Comparison of 1/**F** for l=0 and l= 0.01 cm demonstrates that without stirring the resistance of the unstirred layers can be as high as 80% of the total resistance of membrane. The relative role of unstirred layers must be even higher for biological membranes, having the thickness less than 100 Å. Detailed analysis of the possible role of unstirred layers and fatty acids in regulation of permeability through biomembranes is given in (*5*).

Electrogenic Processes. Measurements of the transmembrane electric potential $\Delta\varphi$ in a gradient of KCl demonstrated that $\Delta\varphi$ is generated on filters impregnated either with oleic and capric acid. Its sign corresponds to K^+ / Cl^- selectivity, i.e. a more nagative potential was in the solution with higher salt concentration. The slope of the electrode function was as high as 50 mV for a 10-fold change in concentration of K^+ in solutions with pH= 5.8 (Figure 6). Essentially the same slope was observed in the case of Teflon porous support, but the slope was much slower if a fatty acid esters were used instead of fatty acids. Membranes had no cation / anion selectivity if pH of both solutions was 2.5.

The difference of transmembrane potentials could be observed also in the experiments with a pH gradient. In this case the increase of KCl concentration in

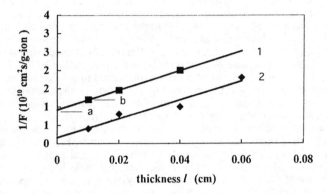

Figure 5. Dependence of 1/F on the Membrane Thickness *l*. 1- without stirring, 2- with stirring.

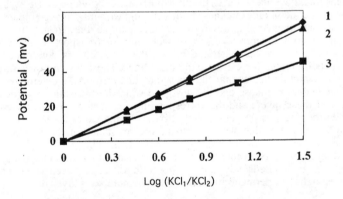

Figure 6. Dependence of the Transmembrane Potential on the Ratio of KCl Concentrations in the Source and Recieving Solutions. 1- Teflon filter impregnated with oleic acid, 2- nitrocellulose filter impregnated with oleic acid, 3- nitrocellulose filter impregnated with the isobutyl ester of lauric acid. (Adapted from ref. 3.).

both solutions resulted in the decrease of the ΔpH-induced transmembrane potential. Also, a slow decrease of potential with time was observed.

If the membrane separated two aqueous solutions in a symmetrical system, its specific resistance decreased from several Mohms x cm^2 to ~ 10-20 kohms x cm^2 when the pH of both solutions was increased from 3 to 6. Again, as it was in the case of the pH-dependence of transmembrane ion transport, the pH dependence resembled a titration curve with an inflection at a pH near the pK of the carboxylic group in water (~5.0). The process was accompanied by a more than 10-fold increase in the electrical capacitance of the membrane *(3)*.

All these effects can be explained by the fact that concurrent with stimulation of the electrically neutral K$^+$-H$^+$ exchange, the anion of the fatty acid traverses through the membrane in a free form having a negative charge. The relative role of both of these processes is determined by the ion exchange and concentrations of ions in the solutions. The higher the K$^+$ or H$^+$ concentration, the lower is the probability that the anion form of carrier will go back through the membrane.

Theoretical Part. Kinetics of Facilitated Transmembrane Transport

The theoretical description of the kinetics of transmembrane transport through a liquid membrane should be based on the principles of solvent extraction kinetics. It should be determined by the processes at both water/membrane interphases and should also involve the intermediate step of diffusion in the membrane. Thus the existence of all these three steps makes the membrane system and its description much more complicated than the relatively simple water/organic phase. However, even the kinetics mechanism in simpler extraction systems is often based on the models dealing only with some limiting situations. As it was pointed out in the beginning of this paper, the kinetics of transmembrane transport is a function both of the kinetics of various chemical reactions occurring in the system and of diffusion of various species that participate in the process. The problem is that the system is not homogeneous, and concentrations of the substances at any point of the system depend on the distance from the membrane surface and are determined by both diffusion and reactions. The solution of a system of differential equations in this case can be a serious problem.

As a rule it is assumed that one of the processes (it could be either a chemical reaction or diffusion in a bulk volume) is much slower than the other and this slow process is the rate-controlling step. These situations are called kinetic and a diffusional regimes, respectively.

Sometimes it is assumed that chemical the reactions can occur in an imaginary two-dimensional region of zero thickness called liquid-liquid interface. Unfortunately we do not know much about the nature of the interface between two liquids. In this case we have to use surface concentrations (per unit of area, using geometrical abstraction) for the processes which occur in a thin, but still having a macroscopic thickness, layer. If the reactions are fast and the rate of extraction is determined by interfacial diffusion, it is often assumed that the resistance for the process is determined by two hypothetical stagnant thin layers of finite thickness located on the aqueous and organic side of the interface. This model of the interface is known as the two-film theory and the layers are called diffusion layers or Nernst layers. To describe experimental results usually it is assumed that the thickness of these layers in water is ~10^{-2} cm with very slow stirring and ~ 10^{-4}cm with intensive stirring *(10)*.

This idea of one rate controlling step may be correct for a series of chemical and diffusion processes, but it could give the wrong interpretation in systems with simultaneous and parallel diffusion and reactions, where the slow step is not really important. When both chemical reactions and film diffusion occur at comparable rates, the extraction is in the so-called mixed diffusional-kinetic regime, and it is a problem to identify the extraction kinetics regime.

In extraction of metal complexes with a fatty acid a metal cation reacts with the carboxylic group to form a neutral salt that is preferentially dissolved in the organic phase. The conventional description of this extraction process supposes an extraction mechanism that is wholly dependent on chemical reactions in the aqueous phase, and requires at least a reasonable water solubility of the extractant (8).

Metal extraction by carboxylic acids is a stepwise process in which the acid partitions between the organic phase and the aqueous phase and ionizes in the latter. Then the anion reacts with the metal cation and the neutral salt partitions between the organic and aqueous phases. Although this description is well-known in extraction, analysis of transport through liquid membranes with the same chemistry is usually based for the sake of simplicity, on the assumption that metal ions in the aqueous phases are present in well-stirred reservoirs and the ion-exchange itself takes place only at the membrane surface. The carrier in this case is able to move only inside the membrane from one of its surfaces to the other. These mechanisms where the carrier is or is not able to leave the membrane could be called, a "big and small carrousel", respectively (6).

The purpose of the next section is to give a short description of the theoretical analysis of the big carrousel mechanism and then to use this model for a quantitative description of a facilitated K^+ transport through a liquid membrane with a fatty acid carrier. A more detailed description of the model and an analysis of transmembrane redox reactions facilitated by quinones is given in (6,7).

In the case of ion-exchange reactions in aqueous solutions, we have:
Source aqueous solution (solution 1)

$$K_1^+ + HA_1 \underset{k_-}{\overset{k_+}{\leftrightarrow}} KA_1 + H_1^+ \tag{1}$$

Receiving aqueous solution (solution 3)

$$K_3^+ + HA_3 \underset{k_-}{\overset{k_+}{\leftrightarrow}} KA_3 + H_3^+ \tag{2}$$

where A^- is an anion of oleic acid. It is assumed that there is no dissociation or ion exchange in the membrane. For fatty acids this assumption is valid if the pH of both solutions is less than 4.

For unstirred aqueous solutions, we can write a set of differential equations for both aqueous phases and membrane phase:
Solution 1

$$\frac{\partial[HA]_1}{\partial t} = D_w[HA]_1'' - k_+[K]_1^+[HA]_1 + k_-[H]_1^+[KA]_1$$
$$\frac{\partial[KA]_1}{\partial t} = D_w[KA]_1'' + k_+[K]_1^+[HA]_1 - k_-[H]_1^+[KA]_1 \tag{3,4}$$

Membrane phase 2

$$\frac{\partial [HA]_2}{\partial t} = D_m [HA]_2''$$

$$\frac{\partial [KA]_2}{\partial t} = D_m [KA]_2''$$

(5,6)

Solution 3

$$\frac{\partial [HA]_3}{\partial t} = D_w [HA]_3'' - k_+ [K]_3^+ [HA]_3 + k_- [H]_3^+ [KA]_3$$

$$\frac{\partial [KA]_3}{\partial t} = D_w [KA]_3'' + k_+ [K]_3^+ [HA]_3 - k_- [H]_3^+ [KA]_3$$

(7,8)

where D_w and D_m are diffusion coefficients of the carrier in water and the membrane, k_+ and k_- are rate constants for the forward and back ion-exchange reactions and are the same in both solutions. The symbol " means a second partial derivative with respect to x. It is assumed that diffusion coefficients for both oleate with K^+ or H^+ are the same.

It is also assumed that the interface does not offer significant resistance to ion transfer, i.e. distribution of the carrier through the membrane/aqueous interface is fast and does not deviate appreciably from equilibrium. In this case for the left and right surfaces of the membrane having a thickness l, we can write:

at $X = -l/2$

$$K_d^{KA} = \frac{[KA]_2}{[KA]_1}$$

$$K_d^{HA} = \frac{[HA]_2}{[HA]_1}$$

(9,10)

and at $X = +l/2$

$$K_d^{KA} = \frac{[KA]_2}{[KA]_3}$$

$$K_d^{HA} = \frac{[HA]_2}{[HA]_3}$$

(11,12)

where $X = 0$ at the middle of the membrane.

In the steady state, the concentrations of both HA and KA are time-independent, which means that for each interface we can write the next four equations:

At $X = -l/2$

$$D_w [HA]_1' = D_m [HA]_2'$$

$$D_w [KA]_1' = D_m [KA]_2'$$

(13,14)

Symbol ' means a first partial derivative with respect to x. At $X = +l/2$

$$D_w[\,HA\,]_3' = D_m[\,HA\,]_2'$$
$$D_w[\,KA\,]_3' = D_m[\,KA\,]_2'$$

(15,16)

Boundary conditions indicating the absence of any flux through the cell walls are :
at $X = \pm(T + \dfrac{l}{2})$

$$[\,HA\,]_1' = 0$$
$$[\,KA\,]_1' = 0$$
$$[\,HA\,]_3' = 0$$
$$[\,KA\,]_3' = 0$$

(17-20)

where T is the thickness of a half-cell.
The equation of the matter conservation law is:

$$\int_{-T-\frac{l}{2}}^{-\frac{l}{2}}([\,HA\,]_1 + [\,KA\,]_1)dx + \int_{-\frac{l}{2}}^{\frac{l}{2}}([\,HA\,]_2 + [\,KA\,]_2)dx + \int_{\frac{l}{2}}^{\frac{l}{2}+T}([\,HA\,]_3 + [\,KA\,]_3)dx = N \quad (21)$$

where N is a content of the carrier per unit area of membrane.

The solution of this problem gives the following equation for the specific steady state rate of transport **F** through the membrane, separating two aqueous solutions with different concentrations of K^+ (6):

$$F = (\frac{k_+[K]_1^+}{k_+[K]_1^+ + k_-[H]_1^+} - \frac{k_+[K]_3^+}{k_+[K]_3^+ + k_-[H]_3^+}) \frac{1}{2T + K_d l} \frac{N}{R_1 + R_2 + R_3}$$

(22)

where R_1 and R_3 are proportional to the transport resistances of the two aqueous phases and R_2 is the membrane resistance. Similar equations can be written for ion transport induced by a pH gradient.

One can see that the flux is equal to the product of three terms. The first one is determined by chemistry, the second - by the geometry of the system and the hydrophobicity of the carrier and the third - by three serial resistances of the source, membrane and receiving solutions. The next three equations describe these resistances:

$$R_{1,3} = \frac{L_{1,3}}{D_w} \coth \frac{T}{L_{1,3}}$$
$$R_2 = \frac{1}{K_d D_m}$$

(23-25)

where

$$L_{1,3} = \sqrt{\frac{D_w}{k_+}} \frac{1}{\sqrt{[K]_{1,3}^+ + \frac{k_-}{k_+}[H]_{1,3}^+}} \qquad (26, 27)$$

While the membrane resistance coincides with the well-known equation for membrane transport without reaction, the equations for each aqueous solution have a correction factor coth T/L. The new parameter L here is determined by the ratio of diffusion and chemical rates and can be called an effective thickness of the reaction layer. If T/L is high, the correction factor is equal to 1. Nevertheless even this situation may be nontrivial in the case of low R_2, where the transmembrane flux is proportional to the square roots of both the diffusion coefficient and rate constants in water.

The opposite situation (T/L is small) corresponds to slow reactions and small distances between the membrane and the chamber walls. In this case, the correction factor can be very large. The rate of the transmembrane processes, then, is not a function of any diffusion coefficient but is determined only by the rates of the chemical processes. This situation could be important for different systems of biological membranes and subcellular structures.

It is noteworthy that earlier the systems with simultaneous transport and reactions of a mobile carrier were discussed in the literature only for the case of different reactions in the membrane, which could be important for gas transport processes (*11,12*).

Implications of the Theory for the Facilitated Transport with Fatty Acids.

1. Analysis of the first term in the equation 22 demonstrates that transmembrane transport could be induced by the difference of either H^+ or K^+ concentrations in the two solutions, i.e. concentration gradient of one ion can be used to provide an uphill transport of another ion. In biology, this type of process is well-known as secondary active transport. At equilibrium, where $F=0$, we have

$$\frac{[K]_1^+}{[K]_3^+} = \frac{[H]_1^+}{[H]_3^+} \qquad (28)$$

In a pH gradient-driven processes, it is necessary to have K^+ in the more alkaline solution. If there were no K^+ in the receiving solution, we would have only a pH-induced transmembrane potential, as determined by the fatty acid anion distribution, and the rate of H^+ transport due to its gradient would be zero.

2. In the experiments with an ion gradient 1/**F** has to be proportional to the concentration of this ion in the receiving solution and to 1/concentration in the source solution. Both of these effects were demonstrated in the experiments with low and approximately constant pH in the source solution and acidification of receiving phase (Figure 7) and also in experiments with a K^+ gradient and varying initial concentrations (Figure 8). The intercepts of these plots give F_{max}. This number has the sense of the maximum rate which can be achieved due to saturation of the carrier in the source solution and negligible rate of transport of the same ion in the opposite direction. F_{max} for oleic acid was 5.5×10^{-10} g-ion/cm^{2}s (Figure 7) and for K oleate it was 2.5×10^{-10} g-ion/cm^{2}s (Figure 8). Using these numbers and the slope, we were able to calculate the K^+/H^+ ion selectivity, which was equal to 1.3×10^{-5} in both experiments.

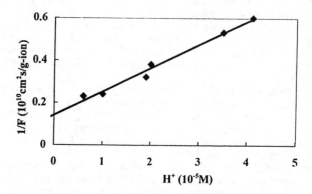

Figure 7. Dependence of $1/\mathbf{F}$ on the H^+-concentration in the Solution with Higher pH. Both solutions had 1M KCl, $pH_{receiving}$=5.3, pH_{source} =2.5, 40°C. The filter was impregnated with oleic acid. (Adapted from ref. 5.).

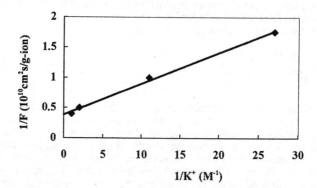

Figure 8. Dependence of $1/\mathbf{F}$ for K^+ Transport on the KCl Concentration in the Source Solution. The receiving solution initially had 0.1 M NaCl and no KCl, $pH_{receiving}$=pH_{source} =6.0, 20°C. The filter was impregnated with oleic acid. (Adapted from ref. 5.).

3. If transport through the membrane is a rate-limiting step ($K_d < 2T/l$ and R_2 is large), the rate of the process should be proportional to K_d, but for carriers with high hydrophobicity F is not a function of K_d. In the experiments performed with bulk liquid membranes it was demonstrated that the fatty acid efficiency increased with their length in the sequence $C_4 \div C_{12}$ but it was essentially the same for lauric (C_{12}) and palmitic (C_{16}) acids (*1*).

4. The experiments performed with impregnated filters allow the membrane thickness to be changed. If $K_d < 2T/l$, according to equation 22 the value of $1/F$ has to be proportional to the membrane thickness. The intercept of this line is determined by resistances of aqueous phases. The slope is determined by the diffusion coefficient in the membrane.

Results presented in Figure 5 correspond to this prediction. In the experiments with pH low enough in source solution and high in the receiving solution, all oleate is in its acidic form at the first surface and in the salt form at the second surface. Taking into account the filter porosity, we have ~ 2.5 M concentration of oleic acid in the membrane at the acidic solution interface. Using this number and the slope of the dependencies in Figure 5, we have a diffusion coefficient of oleic acid in the membrane equal to 1.4×10^{-9} cm^{2}s.

5. As long as the unstirred aqueous solutions play a dominant role as resistances to transmembrane transport, it is important to answer the question which of these layers is rate-limiting and to compare its resistance with the membrane resistance R_2. R_2 can be estimated using the diffusion coefficient in the membrane, its thickness and the partition coefficient of oleic acid, which is 4×10^4 (*5*). In comparison, $\log K$ for octanoic acid distribution from an aqueous phase into decane is 4.3 (*13*). The calculated R_2 value is 180 s/cm. Resistance of the aqueous solutions is 5 times higher (Figure 5) and is 900 s/cm. Then using the ratio k_+/k_- as determined from the Figures 7 and 8 and which is equal to 1.3×10^{-5}, we calculate that $L_1/L_2 = 360$. This means that the transport resistance of the salt solution is much higher than that of the acidic solution.

In this case the equation that describes the aqueous resistance is reduced to:

$$R_1 = \frac{1}{\sqrt{D_w(k_+[K]^+ + k_-[H]^+)}} \qquad (29)$$

Using k_+ and k_- equal to 500 M^{-1} s^{-1} and 5×10^7 M^{-1}s^{-1} (*7*), we find $D_w = 2.5 \times 10^{-9}$ cm^{2}s^{-1} and $L_1 = 2.25 \times 10^{-6}$ cm. The thickness of the reaction layer in the acidic solution is ~1 Å, which is not surprising for this fast reaction. Both D_w and D_m are much lower than typical diffusion coefficients in water, but both of these coefficients were calculated independently. D_m is based on the experimental flux dependence on membrane thickness, while D_w is calculated using its relationship to the transport resistance and reaction constants. The low values for diffusion coefficients in this case cannot be explained only by the influence of the porous support used in these experiments. For example diffusion coefficients of dichlorobenzoquinone (*14*) and different nonelectrolytes (*15*) in similar membranes were $>10^{-7}$ cm^{2}s^{-1}. It seems that an aggregation processes, which usually are not taken into account (*16*) but are important for long chain fatty acids, are responsible for the decrease of the diffusion coefficient of oleic acid both in water and the membrane.

Earlier we demonstrated that the lag period for ion transport induced by fatty acids is sensitive to the fatty acid concentration in the membrane and can be decreased more that 100 times by simple decrease of the carrier content in the membrane. This effect was attributed to the coexistence of volume diffusion and two-dimensional diffusion of fatty acids along the internal surfaces of the pores in nitrocellulose. It was possible to explain this dependence quantitatively assuming local equilibrium between the volume and adsorbed fatty acid and its dimerisation and diffusion coefficient in the membrane $D_m = 2 \times 10^{-9}$ cm^{2}s^{-1} (2).

One might think that as long as the thickness of reaction layer is less then 1000 Å its resistance to transport has to be negligible. In reality the transport resistance in this case without stirring is determined by both chemical and diffusion steps. Their reciprocal dependencies on the linear dimension of the system compensate each other and the transport resistance in this case is not a function of any distance at all. Theoretical analysis of the influence of stirring is discussed in (7).

Acknowledgments. The author would like to thank his former students and colleagues I. S. Osak, L. E. Bromberg and A. V. Mogutov. This paper would never be possible without their help.

Literature Cited

1. Kocherginsky, N. M.; Dolginova, E. A.; Petrov, V. V.; Antonov, V. F.; Moshkovsky, Yu. Sh. *Biophysics* **1980**, *25*, 846-851.
2. Kocherginsky, N. M.; Osak, I. S. *Russ. J. Phys. Chem.* **1986**, *60*, 725-727.
3. Kocherginsky, N. M.; Osak, I. S. *Russ. J. Phys. Chem.* **1987**, *61*, 1018-1020.
4. Kocherginsky, N. M.; Osak, I. S. *Biphysics* **1988**, *33*, 85-89.
5. Kocherginsky, N. M.; Osak, I. S.; Demochkin, V. V.; Rubaylo, V. L. *Biologicheskie Membrani* (Biological Membranes) **1987**, *4*, 838-848, (Russian).
6. Mogutov, A. V.; Kocherginsky, N. M. *J. Membr. Sci.* **1993**, *79*, 273-283.
7. Mogutov, A. V.; Kocherginsky, N. M. *J. Membr. Sci.* **1994**, *86*, 127-135.
8. Cox, M.; Flett, D. S., In *Handbook of Solvent Extraction*; Lo, T. C.; Baird, M. H. I.; Hanson, C., Eds.; Wiley : New York, 1983; pp. 53-89.
9. Stryer, L.; *Biochemistry*, Freeman: New York, 1988; pp. 469-480.
10. Danesi, P. R., In *Principles and Practices of Solvent Extraction*, Rydberg, J.; Musikas, C.; Choppin , G. R., Eds.; Marcel Dekker: New York, 1992; pp. 157-207.
11. Way, J. D.; Noble R. D. In *Membrane Handbook*, Winston Ho, W. S.; Sirkar, K.; Eds., Van Nostrand Reinhold: New York, 1992; pp. 833-865.
12. Cussler, E. L. *Diffusion. Mass Transfer in Fluid Systems*, Cambridge University Press: Cambridge, 1984; pp. 346-412.
13. UIPAC. *Equilibrium Constants of Liquid-Liquid Distribution Reactions.* Part III. Compound Forming Extractants, Solvating Solvents and Inert Solvents; Pergamon: New York, 1974; pp. 4-10.
14. Kocherginsky, N. M.; Goldfeld, M. G.; Osak, I. S. *J. Membr. Sci.* **1989**, *45*, 85-98.
15. Kocherginsky, N. M.; Bromberg, L. E. *Russ. J. Phys. Chem.* **1988**, *62*, 1112-1115.
16. Johnson, P. A.; Babb, A. L. *Chem. Rev.* **1956**, *56*, 387-454.

Chapter 6

Emulsions for Liquid Membrane Extraction: Properties and Peculiarities

E. V. Yurtov[1,2] and M. Yu. Koroleva[1]

[1]Mendeleev University of Chemical Technology and
[2]Scientific and Industrial Enterprise, Ecospectr,
Miusskaya Square 9, Moscow 125047, Russia

Characteristic features of emulsions which are used to extract and concentrate substances from aqueous solutions are examined. The effects of surfactant and carrier concentrations and external and internal phase compositions upon the properties of the extracting emulsions are discussed. Several mathematical models for the rheological curves are considered, and regions of adequacy of the models are determined. An influence of nanodispersion formation on mass transfer through the interface and on the properties of the extracting emulsions for cholesterol is demonstrated.

There is an ever increasing interest in the use of liquid membranes for performing chemical separations. Emulsion liquid membrane (ELM) systems in which the targeted chemical species in an aqueous solution is extracted with a multicomponent emulsion have a variety of applications. These include isolation and concentration of valued or harmful substances in industrial chemistry, separation of substances for determination in analytical chemistry, separation of pollutants in environmental remediation, and detoxification of biological fluids by removal of harmful substances of exogenic and endogenic origins (*1*).

ELM systems are usually prepared by first forming an emulsion between two immiscible phases (*2*). When this emulsion is dispersed in a third (continuous) phase by agitation, the extraction system is produced. The membrane phase is the liquid phase that separates the encapsulated, internal droplets in the emulsion from the external, continuous phase. Although the internal, encapsulated phase and the external, continuous phase are miscible, the membrane phase cannot be miscible with either in order to be stable. In the discussion which follows, the internal and external phases are aqueous solutions which are separated by an oil phase in water-in-oil-in-water (W/O/W) emulsion systems.

To be effective an extracting emulsion must meet a number of requirements. The emulsion must be stable during the period of application. For such stability

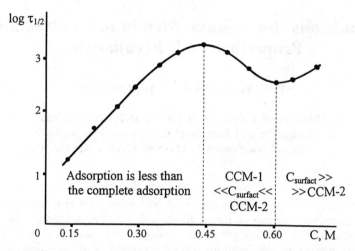

Figure 1. The Half-Separation Period for the Internal Phase at 25 °C Versus the Concentration of Span 80 for a 65% Initial Volume Fraction of Internal Phase.

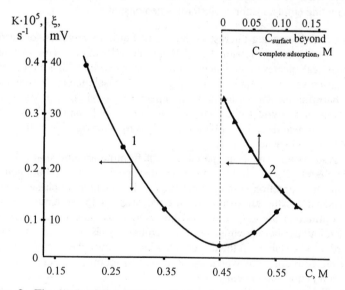

Figure 2. The (1) Rate Constant for Coalescence and (2) ξ-Potential of the Internal Phase Droplets Versus the Concentration of Span 80 in the Liquid Membrane.

minimized transfer of water from the exterior, continuous phase into the internal, encapsulated, aqueous phase of the emulsion is an important factor. It is also beneficial if the emulsion is stable for long periods of storage prior to its application. To provide for high capacity of the extracting emulsion, a concentrated disperse phase should be utilized in which the volume fraction of internal, encapsulated phase may be as high as 60-70%.

Effects of Composition Variables in ELM Systems.

In a W/O/W ELM system, the organic membrane is generally stabilized be a surfactant and often contains a carrier species which provides facilitated transport of the target substance. The properties of the ELM system are influenced by the surfactant and carrier concentrations, as well as the compositions of the external and internal aqueous phases.

Effect of Surfactant Concentration. Raising the surfactant concentration increases the stability of adsorption layer which enhances resistance of the emulsion to coalescence. The radius of the aqueous internal phase droplets in the emulsion initially diminishes with the increase in surfactant concentration and then remains constant. This results from the tendency of the dispersed phase to minimize the interfacial area for a given amount of surfactant. Despite a high degree of dispersion of the internal phase, the droplets coalesce until they attain a surfactant content at the interface which is close to the limiting value of adsorption. When the surfactant is present in an amount sufficient for stabilization of all of the drops of internal phase formed as a result of the dispersion, a maximum stability of the emulsion (Figure 1) and a minimum coalescence rate constant (Figure 2, curve 1) are attained.

Study of the ξ-potential for the drops of internal phase shows that as the surfactant concentration is increased above that needed for saturation of the adsorption layer at the interface in the emulsion, the ξ-potential of the internal phase droplets decreases (Figure 2, curve 2). This lowers the repulsion energy between the disperse phase drops which increases the rate of their coalescence and leads to a decrease in the stability of the emulsion (*3*).

Some increase in the stability of the emulsion as the surfactant concentration exceeds CCM-2 in the liquid membrane (Figure 1) is attributed to formation of a heteroadsorption layer by adsorption of surfactant micelles at the interface or formation of surfactant lamellar layers in the liquid membrane.

Change in the volume fraction of internal phase in the emulsion leads to a variation of the specific interface. For each surfactant concentration, there is a different maximum in the dependence of the emulsion stability on the fraction of the disperse phase (Figure 3). The maximum stability corresponds to saturation of the adsorption layer by the surfactant. For a higher volume fraction of the internal phase at the specified surfactant concentration, the adsorption layer is unsaturated and the stability of the emulsion is diminished. For lower volume fractions of internal phase, the emulsion stability diminishes due to the presence of excess surfactant in the liquid membrane analogous to Figure 1.

As the surfactant concentration is increased, the maximum in emulsion stability shifts toward higher fractions of the internal aqueous phase, but the absolute stability diminishes. Thus to determine the composition of the most stable emulsion, it is

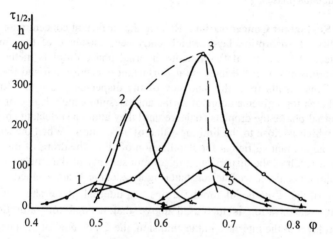

Figure 3. The Half-Separation Period of the Internal Phases at 60 °C Versus
the Initial Fraction of Internal Phase at Span 80 Concentrations of 1) 0.21, 2)
0.35, 3) 0.45, 4) 0.50, and 5) 0.55M.

essential to take into account the interrelated influences of the internal phase and surfactant concentrations.

Effect of Carrier Concentration. The carrier (extractant) also exerts a significant influence on the stability of an extracting emulsion. As a rule, the carrier will be surface-active which will reduce the stability of the emulsion due to competitive adsorption at the interface. However for ELM systems, a high concentration of carrier in the liquid membrane is usually not necessary. For each carrier, the optimal concentration will be determined by the opposing influences of the carrier on the rate of extraction of the target substance and the stability of the extracting emulsion.

Effect of External Aqueous Phase Composition. The properties of the extracting emulsion may be varied by altering the composition of the external, continuous phase which influences its viscosity and polarity. An increase in viscosity can provide a certain kinetic stability for the emulsion as a result of the decrease in rate at which the thickness of a liquid film between two surfaces diminishes (3). This factor can play a significant role in reverse emulsions (*i.e.* oil-in-water-in-oil emulsions).

Effect of Internal Phase Composition. In addition to the factors mentioned above, the stability of the extracting emulsion is influenced significantly by the composition of the internal phase. Among the factors which play the greatest role in determining the stability are the pH and ionic strength of the internal phase and the presence of organic substances (1).

With an increase in the ionic strength of the internal phase, the stability of the emulsion is enhanced. For membrane extraction, this is an important point because it makes possible the enhancement of capacity for the emulsion with respect to the target substrate by increasing the content of a substrate-complexing reagent in the internal phase.

Water Transfer in Extracting Emulsions.

Transmembrane transfer of water from the external (continuous) phase into the internal (encapsulated) phase (*i.e.* swelling of the emulsion) is an undesirable process. Some of the primary factors which determine the rate of water transfer are the type and concentration of surfactant in the liquid membrane. The direction of the transmembrane transfer of water in an extracting emulsion is determined by the sign of the water activity gradients.

Swelling of the emulsion usually does not react equilibrium. Due to the large difference between the osmotic pressures of the external and internal aqueous solutions, the swelling of extracting emulsions may be considerable. Therefore optimization of the residence time of the extracting emulsion in the apparatus to concentrate the target substance to the maximum extent is important.

Rheological Properties of Extracting Emulsions.

Information about the rheological properties of extracting emulsions and their fluidity enables one to optimize the processes of preparation of the extracting emulsion, transportation of the liquid dispersions, and reduction of energy consumption at the various stages of their handling.

Study of the flow curves shows that emulsions with a low content of dispersed phase ($\phi <$ 0.1) appear to be Newtonian fluids. As the dispersed phase content increases ($\phi >$ 0.1), the extracting emulsion becomes a non-Newtonian, pseudo-plastic liquid.

Studies of concentrated emulsions reveal that their rheological properties are not time-dependent. (The rheological studies were performed for a quasi-equilibrium state of the emulsion with respect to the coalescence and sedimentation stability.) There is no hysteresis on the flow curves of the emulsion.

In practice, complete rheological curves for concentrated emulsions often cannot be obtained. This is due to the instability of disperse systems with a high content of dispersed phase at high shear rates. Difficulties with the instrumental techniques employed for such measurements may also arise because the values of the viscosity and the shear stress at the transition from the unperturbed to a completely destroyed structure vary over a wide range. Under certain conditions, incomplete rheological curves must be used for analysis and prediction of the viscosity of emulsions.

A first group of models treat the system as a liquid-like substance with some deviation of its properties from those of a Newtonian fluid (the Ostwald-Weil model)

$$\eta_{eff} = K \gamma^{n-1} \tag{1}$$

where η_{eff} is the effective viscosity of the emulsion, γ is the shear rate, and K is the consistency coefficient, or as a solid-like substance with a yield point (the Bingham model)

$$\eta_{eff} = \eta_{pl} + P_y/\gamma \tag{2}$$

where η_{pl} is the plastic (Bingham) viscosity and P_y is the yield strength.

The important advantage of these models is that they account for the physiochemical nature of the systems.

The Ostwald-Weil model has broader applicability because it is based on treatment of the liquid-like properties of the disperse system. The values of n and K for this model are given in Figure 4. This model adequately describes the experimental data for the region presented in Figure 5. It should be noted that this model is most appropriate for description of properties of emulsions with a low content of the dispersed phase. At high values of ϕ, the region of adequate description is reduced. This makes the model inappropriate for description and prediction of the viscosity of concentrated water-in-oil emulsions.

The second group of models which are based on fitting the mathematical equation with empirical coefficients to adequately describe the experimental data include the:

Steiner model

$$\eta_{eff} = 1/(C + BP^2) \tag{3}$$

Ferry model

$$\eta_{eff} = \eta_0/(1 + CP) \tag{4}$$

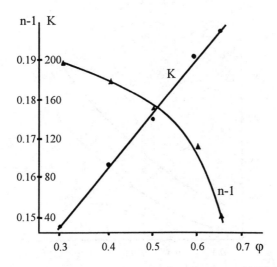

Figure 4. Coefficients (1) K and (2) n for the Ostwald-Weil Equation Versus the Content of Dispersed Phase in the Emulsion.

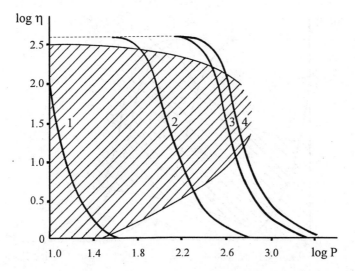

Figure 5. Viscosity Versus the Shear Stress for Emulsions with a Dispersed Phase Content of 1) 0.30, 2) 0.50, 3) 0.60, and 4) 0.65. The Shaded Region Corresponds to the Adequate Description of the Dependence According to the Ostwald-Weil Equation.

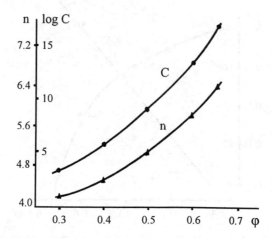

Figure 6. Coefficients (1) C and (2) n for the Haven Equation Versus the Content of Dispersed Phase.

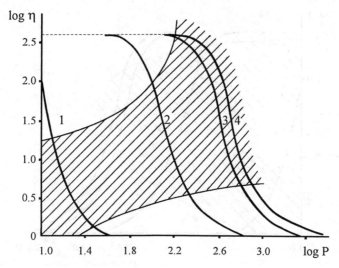

Figure 7. The Shaded Region Corresponds to the Adequate Description of the Dependence According to the Haven Equation. (The symbols are the same as in Figure 5.)

Haven model

$$\eta_{eff} = \eta_0/(1 + CP^n) \tag{5}$$

where η_0 is the viscosity of the unperturbed emulsion, and B and C are empirical coefficients.

Calculations based on the Steiner model do not adequately describe the experimental data.

The Haven model differs from the Ferry model in the value of the exponent n. Calculations show that an adequate description of the experimental data for the emulsion viscosity is possible for the ranges of n and C values given in Figure 6.

The Ferry model with a flow index of n = 1 does not provide a suitable mathematical description of the experimental data. The region of adequate description for the Haven model is presented in Figure 7. It should be noted that as the content of dispersed phase in the emulsion increases, the region of adequate description for the emulsion viscosity expands.

The models proposed for the description of the complete rheological curves are also based on a mathematical approximation (*4*). The power equations are characterized by their similarity to the second group of models which were used to describe the incomplete rheological cures and include the:

Peek-Mak-Lean-Williamson model

$$\eta_{eff} = \eta_\infty + (\eta_0 - \eta_\infty) / (1 + P/P_{av}) \tag{6}$$

Ellis model

$$\eta_{eff} = \eta_0/[1 + (P/P_{1/2})^A] \tag{7}$$

Meter model

$$\eta_{eff} = \eta_0[1 + (P/P_{av})^A (\eta_\infty/\eta_0)]/[1 + (P/P_{av})^A] \tag{8}$$

where η_∞ is the viscosity when the structure of the extracting emulsion has been completely destroyed, P_{av} and $P_{1/2}$ are the average shear stresses at $\eta_{eff} = (\eta_0 + \eta_\infty)/2$ and at $\eta_\infty \ll \eta_0$, respectively, and A is the empirical coefficient.

For the investigated emulsions, $\eta_\infty \ll \eta_0$. Under this condition, the Peeck-Mak-Lean-Williamson equation transforms into the Ferry equation and, therefore, is not suitable for description of the viscosity of extracting emulsions. In the Meter equation, the term $(P/P_{av})^A (\eta_\infty/\eta_0)$ approaches 1 at $\eta_\infty \ll \eta_0$. Thus, Equation 8 is transformed into the Ellis equation. Values of $P_{1/2}$ and the exponential coefficient A for the Ellis model are presented in Figure 8. It should be noted that the value for A is constant and equal to 6 in the equation which describes the rheological curves of the extracting emulsions for the indicated range of dispersed phase content.

The Ellis equation (and consequently the Meter equation) adequately describe the experimental dependence of the emulsion viscosity on the shear stress over the entire range of dispersed phase contents because the values of the Fisher criterion do not exceed the tabular value (Table I). Thus, the Ellis equation allows the variation in emulsion viscosity with ϕ in the indicated range for different values of shear stress to not only be described, but also predicted.

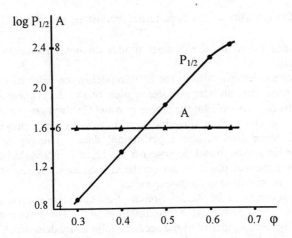

Figure 8. The (1) Average Shear Stress $P_{1/2}$ and (2) Coefficient A for the Ellis Equation Versus the Content of the Dispersed Phase in the Emulsion.

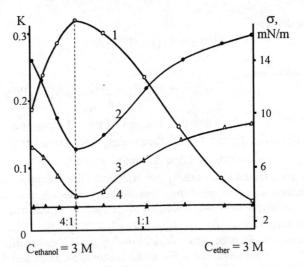

Figure 9. The (1) Coefficient for Cholesterol Extraction and the Interfacial Tension in a Two-Phase Non-Emulsion System at the Surfactant Concentration of (2) 0, (3) 4 X 10^{-7}, (4) 1.2 X 10^{-6} M Versus the Ethanol/Diethyl Ether Molar Ratio.

Table I. Values of the Fisher Criterion for the Emulsion Viscosity Using the Ellis Equation

Region of the rheological curve	Fisher criterion value for dispersed phase content of			
	0.30	0.50	0.60	0.65
$\eta_0 = \eta_\infty$	0.15	0.07	0.07	0.03
$\eta_\infty < \eta < \eta_0$	0.35	1.09	2.16	2.16
$\eta = \eta_\infty$	0.27	3.70	3.90	3.42

The tabular value of the Fisher criterion, $F_{tab} = 4.26$.

Nanodispersion Formation in Extracting Emulsions.

One of the most important phenomena which may occur at the interface in an extracting emulsion is formation of small particles on the order of 10-50 nm (*i.e.* a nanodispersion). Formation of a nanodispersion is promoted by the presence of a surfactant and a co-surfactant in the system (*5*) and by disruption of the equilibrium between the phases of the extracting emulsion (*i.e.* by mass transfer of components of the phases and the solvents through the interface). When microemulsifiers are included in the emulsion to decrease interfacial tension, nanodispersion formation result. The microemulsifiers can diffuse through the interface of the emulsion which results in interface instability and nanodispersion formation. The microemulsifiers can be co-surfactants, such as ethanol and diethyl ether (*6,7*).

Photomicrographs of extracting emulsions were conducted with Pt/C replicas, which were obtained by the freeze-cleavage method. The data indicate an absence of the nanodispersion in the initial emulsion before extraction with formation of 17-25 nm droplets during the course of extraction by the emulsion (*1*).

Such nanodispersions possess the following properties:
a) the size of the droplets is 10-50 nm;
b) the diffusion coefficient of the particles is much smaller than that of the individual surfactant monomer (*i.e.* as the equilibrium shifts to decrease the concentration of the surfactant monomer, there is no decomposition of the nanodispersion droplets); and
c) the system is thermodynamically unstable.

Appropriate concentrations of microemulsifiers are necessary in an extracting emulsion to lower the interfacial tension. Figure 9 compares the dependence of the coefficient for cholesterol extraction by an emulsion (curve 1) and the interfacial tension in a two-phase, non-emulsion system (curves 2-4) on the ethanol to diethyl ether molar ratio. There is a maximum on the cholesterol extraction curve which corresponds to minima in the interfacial tension curves. At this composition the greatest interfacial instability due to diffusion of ethanol and diethyl ether takes place. Consequently the nanodispersion formation is the greatest at the same composition for which the cholesterol extraction coefficient has a maximal value.

During formation of the nanodispersion, the interface of the emulsion increases so the amount of surfactant per unit of interface area decreases. This leads to an increase in the interfacial tension which retards further nanodispersion formation. But at a molar ratio of ethanol to diethyl ether of 4:1, the interfacial tension in such a system is the lowest (Figure 9, curve 3).

Since the capacity of the extracting emulsion increases as a consequence of the increase in degree of adsorption of the target substance (cholesterol, in this example)

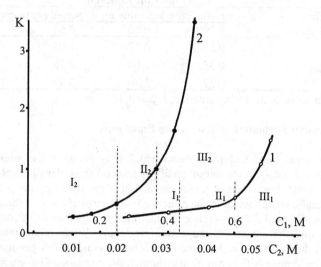

Figure 10. Coefficient for Cholesterol Extraction Versus the Concentration of (1) Monomeric and (2) Polymeric Surfactants.

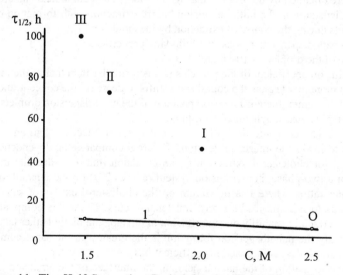

Figure 11. The Half-Separation Period for the Water Phase Versus the Ethanol Concentration (Curve 1). Points: 0 = Without Pre-Treatment; I , II, and III = Pre-Treatment for One, Two, and Three Times, Respectively.

at the interface in the emulsion (*6*), nanodispersion formation exerts a positive influence on the extraction of substances with surface-active properties.

Differing amounts of surfactant can be used for stabilization of the extracting emulsion:

a) less than the amount necessary for complete adsorption at the interface;

b) the amount necessary for complete adsorption at the interface;

c) enough for formation of spherical micelles (more than CCM-1); and

d) enough for formation of non-spherical micelles (more than CCM-2).

Therefore cholesterol molecules can either sorb at the interface or be solubilized in surfactant micelles in the bulk of the liquid membrane during extraction by the multicomponent emulsion. The effect of surfactant type and concentration on the recovery of cholesterol was investigated. Figure 10 shows the dependence of the coefficient for cholesterol extraction from human blood on the surfactant concentration in the emulsion. In Region I, the coefficient for cholesterol extraction increases as the amount of surfactant approaches that needed for complete adsorption on the interface. Enhancement in the coefficient results from an increase in the interface due to reduction in the diameter of the internal phase droplets in the initial emulsion (before extraction). In Region II, the coefficients for cholesterol extraction continue to increase due to an increase in the spherical micelle concentration in the liquid membrane. The sharp enhancement in cholesterol extraction which occurs in Region III is apparently connected with formation of a large amount of nanodispersion and water clusters in the oil phase during the swelling of the emulsion. This has analogy with percolation processes in microemulsion systems. The surfactant molecules may form a lamellar structure in which the interface would increase considerably thereby leading to an increase in cholesterol adsorption.

Droplets of the nanodispersion are situated on the interface in the emulsion and form a structural-mechanical barrier. Therefore nanodispersion formation reduces the coalescence rate for the internal phase droplets and the emulsion stability increases. Thus the formation of nanodispersions may be highly beneficial when it is necessary to have emulsions which remain stable for long periods of time, such as in medicine, in certain analytical reagents, and in the preparation of protective skin creams.

Special conditions for pre-treatment of the initial emulsion to enhance nanodispersion formation have been developed. An additional amount of ethanol (10-25 %) is first incorporated into the internal phase of the extracting emulsion. The resulting emulsion is contacted with an aqueous buffer solution of pH equal to that of the internal aqueous phase and then with doubly distilled water or an aqueous buffer at pH 7.4. During this process some of the ethanol diffused from the internal phase of the emulsion into the external aqueous solution which stimulates nanodispersion formation.

Figure 11 shows the period for half separation of the external aqueous phase in the emulsions as a function of the ethanol concentration with and without this pre-treatment. As can be seen, repeating the pre-treatment three times produces an eight-fold enhancement in stability of the emulsion to coalescence. Although almost all of the nanodispersion is formed during the pre-treatment step, this is not detrimental to the extraction properties of the emulsion.

Table II compares the cholesterol extraction properties and stabilities of emulsions prepared with and without the pre-treatment. These extracting emulsions have the same overall chemical compositions, but different structures. For the untreated emulsion, the nanodispersion forms during cholesterol extraction, and in the

Table II. Comparison of Properties of Extracting Emulsions Prepared With and Without Pre-Treatment

Property	Untreated emulsion	Emulsion after one pre-treatment
Coefficient for cholesterol extraction	0.29 ± 0.03	0.32 ± 0.03
Period for half separation of the internal aqueous phase at 70 °C, hours	6 ± 1	48 ± 1

second type of emulsion the nanodispersion was formed before cholesterol extraction during pre-treatment with the isotonic buffer solution. Although the cholesterol extraction behavior of the untreated and pre-treated emulsions are the same within experimental error, the pre-treated emulsion has much higher stability.

Drops of the nanodispersion can play the role of transfer agents leading to transmembrane transfer of water to the internal phase (*i.e.* swelling of the emulsion) which is undesirable. Thus in addition to the surfactant and carrier molecules, droplets of the nanodispersion may also be a major contributor to water transport from the external aqueous phase into the internal aqueous phase.

Acknowledgments.

The research described in this chapter was made possible in part by Grants NAC000 and NAC300 from the International Science Foundation. The authors with to acknowledge the Dr. Richard A. Bartsch in the preparation of the revised chapter manuscript.

Literature Cited.

1. Yurtov, E.V.; Koroleva, M.Yu. *Russ. Chem. Rev.* **1991**, *60*, 1255-1270.

2. Ho, W.S.W.; Li, N.N. In *Membrane Handbook* ; Ho, W.S.W.; Sirkar, K.K., Eds.;Van Nostrand Reinhold: New York, NY, 1992; Chapt. 36, pp. 597-610.

3. Yurtov, E.V.; Koroleva, M.Yu. *Kolloidn. Zh* **1994**, *56*, 515-518.

4. Skelland, A.H.P. *Non-Newtonian Flow and Heat Transfer*, Wiley: New York, NY, 1967.

5. Eike, H.F. *Interfacial Phenomena in Apolar Media*; Marcel Decker: New York, NY, 1987; pp. 41-92.

6. Yurtov, E.V.; Koroleva, M.Yu. *Kolloidn. Zh.* **1991**, *53*, 86-92.

7. Yurtov, E.V.; Koroleva, M.Yu.; Golubkov, A.S; *et. al. Dokl. Akad. Nauk SSSR* **1988**, *302*, 1164-1166.

Chapter 7

Preparation and Splitting of Emulsions for Liquid Membranes

J. Draxler[1], C. Weiss[1], R. Marr[1], and G. R. Rapaumbya[2]

[1]Institut für Thermische Verfahrenstechnik und Umwelttechnik, Technische Universität Graz, Inffeldgasse 25, A–8010 Graz, Austria
[2]Ecole Centrale Paris, Laboratoire de Chimie Nucléaire et Industrielle, Grande voie des Vignes, F–922295 Chatenay Malabry, France

An apparatus for the preparation of emulsions and various devices for the electrostatic splitting of emulsions were improved by investigating the flow pattern using computational fluid dynamics software. In the case of electrostatic splitting it, an attempt is made to combine the flow pattern with the electric forces acting on the droplets. Although to date only single phase calculations were made, the experimental results show that considerable improvements could already be achieved. In the case of emulsion preparation a much simpler and therefore cheaper device gave the same droplet size distribution. As for the electrostatic splitting, the influence of various geometries and inlet positions was investigated for a given electric field.

The preparation and splitting of emulsions are the key parameters in an emulsion liquid membrane process. A very stable emulsion which avoids any loss of emulsified droplets is a prior condition for the feasibility of the process. However, the more stable the emulsion, the more difficult to split it. So both steps are dependent on each other and have to be optimized, also with regard to cost optimization. In the present work, we try to calculate the flow pattern in the two steps using CFD (computational fluid dynamics) software in order to improve the design of the two steps.

Preparation of Emulsions

There are many suitable devices available, such as high pressure homogenizers or devices using the rotor/stator principle. In the liquid emulsion membrane process often very acidic stripping phases are used in the emulsified phase. Together with the high shear forces in those devices this leads to high corrosion problems which makes the process very expensive. For this reason, a new low pressure homogenizer was developed which met all the requirements for the liquid emulsion membrane process

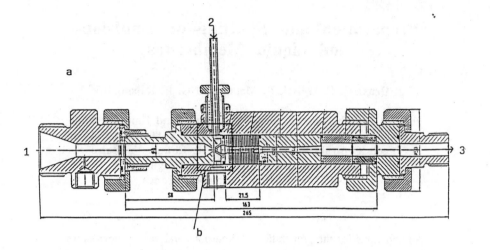

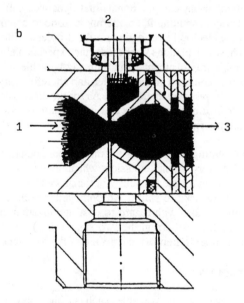

Figure 1. Low-pressure Homogenizer.
a whole homogenizer
b detail of nozzle
1 continuous phase in
2 dispersed phase in
3 emulsion out

(*1*). This homogenizer is shown in Figure 1. It was developed empirically and the objective of the present work was to improve this device by means of CFD.

This type of homogenizer was used in all of our emulsion liquid membrane plants. No corrosion problems were observed after several months of operation. At a pressure difference of only 500 kPa, a mean droplet diameter of 1-2 μm could be obtained.

Although this device worked very satisfactorily, there were two major drawbacks: (1) The radial suction of the dispersed phase by the axial flow of the continuous phase was not sufficient, so that the dispersed phase had to be injected by a pump; and (2) The manufacturing of the end-section with many edges, intended to enhance turbulence and assumed to be the reason for the good performance, is quite complicated and makes the homogenizer expensive.

All the following calculations were made assuming a single phase flow. At first the velocity vector field, the pressure distribution, the mean turbulence fluctuations (local root mean square) and the turbulent kinetic energy dissipation were calculated for the existing type of homogenizer. The calculations were made for a slice of an axial symmetric fluid domain. Figure 2A shows the velocity vector field and Figure 2B the mean turbulence fluctuations.

From these calculations, it could be concluded that the turbulence in the end-section is not as high as expected and that therefore this expensive part of the homogenizer cannot be a dominant factor in the good performance. The highest turbulence fluctuations occur at the periphery of the free liquid jet coming out of the inlet nozzle for the continuous phase. A few millimeters behind this nozzle is a second nozzle and between these two nozzles the radial inlet for the dispersed phase is located. It can be seen that the second nozzle is too small and also that the distance between these two nozzles is too small. The free liquid jet coming out of the first nozzle is broken at the second nozzle, so the pressure in the inlet for the dispersed phase is not low enough to suck in a sufficient amount of the dispersed phase.

Based on these results, an new homogenizer was devised without the end-section with the edges and with improved geometry for the inlets of both phases. The calculations for this new design (Figure 3) show that the free liquid jet is not broken any longer due to a wider second nozzle. Therefore the pressure in the inlet region is much lower so self-suction of the dispersed phase could be expected.

The experimental results confirmed the expectations based on the calculations. Figure 4 shows the dependence of the sucked-in dispersed phase on the flow rate of the continuous phase for the old and new geometries, respectively. It can be seen that the suction of the dispersed phase is much higher with the new geometry, so no extra pump is needed to achieve the desired phase ratio of about 3:1 (continuous : dispersed phase). The drop size distribution was about the same for both devices. Using the following conditions a mean diameter of around 1.5 μm and a Sauter mean diameter of around 2.5 μm could be obtained:

continuous phase:	dispersed phase:
water + 0.15% Na-dodecylsulfate	Shellsol T
flow rate: 100 l/h	flow rate: 30 l/h
pressure difference: 500 kPa	

These results were obtained by pumping the continuous phase only once through the nozzle. Recycling a part of the formed emulsion severals times through the nozzle would give even smaller drops and a narrower droplet size distribution.

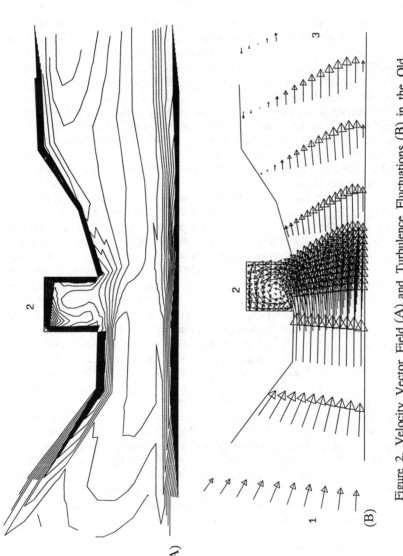

Figure 2. Velocity Vector Field (A) and Turbulence Fluctuations (B) in the Old Device.
Isolines of turbulent energy, maximum = 0.3 m/s, line spacing = 0.04 m/s

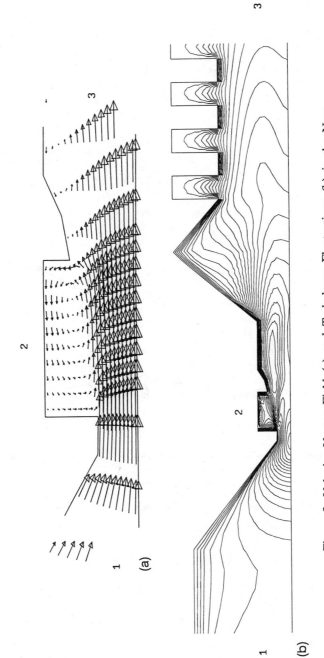

Figure 3. Velocity Vector Field (a) and Turbulence Fluctuations (b) in the New Device.
Isolines of turbulent energy, maximum = 1.3 m/s, line spacing = 0.04 m/s

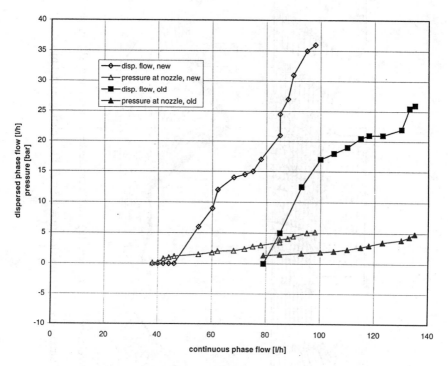

Figure 4. Flow Rate of the Dispersed Phase vs. the Continuous Phase.

In our further work we will try to account for the dispersed phase and calculate a two phase flow. The goal will be to predict a mean droplet size for given operating conditions and physical properties.

Splitting of Emulsions

For emulsion liquid membranes mostly water-in-oil-emulsions are used. For this type of emulsions the electrostatic coalescence is the method of choice. This is a well known method used in crude oil desalting. However, the emulsions in the liquid membrane process are much more stable, therefore this process had to be improved. These improvements concerned mainly electrical parameters, but also the flow regimes in order to avoid the formation of spongy emulsions. Although a myriad of different emulsions has been split, not only for the liquid membranes process but also for many other applications, no general design criteria could be found. The efficiency of the splitting depends very much on the physical properties of the components and on the phase ratio. Sometimes, a high frequency, a homogenous electric field, a countercurrent flow, or a high temperature is favourable; but also the contrary might be true.

In our present work we try to combine existing correlations for droplet-droplet coalescence efficiency with the flow pattern in different splitting devices with the main objective to compare different splitting designs for a given emulsion.

As a starting point the following conditions were chosen:

- An emulsion consisting of 99.5% continuous organic phase with a density of 934 kg/m³ and a viscosity of 4.6 mPa.s at 20°C and 0.5% tap water. This low water content was chosen because initially all of the calculations were based on a single phase flow and this small amount of dispersed phase was assumed to be negligible.
- Apparatus: Parallel plate coalescer with 7 electrodes and a height of 0.26m, a length of 0.27 m and a width of 0.17 m. This type was chosen because existing gravity settlers could be easily equipped with such electrodes and because of the homogenous field between the electrodes.
- Voltage: 5 kV/cm
- Frequency: 10 kHz
- Throughputs: 50 l/h and 100 l/h, respectively.

In a first step the velocity pattern was calculated for different inlet heights and different throughputs. These velocity vectors are shown in Figure 5. Now for each cell the droplet diameter was calculated when the droplet starts to settle (one droplet in each cell). This point is given when the upward velocity of the continuous phase is balanced by the sinking velocity of the droplets according to Stokes law:

$$v_c = v_{drop} = \frac{\Delta\rho.g.d^2}{18\eta_c}, \quad with \quad d = \sqrt{\frac{18\eta_c.v_c}{\Delta\rho.g}} \tag{1}$$

where v_c and v_{drop} are the velocity of the continuous phase and the droplet velocity, respectively, $\Delta\rho$ the density difference between the two phases, g the acceleration due to gravity, d the droplet diameter, and η_c the viscosity of the continuous phase.

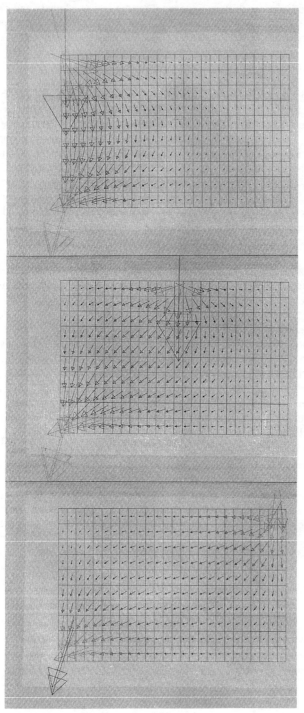

Figure 5. Velocity Vector Field in a Parallel Plate Coalescer.

These droplet size distributions are shown in Figure 6. It can be seen that the average diameter of the droplet when they start to sink is around 165 µm with the inlet at the bottom (a). This value is shifted to around 100 µm and 35 µm with the inlet in the middle (b) and at the top (c), respectively. These differences are due to the velocity pattern of the continuous phase. When the inlet is at the bottom, the continuous phase flows up and the droplets must flow down countercurrently. When the inlet is at the top, the continuous phase flows partially down and this cocurrent movement promotes the settling of small droplets. At a higher flow rate (100 l/h, inlet at bottom) the droplet size is shifted to higher values (Figure 6d). According to these numerical results, settling starts earlier with the inlet at the top than at the bottom.

To calculate the time till settling starts we need a correlation of the droplet growth in dependance of the electric field. There are several such correlations available in literature. We used the relation of Williams and Bailey (2), which gives the growth pattern of a droplet d(t) for the assumption that droplets coalesce only in pairs:

$$d(t) = d_0 \cdot \exp\left[\frac{8\varepsilon_c \cdot E^2}{\eta_c}\left(\frac{3x_d}{4\pi}\right)^{5/3} \cdot t\right]$$ (2)

where d(t) is the droplet diameter at time t, d_0 the initial droplet diameter, ε_c the dielectric constant of the continuous phase, E the electric field strength, and x_d the dispersed phase hold-up.

With equation (2) the times were calculated to reach the droplet diameters shown in Figure 6. The initial condition for the droplet diameter was $d_0 = 1$ µm at t = 0. After 30 seconds, 80% of the droplets start settling with the inlet at the top. With the inlet in the middle it takes 35 seconds and 45 seconds with the inlet at the bottom, respectively. However, the time till 99% of the droplets started settling is about the same at each inlet, because the maximum diameter of the droplets at each inlet is about the same, as can be seen in Figure 6a-c.

Having the droplet diameter at each cell and the time to reach this diameter, an estimation of the overall settling time was made by use of Stokes´equation and the assumption of stagnant continuous phase. The results are presented in Figure 7A. We can see that with the inlet at the bottom around 80% are settled within 15 min and 98% within one hour, respectively. The results are much worse with the inlet in the middle or the top (a,b,c,d correspond to Figure 6), because smaller drops need a longer time for settling.

Despite the assumption of a stagnant continuous phase (no upward velocity of the continuous phase ⇒ maximum sinking velocity of droplets) the experimental results were even better than the calculated ones. At stationary conditions a residence time of around 10 minutes was sufficent to settle 99% of the droplets with the inlet at the bottom. The results were slightly worse with inlet in the middle (95%) and at the top (90%). The reasons for this could be: (1) During sedimentation the droplets still grow, which could not be accounted for; and (2) There are some other important parameters which influence the separation efficiency that are not accounted for in the equation of Williams and Bailey. For instance we noticed that under given conditions splitting is much more effective at a frequency of 10 kHz than at 50 Hz. Since any reliable correlations for the frequency dependence do not exist, we need at least one

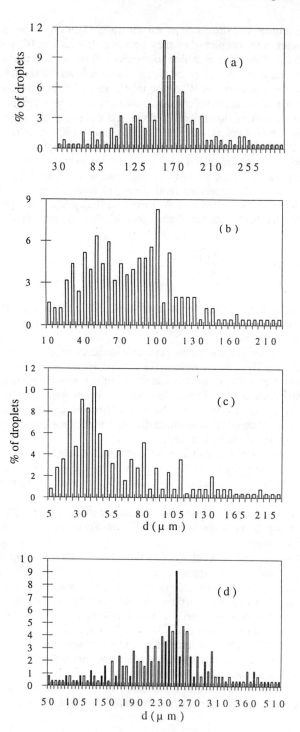

Figure 6. Droplet Size Distribution at the Beginning of Settling.

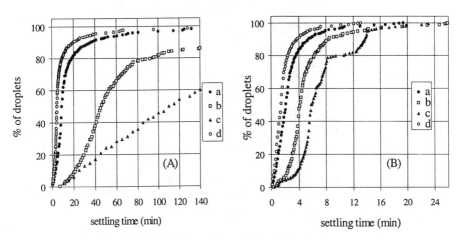

Figure 7. Settling Time for a Parallel Plate Coalescer.

adjustable parameter. Therefore the calculations were repeated using a correlation suggested by Cho (3) for the sinking velocity of the droplets:

$$v_{drop} = \eta \frac{6.6\, \varepsilon_c.d_0.E^2}{12\eta_c} \tag{3}$$

In this equation the hydrodynamic drag (for small droplets expressed by Stokes´law) is made equal to the electric force. η is an adjustable parameter. In the first experiment with the inlet in the middle, this parameter was determined (0.01) and used for subsequent calculations. The results are shown in Figure 7B. Now the difference in the position of the inlet is smaller and fits much better with the experimental results.

The same calculations were also made for a cylindrical device described in (4). Calculation of the splitting efficiency for a different thickness of the central electrode at a constant outer electrode was attempted. The flow pattern is not changed very much by the thickness of the electrode. Therefore the necessary droplet diameter for settling is nearly the same. However, the electric field strength at a given voltage is changed considerably. Again using the correlation of Williams and Bailey with local electric field strength in each cell, we get the results shown in Figure 8. At the high electric field with the thicker electrode, the droplet diameter necessary for settling is reached almost immediately.

Using Cho´s relation with an efficiency of 0.01 the settling time was calculated (Figure 8a). It can be seen that settling is slowest with an electrode of 2 mm and highest with the electrode of 12 mm. The experimental results (Figure 8b) show an acceptable agreement only when Cho´s relation is used for the calculations.

The results prove that it is possible to calculate and optimize an electrostatic splitting device by combining the forces acting on the droplets in an electrical field to the flow pattern. To be able to account for higher phase ratios, the calculations must be extended to a two phase flow, which was not done yet. Even though the results look very promising, one should bear in mind that there are still some parameters

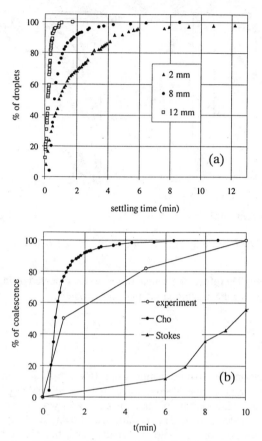

Figure 8. Settling Time for a Cylindrical Coalescer.

whose influence is not fully understood. Therefore one will need at least one adjustable parameter which has to be determined experimentally.

Acknowledgement

The authors would like to thank KINEMATICA AG (Luzern, CH) for providing the homogenizers (trade name ATOMIX-SND) and the European Commission for a HCM-grant for G.R.Rapaumbya.

References

1. Marr, R.; Draxler, J. *Aachener Membran Kolloquium 1989*; VDI-Verlag: Düsseldorf; pp. 221-233.
2. Williams, T. M.; Bailey A. G. *Oxford Inst. Phys. Conf. Ser.* **1983**, *66*, 39-44.
3. Cho, A. Y. H. *J.Appl.Phys.*, **1964**, *35*, 2561-2565.
4. Draxler, J.; Marr, R. In *Solvent Extraction 1990*; Sekine T., Ed.; Elsevier: New York 1992, Part A, pp. 37-48.

Chapter 8

Mathematical Modeling of Carrier-Facilitated Transport in Emulsion Liquid Membranes

Ching-Rong Huang, Ken-Chao Wang, and Ding-Wei Zhou

Department of Chemical Engineering, Chemistry, and Environmental Science, New Jersey Institute of Technology, University Heights, Newark, NJ 07102

Mass transfer in emulsion liquid membrane (ELM) systems has been modeled by six differential and algebraic equations. Our model takes into account the following: mass transfer of the solute across the film between the external phase and the membrane phase; chemical equilibrium of the extraction reaction at the external phase-membrane interface; simultaneous diffusion of the solute-carrier complex inside globules of the membrane phase and stripping of the complex at the membrane-internal phase interface; and chemical equilibrium of the stripping reaction at the membrane-internal phase interface. Unlike previous ELM models from which solutions were obtained quasi-analytically or numerically, the solution of our model was solved analytically. Arsenic removal from water was chosen as our experimental study. Experimental data for the arsenic concentration in the external phase versus time were obtained. From our analytical solution with parameters estimated independently, we were able to obtain an excellent prediction of the experimental data.

Since Norman N. Li (1) introduced the emulsion liquid membrane in 1968, many publications have appeared both on experimental work and on theoretical modeling of such separations. Most of the theoretical work was concentrated on mass transfer in the ELM process in a mixing vessel where the water-in-oil emulsion is mixed with an external aqueous phase and globules of emulsion are formed and suspended in the external phase. The solute, after a series of mass transfer steps, is transferred from the external aqueous phase to the internal aqueous receiving phase through the oil membrane phase. Many mathematical

0097–6156/96/0642–0115$15.00/0

models were developed to describe and represent this mass transfer operation. The first model, the advancing front model, was developed by Ho *et al.* *(2)*. In this model, the solute was assumed to react instaneously and irreversibly with the reagent in the internal phase. The solute concentrations in the membrane phase and in the internal phase were treated as a single term. Fales and Stroeve extended the advancing front model to include external mass transfer resistance *(3)*. The model was further modified independently by Teramoto *et al.* *(4)*, and by Bunge and Noble *(5)* by considering the reversibility of chemical reactions. However, their resulting equations were too complicated to develop even quasi-analytical solutions. Lorbach and Marr simplified the modeling equations by assuming constant summation of free and complexed carrier concentrations *(6)*. In a recently published book by Ho and Sirkar *(7)*, an extensive review is given of the experimental and theoretical work on ELM.

We starting working on modeling of the ELM process in 1984 *(8)*. Our model involves the following steps: mass transfer of the solute from the bulk external phase to the external phase-membrane interface; an equilibrium reaction between the solute and the carrier to form the solute-carrier complex at the interface; mass transfer by diffusion of the solute-carrier complex in the membrane phase to the membrane-internal phase interface; another equilibrium reaction of the solute-carrier complex to release the solute at the membrane-internal phase interface into the internal phase. Assumptions used in our model include: a uniform radius of the globules; negligible leakage of the internal phase into the external phase; and negligible mass transfer resistance from the membrane-internal interface into the internal phase. From the model, simultaneous partial differential equations and algebraic equations were formulated. Analytical solutions were obtained by Laplace transform. From these solutions, we were able to predict theoretically: 1) the concentration of the solute in the external phase as a function of time; 2) the concentration profile in a globule at different times; and 3) the change in surface concentration of the solute at the external phase-membrane interface with time.

The features of our model are the following: 1) The most significant feature of our model is that an analytical solution was obtained. Previous models gave solutions in quasi-analytical or numerical forms. 2) Most models predicts only the external phase concentration of the solute versus time. Our model allows calculation of the distribution of solute in the external, membrane and internal phases at any time. 3) Two dimensionless groups with physical significance were developed from our model to characterize the ELM systems. One dimensionless group, G, provides the ratio of mass transfer resistance between the film of external phase and the membrane phase. This group is the reciprocal of the Biot number. The other dimensionless group, B, indicates the capability of reduction of the solute concentration in the external phase in an ELM system. A larger value of this dimensionless group gives greater percent reduction of the solute concentration in the external phase.

Mathematical Equations Governing the ELM System.

<u>External Phase</u>

$$Ve\frac{dCe}{dt} = -N(4\pi R^2)De(\frac{\partial Cm}{\partial r})_{r=R} \tag{1}$$

$$-N(4\pi R^2)De(\frac{\partial Cm}{\partial r})|_{r=R} = -N(4\pi R^2)k(Ce-Ce^*) \tag{2}$$

Initial Condition (I.C.): Ce = Ceo, when t = 0
where: Ve = volume of the external phase (liter)
      Ce = concentration of solute in the external phase (mol/liter)
      N = total number of globules
      R = radius of the globules (m)
      De = effective diffusivity of the solute-carrier complex in the
          globule (m^2/sec)
      Cm = concentration of the complex in the membrane phase (mol/liter)
      k = external mass transfer coefficient (m/sec)
      $k(Ce-Ce^*)$ is the flux for mass transfer across the external resistance film
      $De(\frac{\partial Cm}{\partial r})|_{r=R}$ is the flux for mass diffusion inside the globules
At the external phase-membrane interface, chemical equilibrium is reached. Thus,

$$Cm^* = pCe^* \quad \text{and } Ce^* \neq Ce \qquad \text{when } r = R \tag{3}$$

where: Cm* = concentration of the solute-carrier complex in the membrane phase
        at the interface
      Ce* = concentration of the solute in the external phase at the interface
      p = the extraction partition function at the external phase-membrane
        interface

<u>Membrane Phase</u>
A spherical shell material balance is taken inside the globule membrane:

$$Vm*\frac{\partial Cm}{\partial t} = (Vi+Vm)*De*[\frac{1}{r^2}\frac{\partial}{\partial r}(r^2\frac{\partial Cm}{\partial r})] - Vi*Rx \tag{4}$$

Initial conditions (I.C.):   Cm = 0 for all r, when t = 0
Boundary Conditions (B.C.): Cm = finite, when r = 0; Cm = Cm*, when r = R
where: Vm = volume of the membrane phase (liter),
      Vi = volume of the internal phase (liter)
      Vm+Vi = total volume of the emulsion globules (liter)
      Rx = rate of release of the solute into the internal phase from the
      membrane-internal phase interface per unit volume of the internal
      phase (mol/liter/sec.)

Internal Phase

$$Rx = \frac{\partial Ci(r,t)}{\partial t} \tag{5}$$

Initial conditions (I.C.): $Ci = 0$, when $t = 0$
where: Ci = concentration of the solute in the internal stripping phase

At the membrane-internal phase interface, chemical equilibrium is reached.
Thus, at any r

$$Ci = qCm \tag{6}$$

where: q = the stripping partition function at the internal phase-membrane
          interface
          The partition functions, p and q, represent equilibrium chemical reactions
for formation of the solute-carrier complex at the external phase-membrane
interface and decomposition of the solute-carrier complex at the membrane-
internal phase interface, respectively.
          We have four partial differential equations and two equilibrium equations.
Laplace transform is applied to solve four variables: $Ce(t)$, $C^*e(t)$, $Cm(r,t)$ and
$Ci(r,t)$.
          These equations are combined and changed into dimensionless form as
follows:

$$\frac{dUe}{d\tau} = -Ko(Ue - Ue^*) \tag{7}$$

$$G(\frac{\partial Um}{\partial v})\big|_{v=1} = (Ue - Ue^*) \tag{8}$$

Equilibrium: $U^*e = U^*m$ $Ue \neq U^*e$
I.C. $Ue = 1$, when $\tau = 0$

$$\omega \frac{\partial Um}{\partial \tau} = \frac{1}{v^2} \frac{\partial}{\partial v}(v^2 \frac{\partial Um}{\partial v}) \tag{9}$$

I.C. $Um = 0$,   when $\tau = 0$
B.C. Um = finite   when $v = 0$; $Um = Um^*$ when $v = 1$

where: $Ue = \dfrac{Ce}{Ceo}$ $\tau = \dfrac{Det}{R^2}$ $f = \dfrac{3f'}{1-f'}$ $f' = \dfrac{Vi+Vm}{Vi+Vm+Ve}$

$\omega = 1 - \dfrac{Vi}{Vi+Vm}(1+q)$ $v = \dfrac{r}{R}$ $Um = \dfrac{Cm}{pCeo}$ $e = \dfrac{Vi}{Vm+Vi}$

$Ko = \dfrac{Rkf}{De}$ $G = \dfrac{pDe}{Rk}$ $Ue^* = \dfrac{Ce^*}{Ceo}$ $Um^* = \dfrac{Cm^*}{pCeo}$

Results and Discussion.

Laplace transform is taken for all these dimensionless equations, together with their initial and boundary conditions. The analytical solutions can be found by taking the inverse Laplace transform:

$$\frac{Ce}{Ceo} = \frac{3}{B+3} + \sum_{n=1}^{\infty} \frac{2B}{3B+B^2+bn^2+Gbn^2(Gbn^2-2B-1)} \exp(-\frac{bn^2}{\omega}\tau) \tag{10}$$

$$\frac{C^*e}{Ceo} = \frac{3}{B+3} + \sum_{n=1}^{\infty} \frac{2(B-Gbn^2)}{3B+B^2+bn^2+Gbn^2(Gbn^2-2B-1)} \exp(-\frac{bn^2}{\omega}\tau) \tag{11}$$

$$\frac{Cm}{pCeo} = \frac{3}{B+3} + \sum_{n=1}^{\infty} \frac{2(B-Gbn^2)(\frac{\sin(bn*v)}{v*\sin(bn)})}{3B+B^2+bn^2+Gbn^2(Gbn^2-2B-1)} \exp(-\frac{bn^2}{\omega}\tau) \tag{12}$$

$$\tan(bn) = \frac{bn*(B-Gbn^2)}{B+bn^2(1-G)} \tag{13}$$

where bn = eigen values defined by Equation (13), n=1,2,3···
Dimensionless groups are involved in the solution equations. They are

$$B = pf\omega, \qquad G = \frac{pDe}{Rk} = \frac{1}{Biot} \qquad tau = \frac{\tau}{\omega} = \frac{tDe}{\omega R^2}$$

Experimental Measurements.   Removal of arsenic from metallurgical wastewater by emulsion liquid membrane was studied experimentally *(9)*. One set of experimental data is used here to verify the mathematical model. The experiment on emulsion liquid membrane removal of arsenic was conducted as follows.

The arsenic solution (feed) was prepared by dissolving arsenic trioxide (As_2O_3) in sodium hydroxide solution. The pH was adjusted by adding sulfuric acid solution and the solution was then diluted to approximately 100 ppm. The emulsion consisted of the membrane phase and the internal phase. The membrane phase was formulated with a diluent, such as heptane, a carrier, such as 2-ethylhexyl alcohol (2EHA), and a surfactant, such as ECA4360J (Exxon Chemical Company). The emulsion was prepared by adding the internal aqueous phase (1 N NaOH solution) to the formulated membrane phase and then emulsifying with Warring Blender for 30 minutes at 10,000 rpm, and cooling down to room temperature. The emulsion was freshly prepared before each permeation experiment.

The prepared emulsion was dispersed in an agitated vessel with the feed arsenic solution in a volume ratio of 1/5. The agitation speed was controlled at 300 rpm as monitored by a digital stroboscope (Cole-Parmer). The pH of the external aqueous phase was measured by a pH meter (PHCN-31, OMEGA), and samples were removed periodically for further separation, dilution and analysis. Inductively Coupled Plasma-Mass Spectrometry (ICP-MS, VG PLasma Quad, VG Elemental Limited) was used for quantitative analysis. A set of experimental conditions is listed in Table I under which optimal arsenic removal efficiency was obtained.

Table I: One Set of Experimental Conditions of Arsenic Removal by ELM

Experimental Conditions
Room temperature, agitation speed: 300 rpm
External phase: 500 ml, Ceo = 5.51 ppm (initially) , [H$_2$SO$_4$]=0.2M
Membrane phase: 90 ml, 10 vol% 2EHA, 2 vol% ECA4360J, remainder: heptane
Internal phase: 10 ml of NaOH (2.0 N)

Equilibrium data were obtained by independent extraction and stripping experiments conducted in separatory funnels. Extraction was conducted with O/A = 2:1 with an organic phase formulated as 90 vol% of heptane and 10 vol% of 2EHA and an aqueous phase which contained 100 ppm of arsenic in 0.2 M sulfuric acid solution. Stripping was conducted with A/O = 1:1 with a stripping phase of 2.0 N NaOH solution. Both extraction and stripping were conducted in closed vessels with magnetic stirring for 24 hours.

Evaluation of Parameters. An important feature of our model is that all of the parameters can be evaluated independently. Thus, De, the effective diffusivity of the solute-carrier complex in the membrane phase, is evaluated by the Jefferson-Witzell-Sibbert equation *(10)*, Di and Dm are determined by the Wilke-Chang equation *(11)*, the external mass transfer coefficient k is determined by the Skelland and Lee equation *(12)*, and the Sauter mean radius R is determined by the same method as Ohtake *(13)*. The two partition function values p and q are obtained from the extraction and stripping experiments conducted in the separatory funnels. For the ELM removal of arsenic, Table II gives the predicted parameters by these methods under the experimental conditions listed on Table I.

Table II: List of Model Parameters for Arsenic System

Model Parameters	Value
Di=Diffusivity of solute in the external phase (m^2/sec)	$1.16*10^{-9}$
Dm=Diff. of complex in membrane phase (m^2/sec)	$1.18*10^{-9}$
De=Eff. Diff. of complex in membrane phase (m^2/sec)	$8.73*10^{-10}$
f'=emulsion volume fraction (Vi+Vm/Vtot)	$1.67*10^{-1}$
e=sink phase volume fraction (Vi/(Vi+Vm))	0.1
R=radius of the globule (m)	$5.8*10^{-4}$
p=external distribution coefficient	0.22
q=internal distribution coefficient	$7.0*10^3$
k=external mass transfer coefficient (m/sec)	$7.54*10^{-6}$

Prediction of the Experimental Data for Arsenic Removal by ELM.
Experimental data for the time dependence of the arsenic concentration in the external phase is shown in Figure 1. The theoretical curve predicted from the model is plotted on the same figure. The plot demonstrates that the theoretically predicted curve gives an excellent representation of the experimental data.

Engineering Analysis of the ELM System. This experimentally verified mathematical model can be used for engineering analysis of the ELM system. The effect of the two dimensionless groups (B and G) on the permeation rate and removal efficiency are discussed below. Such discussion strengthens our understanding of the ELM operations and facilitates determination of the optimal design of experiments for removal of other species by ELM processes.

Steady State Solution. The solutions for Equation 10, 11 and 12 have a non-series term plus a summation of a series. The non-series term of $3/(3+B)$ is the steady state solution of the partial differential equations when time approaches infinite. As mentioned before that dimensionless group B could be considered as a indicator of the emulsion capacity. Thus, the bigger the B number, the smaller is Ce/Ceo, which implies an enhanced removal efficiency. This steady state solution can also be obtained by material balance of the emulsion liquid membrane system as follows:

$$Ceo*Ve = Ce*Ve + Cm*Vm + Ci*Vi \qquad (14)$$

with two equilibrium equations: $Cm=pCe$ and $Ci=qCm$
The steady state solution becomes:

$$\frac{Ce|_{t=\infty}}{Ceo} = \frac{Ve}{Ve + p*(Vm + q*Vi)} = \frac{3}{B+3} \qquad (15)$$

Therefore, a larger B value gives better performance of the ELM system. It is apparent that a good choice of the extractant and extraction conditions will give a large extraction partition function, p, and a good choice of stripping reagent and stripping conditions will give a large stripping partition function, q. Together with a large emulsion volume ratio f and a large internal volume ratio e, the B value can be increased. However, the two volume ratios cannot be selected arbitrarily because of the stability of the emulsion.

Effect of B Number on ELM Performance. The effect of B, the emulsion capability dimensionless group, on the external phase concentration versus time is shown in Figure 2 for a fixed Biot number of 22.8. Results for the three curves obtained with different B number shows that the curve with the largest value of B gave the best ELM performance in solute removal for this group.

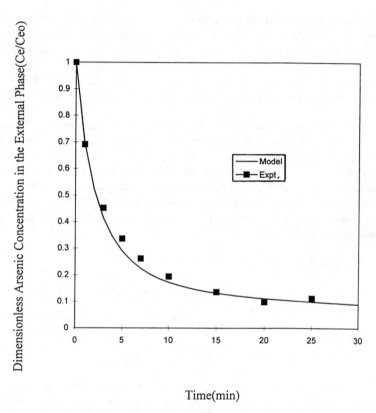

Time(min)

Figure 1. Prediction of the Arsenic ELM Experimental Data by the Model
(B=92.5 and Biot=22.8)
Experimental conditions: External phase was 500 ml of 5.51 ppm arsenic
solution, 0.2 M H_2SO_4; Emulsion phase was 100 ml of fresh emulsion, which
contains 90 ml of organic phase (10 vol% 2EHA, 2 vol% ECA4360J, remainder
was heptane) and 10 ml of internal phase (2 N NaOH solution); Room
temperature and 300 rpm of agitation speed.

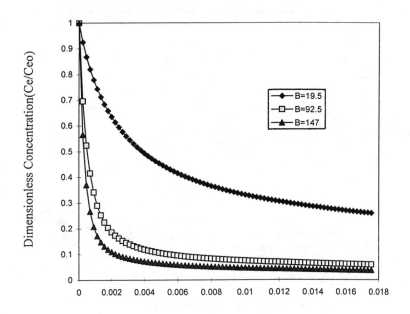

Dimensionless Time[Tau(De*t/w/R^2)]

Figure 2. Effect of the B Number on ELM Performance (Biot=22.8)

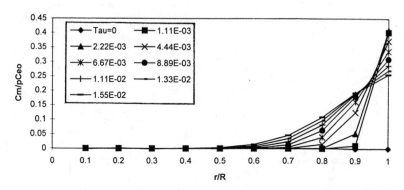

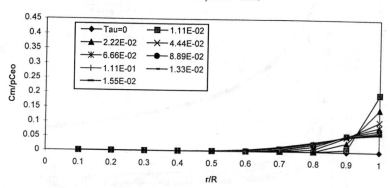

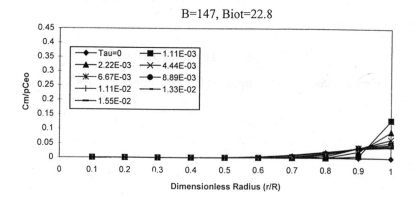

Figure 3. Effect of the B Number on the Solute-Carrier Complex Distribution
         in the Emulsion Globule

It was interesting to observe theoretically the effect of B on the concentration profile in a globule at different times. In Figure 3, plots of the globule concentration profiles at three B values of 19.5, 92.5 and 147 are presented. The results show that at a low value of B=19.5 with a low stripping capability of the internal phase keeps the solute (in the form of the solute-carrier complex) concentration high in the membrane phase. Therefore, the solute-carrier complex in the membrane penetrates deeper toward the center of the globule. On the other hand, at the high value of B=147, most of the solute is stripped into the internal phase and which leaves a small amount in the membrane phase. Since the membrane concentration of the solute-carrier complex is a function of time, the concentration profiles for a globule are plotted at different times from tau=0 to tau=$1.55*10^{-2}$.

Effect of the Biot Number on ELM Performance.   The Biot number, the reciprocal of the G number, represents the ratio of the internal diffusion transfer resistance to the external mass transfer resistance. The case of an infinite Biot number (G=0) represents no external film resistance.

Figure 4 presents the effect of Biot number on ELM performance as calculated from the model. At the fixed value of B=105, the dimensionless external phase concentration is plotted versus time at three different values of the Biot number, namely 7.01, 22.18 and 49.60. It is observed that the smaller the Biot number, the slower is the permeation rate. It is also noted that the curve for an infinite Biot number is exactly the same as the curve for Biot=49.6. This means that a critical Biot number exists so that the external mass transfer resistance may be neglected, if the operative Biot number is greater this critical Biot number.

In Figure 5, the external phase concentration of the solute and the external phase-membrane interface concentration of the solute are plotted simultaneously versus time at a constant B=105 and different values of the Biot number. It is seen that a larger film resistance (small Biot number) produces greater difference in the two solute concentrations across the film.

Conclusion.

The mass transfer of an emulsion liquid membrane process has been modeled mathematically. An analytical solution which allows prediction of concentrations of solutes in the external phase, membrane phase, and external phase-membrane interface was obtained. Experimentally, arsenic was selected as a solute in the external phase to be removed by the ELM process. Our model gives an excellent representation of the experimental data for the external concentration of the arsenic versus time. In addition, the model predicts the concentration distribution in the membrane phase at any time. Thus, the overall distribution of solutes in three phases (external, membrane and internal) at any time of the ELM process can be evaluated. From the model, it was found that the ELM process was characterize by two dimensionless groups. One group for the transport phenomena governs the rate of mass transfer or the Biot number. The other group includes the

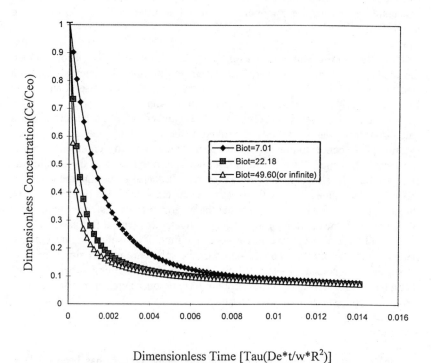

Dimensionless Time [Tau(De*t/w*R^2)]

Figure 4. Effect of the Biot Number on ELM Performance (B=105)

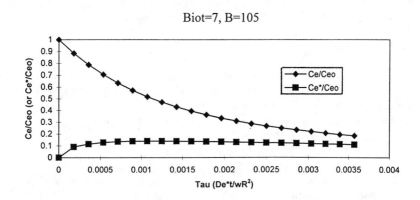

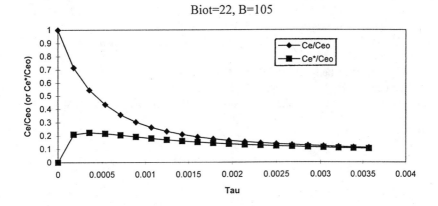

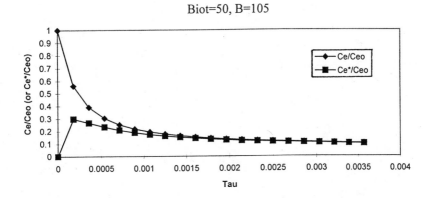

Figure 5. Comparison of External and Interfacial Concentrations

equilibrium constants for extraction and stripping reactions at both interfaces and expresses the limit or the capability of the separation process.

Literature Cited

1. Li, N. N. *US Patent,* 3,410,794 (Nov.12,1968).
2. Ho, W. S.; Hatton, T. A.; Lightfoot, E. N.; Li, N. N. *AIChE J.* **1982**, *28,* 662.
3. Fales, J. L.; Strove, P. *J. Membr. Sci.* **1984**, *21,* 35.
4. Teramoto, M.; Takihana, H.; Shibutani, M.; Yuasa, T.; Hara, N. *Sep. Sci. Technol.* **1983**, *18,* 397.
5. Bunge, A. L.; Noble, R. D. *J. Membr. Sci.* **1984**, *21,* 55.
6. Lorbach , D; Marr, R. J. *Chem. Eng. Process.* **1987**, *21,* 83.
7. Ho, W. S.; Li, N. N. In *Membrane Handbook*; Ho, W. S., Sirkar, K. K., Ed.; Van Nostrand Reinhold: New York, NY, 1992; pp. 597-611.
8. Wang, G.C. *Ph.D. Dissertation*, New Jersey Institute of Technology, 1984.
9. Huang, C.R.; D.W. Zhou; W.S. Ho; N.N. Li. presented at *1995 AIChE Spring National Meeting*, March 19-23, Houston, Texas.
10. Jefferson, T. B.; Witzell, O. W.; Sibbett, W. L. *Ing. Eng. Chem.* **1958**, *50,* 1589.
11. Wilke, C. R.; Chang, P. *AIChE J.* **1955**, *1,* 264.
12. Skelland, A. H. P.; Lee, J. M. *AIChE. J.* **1981**, *27,* 99.
13. Ohtake, T.; Hano, T.; Takagi, K.; Nakashio, F. *J. Chem. Eng. Jpn.* **1987**, *20,* 443.

CARRIER DESIGN, SYNTHESIS, AND EVALUATION

Chapter 9

Design of Macrocyclic Carriers for Liquid Membranes

Randall T. Peterson and John D. Lamb

Department of Chemistry and Biochemistry, Benson Science Building, Brigham Young University, Provo, UT 84602–5700

When used as carriers, macrocyclic multidentate ligands may endow liquid membrane systems with excellent cation selectivity. This selectivity, along with several other membrane characteristics, is determined primarily by the stability of the macrocyclic ligand/cation complex. The factors which determine complex stability including donor atom number and type, chelate and macrocyclic ring size, pendant group type, and ligand preorganization are discussed in an attempt to develop a more rational approach to macrocyclic carrier design. Tools which offer aid in the design of such carriers are also briefly reviewed.

Multidentate macrocyclic ligands are frequently employed as metal ion carriers in liquid membrane systems. Their ability to selectively and reversibly bind metal cations may enable a liquid membrane in which they are incorporated to perform difficult separations. A wide variety of macrocyclic carriers exist, and choosing the appropriate carrier for a given separation is crucial to the proper functioning of a liquid membrane. Frequently this choice is made by trial and error. However, with more information becoming available, rational design of macrocyclic carriers with the desired properties can hopefully be developed. In this chapter, we examine several structural features of macrocyclic multidentate ligands which influence carrier effectiveness and which provide the basis for rational ligand design. We also review two emergent technologies which are making rational ligand design even more feasible.

Complex Stability

Perhaps no characteristic of macrocyclic ligands has been more thoroughly studied than metal ion complex stability. Exhaustive reviews (*1,2* and references therein) have been dedicated to the compilation and discussion of binding constants and other thermodynamic data for cation binding with a large number of macrocyclic ligands. This wealth of information is a great resource to the separation scientist because complex

0097–6156/96/0642–0130$15.00/0

stability is inseparably connected to carrier effectiveness in membrane separations, as well as performance in other separation systems. With few exceptions *(3,4)*, and specifically for membranes, separation selectivity is governed principally by binding constants. Moreover, transport efficiency in membranes is also dictated largely by complex stability, though the relationship is not as straightforward nor as exclusive as it is with selectivity.

We have demonstrated the relationship of metal ion flux across bulk chloroform membranes upon log K values in methanol for several crown ethers and various cations *(5)*. Figure 1 shows that for transport of alkali and alkaline earth metals by a variety of macrocyclic ligands, flux is negligible when log K values are less than 4 or greater than 10. Significant transport occurs when log K values lie between 4 and 10 with maximum transport occurring at log K values of 6-7. The work of others, most notably Kirch and Lehn *(6)*, has confirmed this phenomenon both theoretically and empirically. Thus, when designing macrocyclic carriers for membrane separation systems of this type, stability constants must be carefully manipulated so as to be neither too low nor too high. In this one system studied, for example, the optimal log K value was approximately 7. Carriers must be capable of readily binding cations from the source phase while retaining the ability to release them into the receiving phase unless a stripping agent is used in the receiving phase *(5)*. Stability constants which are either too high or too low can compromise one or the other of these delicately balanced requirements.

Complex stability for macrocyclic carriers is determined by a number of factors including donor atom type, macrocyclic ring size, chelate ring size, pendant group characteristics, steric considerations, and degree of preorganization. Because complex stability is so crucial to carrier effectiveness, we examine a few of these factors below.

Donor Atoms. Few structural features influence complex stability more dramatically than the donor atoms themselves. Simple substitution of donor atom types in otherwise identical macrocyclic ligands can produce dramatic changes in stability constants, even to the extent that selectivities are reversed *(1)*. Several attempts have been made using both empirical data and theoretical principles to establish rules for predicting the influence of new donor atom combinations on complex stabilities *(7-14)*. Unfortunately, ligands do not always adhere to these rules, and the success of these models in predicting actual cation-macrocycle affinity seems to be proportional to the complexity of the model. However, there are a few basic rules that should be observed when selecting donor atoms during ligand design.

Atom Hardness. Figure 2 groups acids and bases by hardness and softness as described by Pearson *(15)*. Generally, hard metal ions (*e.g.* the small alkali metals and alkaline earths) are complexed more strongly by hard donor atoms (*e.g.* O) and soft metal ions (*e.g.* Cu^+, Ag^+, Tl^+, Hg^{2+}) more strongly by soft donor atoms (*e.g.* R-S-R). There are many exceptions to this rule, and it is limited in its applicability. Nevertheless, it serves as a good starting place when selecting donor atoms.

Neutral Oxygen Donor Atoms. In a thorough review, Hancock and Martell *(13)* recently concluded that the addition of neutral oxygen donors to a ligand serves to increase its selectivity for large cations over smaller cations. This trend is displayed in

CHEMICAL SEPARATIONS WITH LIQUID MEMBRANES

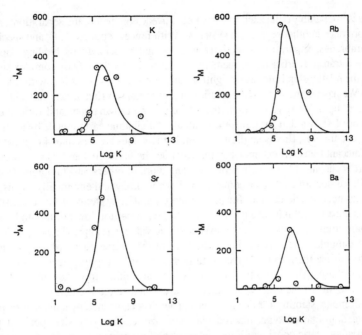

Figure 1. Effect of Complex Stability [log K (Methanol)] on Metal Ion Flux Through Chloroform Bulk Liquid Membrane [J_M (x $10^7/24$ hr)]. Both data points and calculated curves are shown. (Reproduced with permission from ref. 5. Copyright 1981 Wiley.)

Acids

Hard	Soft
H^+, Li^+, Na^+, K^+	Cu^+, Ag^+, Au^+, Tl^+, Hg^+
Be^{2+}, Mg^{2+}, Ca^{2+}, Sr^{2+}, Ba^{2+}	Pd^{2+}, Cd^{2+}, Pt^{2+}, Hg^{2+}
Al^{3+}, Sc^{3+}, Ga^{3+}, In^{3+}, La^{3+}	CH_3^+Hg, $Co(CN)_5^{2-}$, Pt^{4+}
Gd^{3+}, Lu^{3+}, Cr^{3+}, Co^{3+}, Fe^{3+}, As^{3+}	Te^{4+}, Br^+, I^+
Si^{4+}, Ti^{4+}, Zr^{4+}, Hf^{4+}, Th^{4+}, U^{4+}	
Pu^{4+}, Ce^{4+}, WO^{4+}, Sn^{4+}	
UO^{2+}, VO^{2+}, MoO^{3+}	

Borderline

Fe^{2+}, Co^{2+}, Ni^{2+}, Cu^{2+}, Zn^{2+}, Pb^{2+},
Sn^{2+}, Sb^{3+}, Bi^{3+}, Rh^{3+}, Ir^{3+}, $B(CH_3)_3$

Bases

Hard	Soft
H_2O, OH^-, F^-, $CH_3CO_2^-$, PO_4^{3-}	R_2S, RSH, RS^-, I^-, SCN
SO_4^{2-}, Cl^-, CO_3^{2-}, ClO_4^-, NO_3^-	$S_2O_3^{2-}$, R_3P, R_3As, $(RO)_3P$
ROH, RO^-, R_2O, NH_3, RNH_2,	CN^-, RNC, CO, C_2H_4, H^-, R^-
NH_2NH_2	

Borderline

$C_6H_5NH_2$, C_5H_5N, N_3^-, Br^-, NO_2^-, N_2, SO_3^{2-}

Figure 2. Classification of Acids and Bases by Hardness and Softness. (Reproduced with permission from ref. 15. Copyright 1963 American Chemical Society.)

Figure 3 which shows a linear relationship between the metal ion radius and the change in log K upon addition of neutral oxygen donors. This phenomenon also explains why the addition of four hydroxyethyl groups to the macrocycle 18-aneN$_4$O$_2$ reverses its selectivity from zinc over lead to lead over zinc *(13)*.

Negative Oxygen Donor Atoms. The complexing strength of a negative oxygen donor is closely related to its proton basicity and the acidity of the metal ion. Highly basic oxygen donors tend to bind most strongly to highly acidic cations (*e.g.* Mg^{2+} and Ca^{2+}), so selectivity for acidic metal ions can be enhanced by increasing both the number and basicity of the negative oxygen donors through the addition of electron releasing groups *(13)*.

Saturated Nitrogen Donor Atoms. Steric and related affects aside, Ca^{2+}, Sr^{2+}, Ba^{2+}, and the alkali metal ions prefer oxygen donor atoms to nitrogen donor atoms while other metal ions (*e.g.* Ag$^+$, Hg^{2+}) bind more readily to nitrogen donor atoms than to oxygen *(13)*. This fact is closely tied to the difference in their affinities for the ammonia ligand. However, due to the steric strain frequently present in macrocycle-cation binding, this rule is not highly reliable in predicting complex stabilities.

Unsaturated Nitrogen Donor Atoms. Unsaturated nitrogen donors have nitrogen atoms which are double bonded to an adjacent atom. As a result, they can participate in π bonding with metal ions. The result greatly stabilizes complexes with Os^{3+} and Ru^{2+} or other transition metal cations *(13)*. Another advantage of unsaturated nitrogen donors is the resultant structural rigidity they possess. This preorganization can greatly increase complex stability and is discussed further in a later section.

Larger Donor Atoms. Larger donor atoms, such as S, P, Se, and As, are less able to coordinate with small metal ions which prefer tight solvation spheres. As a result, they exhibit greater selectivity for larger, softer, weakly-solvated cations such as Ag$^+$ and Hg^{2+} *(13)*.

Categories of Cations. Hegetschweiler, *et al.* *(7)* recently proposed that metal ions be placed into one of five categories depending upon their electronic configurations. This categorization can then be used to determine whether each cation would be most readily bound by oxygen donors, nitrogen donors, or a combination of the two. Their system is based upon results from metal ion complexation by 1,3,5-triamino-1,3,5-trideoxy-*cis*-inositol (taci). The five categories are as follows:

　　　1) Group 1 contains metal ions with a d^0 electronic configuration and z ≤ +2. Examples include Mg^{2+}, Ca^{2+}, Sr^{2+}, and Ba^{2+}. Ions of this group form complexes most readily with neutral oxygen donors. Frequently, three oxygens from the ligand participate in binding with water molecules to complete the coordination sphere. Complexes are rather weak.

　　　2) Group 2 consists of hard trivalent transition metal ions with a d^q electronic configuration, $1 \leq q \leq 5$, and d^0 metal ions with z ≥ 3. Examples are

Al^{3+}, Cr^{3+}, and Ti^{4+}. The coordination number for ions of this group is always 6, and they form stable complexes with alkoxo groups.

3) Group 3 contains divalent transition metal ions with a d^q electronic configuration, $1 \leq q \leq 9$, divalent d^{10} cations, and large d^{10} trivalent metal ions. Examples include Ni^{2+}, Cu^{2+}, Zn^{2+}, Cd^{2+}, and Tl^{3+}. They prefer complexes having a MN_6 coordination sphere.

4) Group 4 is made up of small trivalent d^{10} metal ions and some of the trivalent transition metal ions with d^{10} electronic configurations, $1 \leq q \leq 9$. Examples are Cr^{3+}, Fe^{3+}, and Ga^{3+}. These cations share properties with cations from both Groups 2 and 3 and prefer complexes with a MN_3O_3 coordination sphere.

5) Group 5 contains large ions with a $d^{10} s^2$ electronic configuration, such as Pb^{2+} and Bi^{3+}. They form trinuclear species with taci by binding one amino group and two alkoxo groups. The coordination number of metal ions in this group can vary considerably.

While this work did not involve the use of macrocyclic ligands, the classification system used reveals several principles which are pertinent to macrocycle-metal ion binding. Further study is required to determine if this system can be extended to the rational design of macrocyclic ligands.

Ring Size. Shortly after Pedersen's discovery of macrocycles *(16)*, the size-match selectivity theory was proposed. It suggests that macrocycle-cation binding is dictated by the closeness of fit between ionic radii and the breadth of the inner macrocyclic cavity *(5,11,17)*. This concept has been cited for most of macrocyclic history, and continues to be used to explain empirical phenomena, though current research shows it to be largely coincidental *(8,10,13,18-20)*.

One shortcoming of the size-match selectivity theory is its inability to explain selectivities in the gas phase. The binding constants of 18-crown-6 with the alkali metals in the gas phase decrease in the order $Li^+ >> Na^+ >> K^+ >> Rb^+ >> Cs^+$ *(18)*. This trend has no apparent relationship to macrocycle cavity size. Another shortcoming is the observation that open-chain ligands, with no fixed cavity size, exhibit selectivity patterns similar to those of their macrocyclic counterparts *(13)*. These results greatly weaken the argument that the fixed cavity size determines selectivity. Furthermore, X-ray crystallographic data show that some macrocyclic ligands fold to complex with cations which do not match their cavity sizes *(13)*. In short, most macrocycles are much too flexible to be limited to the restrictions of the size-match selectivity theory, except in the case of small, rigid, highly preorganized macrocycles (*e.g.* 9-aneN_2O) which do exhibit size selectivity *(19)*.

As expressed by Hancock, *et al.*, the effect of macrocyclic ring size can be better described by examining the chelate ring size *(19)*. Abundant data suggest that an increase in the chelate ring size decreases complex stability for larger metal ions and increases complex stability for smaller metal ions. This behavior is somewhat counter-intuitive without the aid of molecular modelling calculations. As the radius of the metal cation increases, so does its coordination number. This in turn decreases the ligand-cation-ligand bond angles, resulting in increased strain and decreased complex stability.

Therefore, close attention to chelate ring size becomes an important factor in designing novel ligands while helping to explain the frailties in the size-match selectivity theory.

Adam, *et al. (10)* have recently proposed a model that relates the macrocycle ring size to the chelate ring size and use this model to describe what they call "enhanced discrimination." As the ring size was increased or decreased in a systematic fashion, they observed that the actual coordination structure remained essentially the same but was accompanied by an increase in ring strain. However, at certain threshold sizes, the strain became too great, resulting in "a major change in the structure of adjacent complexes in the series." This dislocation event was shown to be crucial to the selectivities of various macrocyclic ligands.

Pendant Donor Groups. Due to the ease with which pendant donor groups can be added to macrocyclic rings, hundreds of "armed" macrocyclic ligands have been prepared *(21,22)*. In general, the donor atoms in these pendant groups influence complex stability according to the rules discussed previously. There are, nonetheless, a few points that need to be considered when designing macrocycles with pendant groups.

Addition of pendant donor groups increases the coordination number of the ligand-metal complex. Hence, consideration of the ability of the target ion to accommodate the increased coordination number is necessary. For example, Hancock and Martell *(13)* review the effect of addition of four N-acetate pendant groups to tetraazamacrocycles. Metals such as Ca^{2+} that can accommodate a coordination number of eight experience a dramatic increase in complex stability, whereas metals that cannot accommodate the increased coordination (*e.g.* Cu^{2+}) experience a decrease in complex stability.

Pendant donor groups display a wide range of steric efficiencies. Steric efficiency describes the amount of steric crowding caused by bringing a pendant donor atom into the coordination sphere. For example, when a primary amine complexes with a metal ion, it brings its two associated hydrogen atoms into close proximity with the metal ion. This crowding causes low steric efficiency. In contrast, carboxylate donor groups are essentially planar. The donor oxygen carries no associated hydrogen atoms and can approach the metal ion without crowding. The high steric efficiency of carboxylate is partially responsible for the dramatic increase in complex stability that the addition of carboxylate groups can produce. For example, the log K value for 12-aneN$_4$ complexation with Ca^{2+} increases from 3.1 to 16.5 upon addition of four carboxylate pendant groups *(13)*.

We have recently synthesized and characterized a series of macrocyclic ligands whose pendant groups give them unique selectivities *(23)*. Each of the five crown ethers studied possesses the same pendant groups which differ only in the point of attachment and the substitution of a methoxyl group for a hydroxyl group in two cases. Nevertheless, these simple changes were sufficient to completely reverse the selectivities of these macrocycles in solvent extraction of alkaline earth metal ions. For example, the ligand shown in Figure 4A exhibits selectivity for the alkaline earth cations in the order $Mg^{2+} > Ca^{2+} > Sr^{2+} > Ba^{2+}$, while the modified ligand depicted in Figure 4B exhibits the selectivity $Ba^{2+} > Sr^{2+} > Ca^{2+} > Mg^{2+}$. This ability to completely reverse cation selectivity

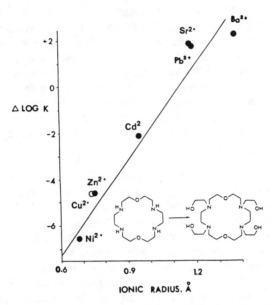

Figure 3. Effect on Stability of Complexes of 18-aneN$_2$O$_4$ of Adding Neutral Oxygen Donors (N-Hydroxyethyl groups). The change in complex stability, Δlog K, upon addition of oxygen donors is plotted against the ionic radii of the cations. (Reproduced with permission from ref. 13. Copyright 1989 American Chemical Society.)

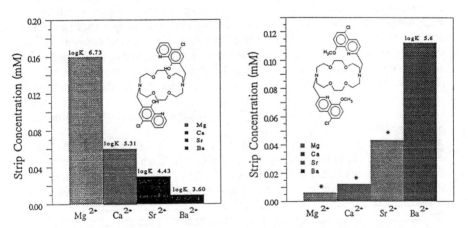

Figure 4. Reversal of Solvent Extraction Selectivity by Altering Pendant Groups. The concentration of metal ions in the strip solution after extraction of the 0.1 M alkaline earth nitrate with a 1.0 mM solution of the macrocyclic ligand in chloroform followed by stripping of the organic phase with 0.1 M HNO$_3$ is shown. Log K values were determined by calorimetry in methanol. (*) indicates insufficient heat to measure log K.

without altering the structure of the macrocyclic ring is another mark against the size match selectivity theory and demonstrates the power of using pendant donor groups to control macrocycle selectivity.

Preorganization. Preorganization refers to the degree of similarity between a ligand in its bound and unbound conformations. A highly preorganized ligand will undergo less conformational change upon binding a cation than a ligand with less preorganization. Preorganization may affect the membrane transport of metal ions in a variety of ways by altering complex stability, selectivity, and complexation kinetics.

Preorganization is intimately linked to entropy *(9, 24-26)*. One determinant of the strength of a macrocycle-cation bond is the difference between the entropies of the bound and unbound macrocyclic ligand. Leach and Lewis *(26)* explain that rigid ligands suffer a smaller loss of conformational entropy during binding than a less rigid counterpart. Because a preorganized ligand loses less entropy during binding, its stability constant is usually larger. Leach and Lewis work primarily with biological systems, but the principles may be extended to the present application. To design ligands with improved substrate affinities, these authors worked toward increased preorganization using a "ring-bracing approach to computer-assisted ligand design." This approach involved the use of bridging groups to give the ligands a higher degree of preorganization.

A lower loss of entropy in preorganized macrocycles is one reason for their elevated stability constants. Another reason is the relatively low solvation energy of the ligand. X-ray crystal structures of spherands, the most highly preorganized macrocycles, reveals an absence of solvent in their ligand cavities *(13)*. Highly preorganized tetraazamacrocycles are also known to exclude solvent molecules to some degree from their cavities. In the absence of solvent molecules in their interiors, preorganized macrocycles need not undergo the energy-expensive process of desolvation before binding a cation. With hydrogen bonds typically measuring about 7 kcal-mol^{-1}, the energy cost of desolvation of unorganized macrocycles is considerable. Therefore, preorganized macrocyclic ligands enjoy a marked advantage in complex stability.

Selectivity is also dictated to some degree by preorganization. As discussed above, large, flexible macrocycles do not faithfully follow the size-match selectivity principle, but preorganized macrocyclic ligands tend to adhere more closely *(8,13)*. Molecular mechanics is usually successful in predicting which cations will bind most readily to preorganized ligands on the basis of cationic radius and cavity size. Ligands which exhibit a high level of preorganization include spherands, small cryptands, porphyrins, corrins, sepulchrates, sarcophaginates, and small crown ethers *(13)*.

While preorganization can be an effective means of increasing complex stability, its use in designing carriers for liquid membrane systems is limited. Hancock and Martell *(13)* review two problems associated with preorganization. First, it is difficult to design preorganized ligands with large cavities. Small ligands generally display high levels of preorganization, but larger ligands are almost always too flexible to derive the benefits of preorganization. "Bridging groups" that lend some preorganization to larger ligands have proven successful. Design of preorganized ligands with larger cavities will almost certainly rely upon aromatic rings and pyridyl groups for increasing preorganization, as

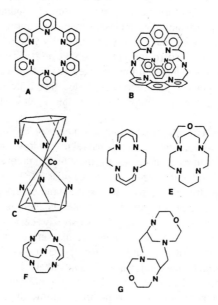

Figure 5. Examples of highly preorganized ligands. (Reproduced with permission from ref. 13. Copyright 1989 American Chemical Society.)

in the case of sexipyridine (Figure 5A). The ring-bracing strategies of Leach and Lewis *(26)* and Wade, *et al. (25) (e.g.* addition of bridging groups) may also prove to be rewarding in the search for larger, preorganized macrocycles (*e.g.* Figures 5 B-G). The second limitation is the decreased lability of the cation-macrocyclic ligand bonds of highly preorganized ligands. As a result of this feature, preorganized macrocycles are good candidates for applications which require stable, strong complexes, but their use is not as practical in membrane systems which require more dynamic interactions.

Emergent Tools for Improved Ligand Design

We owe much of our understanding of ligand design principles to the established fields of solvent extraction, calorimetry, X-ray crystallography, potentiometry and others, along with a large body of empirical data obtained from studies of metal ion transport in membrane systems. While these means of acquiring data will not soon be replaced, emergent tools are offering their strength to the cause of improved ligand design. Two of these emergent tools are gas-phase studies of macrocycle-cation interactions and computer-based molecular modeling.

Gas-Phase Studies. Gas-phase chemistry involves the study of interactions in the gas phase, without the complicating effects of solvent. Recent research has yielded a few papers which elucidate the interaction of several macrocyclic ligands (including 18-crown-6, dicyclohexano-18-crown-6, and valinomycin) with cations in the gas phase *(27-30)*. Cation selectivities determined in the gas phase vary significantly from those observed in the presence of a solvent. Because all membrane systems function in the presence of a solvent, not in the gas phase, the value to membrane separations of data determined by gas-phase study is indirect. By comparing gas-phase results with results from traditional methods of study, the proportion of the interaction which is due to solvent and the proportion due to the macrocycle itself can be separated. Then, as more information becomes available, the solvent effects and macrocycle effects can be treated in combination to predict the outcome of new systems. Because many computer-based molecular modelling programs ignore the influence of solvent, they are similar to gas-phase interactions. Thus, data from gas-phase studies may be used to test the viability of computer-based theoretical calculations, which may in turn find practical application to predict behavior in solution.

In truth, too little gas-phase data are as yet available to contribute significantly to the vast bank of information available right now from traditional sources. However, as the amount of gas-phase information increases, it may shed new light on the science of rational ligand design.

Molecular Modelling. Computer-based molecular modelling (MM) is a tool of considerable power in predicting the formation kinetics and thermodynamic stability of macrocycle-cation complexes. Many papers have already described the successful employment of MM in their study of macrocyclic ligands *(31-33)*. Several MM software packages are now available commercially, making this tool accessible to most separation scientists. While MM holds tremendous potential, caution is warranted when attempting to apply MM results to ligand design. Shirts and Stolworthy *(31)* show that even small variations in the parameters such as charge distribution used in the calculations, even variations that are within experimental error, can result in dramatic alterations in calculation results.

Acknowledgement. The authors express appreciation to the United States Department of Energy, Office of Basic Energy Sciences who funded this work through Grant No. DE-FG03-95-ER14506.

Literature Cited

1. Izatt, R. M.; Pawlak, K.; Bradshaw, J. S.; Bruening, R. L. *Chem. Rev.* **1991**, *91*, 1721-2085; Izatt, R. M.; Bradshaw, J. S.; Nielsen, S. A.; Lamb, J. D.; Christensen, J. J.; Sen, D. *Chem. Rev.* **1985**, *85*, 271-331.
2. *Handbook of Metal Ligand Heats;* Christensen, J. J.; Izatt, R. M., Eds., Dekker: New York, N.Y., 1990.
3. Olsher, U.; Hankins, M. G.; Kim, Y. D.; Bartsch, R. A. *J. Am. Chem. Soc.* **1993**, *115*, 3370-3371.
4. Izatt, R. M.; Bruening, R. L.; Christensen, J. J. In *Liquid Membranes: Theory and Applications*; Noble, R. D.; Way, J. D., Eds.; ACS Symposium Series 347; American Chemical Society: Washington, DC, **1987**; pp 98-108.
5. Lamb, J. D.; Izatt, R. M.; Christensen, J. J. In *Progress in Macrocyclic Chemistry*, Vol. 2; Izatt, R. M.; Christensen, J. J., Eds.; Wiley: New York, N.Y., **1981**; pp 41-90.
6. Kirch, M.; Lehn, J. M. *Angew Chem. Int. Ed. Engl.* **1975**, *14*, 555.
7. Hegetschweiler, K.; Hancock, R. D.; Ghisletta, M.; Kradolfer, T.; Gramlich, V.; Schmalle, H. W. *Inorg. Chem.* **1993**, *32*, 5273-5284.
8. Hancock, R. D.; Bhavan, R.; Wade, P. W.; Boeyens, J. C. A.; Dobson, S. M. *Inorg. Chem.* **1989**, *28*, 187-194.
9. Jarvis, N. V.; Hancock, R. D. *Radiochim. Acta* **1994**, *64*, 15-22.
10. Adam, K. R.; Arshad, S. P. H.; Baldwin, D. S.; Duckworth, P. A.; Leong, A. J.; Lindoy, L. F.; BcCool, B. J.; McPartlin, M.; Bailor, B. A.; Tasker, P. A. *Inorg. Chem.* **1994**, *33*, 1194-1200.
11. Lamb, J. D.; Izatt, R. M.; Swain, C. S.; Christensen, J. J. *J. Am. Chem. Soc.* **1980**, *102*, 475.
12. Bradshaw, J. S.; Maas, G. E.; Izatt, R. M.; Lamb, J. D.; Christensen, J. J. *Tetrahedron Lett.* **1979**, *7*, 635-638.
13. Hancock, R. D.; Martell, A. E. *Chem. Rev.* **1989**, *89*, 1875-1914.
14. Wolf, R. E.; Hartman, J. R.; Torey, J. M. E.; Foxman, B. M.; Cooper, S. R. *J. Am. Chem. Soc.* **1987**, *109*, 4328-4335.
15. Pearson, R. G. *J. Am. Chem. Soc.* **1963**, *85*, 3533.
16. Pedersen, C. J. *J. Am. Chem. Soc.* **1967**, 89, 7017.
17. Lamb, J. D.; Izatt, R. M.; Garrick, D. G.; Bradshaw, J. S.; Christensen, J. J. *J. Membr. Sci.* **1981**, *9*, 83-107.
18. Hay, B. P.; Rustad, J. R.; Hostetler, C. J. *J. Am. Chem. Soc.* **1993**, *115*, 11158-11164.

19.   Hancock, R. D.; Wade, P. W.; Ngwenya, M. P.; de Sousa, A. S.; Damu, K. V. *Inorg. Chem.* **1990**, *29*, 1968-1974.

20.   Walkowiak, W.; Kang, S. I.; Stewart, L. E.; Ndip, G.; Bartsch, R. A. *Anal. Chem.* **1990**, *62*, 2022-2026.

21.   Brown, P. R.; Bartsch, R. A. In *Inclusion Aspects of Membrane Chemistry*; Osa, T.; Atwood, J. L., Eds.; Topics in Inclusion Science; Kluwer Academic: Boston, MA **1991**; pp 1-57.

22.   Tsukube, H. In *Liquid Membranes: Chemical Applications;* Araki, T.; Tsukube, H., Eds.; CRC Press: Boca Raton, FL **1990**; pp 51-75.

23.   Zhang, X. X.; Bordunov, A. B.; Bradshaw, J. S.; Dalley, N. K.; Kou, X. L.; Izatt, R. M. *J. Am. Chem. Soc.* **1995**, *117*, 11507-11511; Bordunov, A. B.; Bradshaw, J. S.; Zhang, X. X.; Dalley, N. K.; Kou, X. L.; Izatt, R. M. *J. Am. Chem. Soc.*, Submitted.

24.   Kaplan, W. A.; Scott, R. A.; Suslick, K. S. *J. Am. Chem. Soc.* **1990**, *112*, 1283-1285.

25.   Wade, P. W.; Hancock, R. D.; Boeyens, J. C. A.; Dobson, S. M. *J. Chem. Soc. Dalton Trans.* **1990**, 483-488.

26.   Leach, A. R.; Lewis, R. A. *J. Comp. Chem.* **1994**, *15*, 233-240.

27.   Wong, P. S. H.; Antonio, B. J.; Dearden, D. V. *J. Am. Soc. Mass Spectrom.* **1994**, *5*, 632-637.

28.   Chu, I.; Dearden, D. V.; Bradshaw, J. S.; Huszthy, P.; Izatt, R. M. *J. Am. Chem. Soc.* **1993**, *115*, 4318-4320.

29.   Chu, I.-H.; Dearden, D. V. *J. Am. Chem. Soc.* **1995**, *117*, 8197-8203.

30.   Zhang, H.; Chu, I.; Leming, S.; Dearden, D. V. *J. Am. Chem. Soc.* **1991**, *113*, 7415-7417.

31.   Shirts, R. B.; Stolworthy, L. D. *J. Incl. Phenom.* **1995**, *20*, 297-321.

32.   Pletnev, I. V. *Can. J. Chem.* **1994**, *72*, 1404-1411.

33.   Hay, B. P.; Rustad, J. R. *J. Am. Chem. Soc.* **1994**, *116*, 6316-6326.

Chapter 10

New Tactics in Design of Ion-Specific Carriers: Approaches Based on Computer Chemistry and Rare Earth Complex Chemistry

Hiroshi Tsukube

**Department of Chemistry, Faculty of Science,
Osaka City University, Sugimoto, Sumiyoshi-ku, Osaka 558, Japan**

New approaches for development of specific carriers for use in liquid membrane are described: (i) computer-aided design of cation-specific carriers and (ii) functionalization of rare earth complexes as anion carriers. A new series of Li(I) and Ag(I) ion-specific carriers are successfully designed using MM2, MNDO and density functional calculations. Computer chemistry provides a rational basis for design and characterization of cation-specific carriers of armed crown ether- and podand-types. Lipophilic lanthanide tris(β-diketonates) are shown to be a new class of membrane carriers. They form 1:1 complexes with anionic guests and mediate transport of amino acid derivatives. Since these complexes exhibit different anion transport properties from those of crown ethers, further applications of rare earth complexes offer promising possibilities in the development of specific anion carriers for liquid membrane systems.

Liquid membranes incorporating host species can effect rapid and selective transport of a variety of guest species and have wide applications in sensing and separation processes *(1)*. As observed in biomembrane transport systems, host molecules recognize guest species of interest and selectively transport them. Valinomycin is an example of such biological carriers. It specifically mediates biological transport of K(I) cation, but rarely carries other substrates such as organic ammonium cations. In contrast, the synthetic ionophore 18-crown-6 and other crown ethers are known to transport alkali, alkaline earth, heavy and transition metal cations as well as organic substrates. This means that synthetic carriers can mediate artificial transport of various guests, while naturally occurring carriers transport only limited kinds of biological guests such as K(I) cation and biogenetic amines. Since the carrier dramatically improves transport performance, development of new, specific carriers is the most exciting task in liquid membrane science and technology *(2, 3)*. This chapter describes new tactics in carrier chemistry which have recently been established in our laboratory. They provide new strategies and concepts in the design of powerful carriers which are specific for some metal cations and organic guests. Our discussion here focuses on (i) computer-aided design of organic carriers specific for Li(I) and Ag(I) cations and (ii)

0097–6156/96/0642–0142$15.00/0

novel transport functions of rare earth complexes toward amino acid anions. Although carrier design is still empirical and challenging, the tactics described below can make it possible to determine the essential features of excellent new carriers.

Computer-Aided Design of Metal Cation-Specific Carriers

Molecular design of a new, specific carrier is a short way to create practical liquid membrane separations, although this usually includes empirical "trial-and-error" exercises. Computer chemistry is a promising methodology in the design of specific carriers and a rational basis for carrier synthesis *(4, 5)*. Although there are three kinds of computational methods, empirical, semi-empirical and non-empirical calculations, bench chemists have limited themselves to the use of empirical methods such as MM2 and have rarely employed semi-empirical or non-empirical calculations in carrier chemistry. However such calculations can be currently performed using personal computers and we successfully applied some of them to develop metal-specific carriers.

Design of Ag⁺ Ion-Specific Carriers Based on MM2 Calculation.
Molecular mechanics calculation (MM) is the computational method most familiar to bench chemists *(6, 7)*. Many kinds of software are on the market and are successfully used to address the conformational properties of various organic molecules. Since the programs have some limitations in dealing with weak interactions, crown ether-alkali metal complexes and related systems have hardly been examined. We applied the "extended MM2 program" (CAChe Scientific, version 3.0) to design Ag(I) ion-specific carriers *(8)*. Because interaction between Ag(I) and donor atoms of the carrier is adequately strong, the MM2 calculation is applicable in such a case.

Highly selective carriers for Ag(I) ion are of commercial interest. Ag(I) ion occurs in nature together with Pb(II) and other metal cations, and some stable complexes have potential in cancer radioimmunotherapy *(9)*. Furthermore, coordination chemistry of Ag(I) complexes is interesting from the standpoint of pure chemistry. Ag(I) usually forms linear bidentate complexes but occasionally forms unique tridentate complexes *(10, 11)*. Since tridentate coordination chemistry has rarely been observed in other metal complex systems, Ag(I) ion-specific carriers can be prepared if donor atoms can be organized in a tridentate fashion. We chose a podand-type carrier which had three pyridine moieties as potential binding sites for Ag(I) ion (Figure 1). Podand-type carriers are usually regarded as poor ligands when compared with macrocyclic carriers, but they have great advantages of structural versatility and variation *(12)*. Many computational modeling experiments were performed to optimize the structure of the tridentate podand and we ultimately decided to link three pyridine moieties through two ester units. The optimized structure of the Ag(I) complex with the podand **2** is shown in Figure 1. When $-CH_2-$ units were inserted between the ester and terminal pyridine units, the three pyridine rings were suggested to cooperatively coordinate with Ag(I) ion in the tridentate fashion. We actually prepared a series of pyridine podands **1-3** (Figure 2) and compared their "real" Ag(I) ion selectivity with that predicted by computational method. Tridentate Ag(I) cation binding of podand **2** was confirmed by 13 C NMR binding experiments. Addition of silver perchlorate caused significant spectral changes of podand **2** upon 1:1 complexation. Since carbon signals for the three pyridine rings shifted greatly, this podand was demonstrated to form a tridentate complex. Podand **3** with two pyridine rings and one benzene ring also formed a 1:1 complex with Ag(I) ion. The 13 C NMR signals shifted only for two pyridine rings, suggesting this forms a bidentate Ag(I) complex as reported in many Ag(I) complexes. These spectroscopic observations clearly indicated that MM2 calculation provided results consistent with experiment and was useful in designing a Ag(I) ion specific carrier.

Transport properties of these podand carriers were examined in a U-tube liquid membrane system (Figure 3) *(13)*. When a neutral podand is dissolved in Membrane,

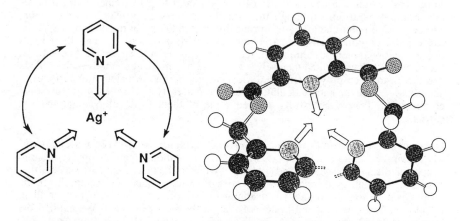

Figure 1. Optimized structure of podand **1** by extended MM2. Two substituents and the Ag$^+$ ion are omitted for simplified illustration (Adapted from ref. 8).

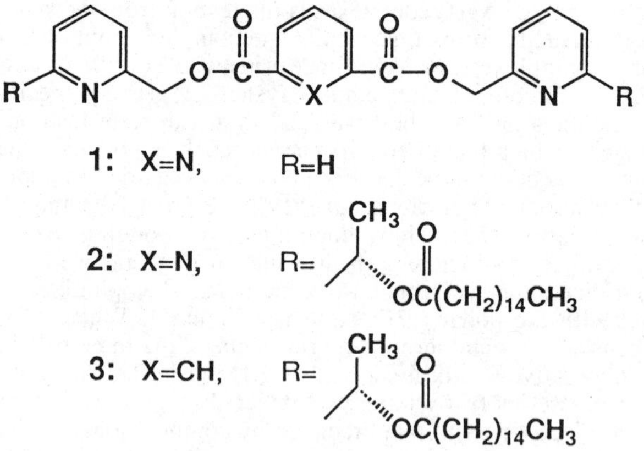

1: X=N,   R=H

2: X=N,   R=

3: X=CH,   R=

Figure 2. Pyridine podands as Ag$^+$ ion-specific carriers.

it selectively binds Ag(I) ion at the interface between Aq. I and Membrane and smoothly extracts it into Membrane. The resulting cationic complex moves across Membrane together with the counter-anion and decomposes at the interface between Membrane and Aq. II. As a result, Ag(I) ion and the counter-anion are transported by the podand carrier across the membrane. Tridentate pyridine podand 2 with bulky substituents exhibited perfect selectivity and high efficiency for Ag(I) ion in competitive cation transport experiments. It predominantly transported Ag(I) ion in the presence of equimolar Pb(II), Cu(II), Ni(II), Co(II) and Zn(II) cations. Tridentate pyridine podand 1 with no substituents also transported Ag(I) ion but its efficiency was much lower than that of podand 2. Podand 3 with two pyridine rings and one benzene ring formed the bidentate complex with Ag(I) ion and hardly transported it. Therefore, a tridentate arrangement of the three pyridine moieties in the podand skeleton is required for the Ag(I) ion-specific and efficient transport. Liquid-liquid extraction experiments demonstrated that tridentate pyridine podand 2 specifically bound Ag(I) ion, while Pb(II), Cu(II), Ni(II), Co(II), Zn(II), Li(I), Na(I) and K(I) ions were not extracted at all. The less lipophilic podand 1 formed a considerable amount of insoluble material. Thus, a combination of three pyridine moieties and bulky/lipophilic substituents can prevent high-order complexation, enhance solubility of the Ag(I) complex and offer highly selective/effective extraction and transport of Ag(I) ion.

Design of Li⁺ Ion-Specific Carriers Based on Semi-Empirical and Non-Empirical Calculations. Li(I) ion-specific carriers are important tools for analysis and separation in biological and environmental systems *(14)*. Crown ether derivatives are undoubtedly good candidates as effective carriers. In particular, 13-crown-4 and 14-crown-4 derivatives were reported to exhibit excellent carrier activities for Li(I) ion, though the diameter of Li(I) ion is formally size-fitted to the 12-crown-4 ring. We first confirmed the size-fitting of Li(I) ion with 12-crown-4 ring using MNDO calculation *(15)*. Figure 4 shows optimized structures of two different types of 12-crown-4 · Li(I) complexes which have D_{2d} and C_{4v} symmetries. The Li(I) ion is completely accommodated in the cavity of the 12-crown-4 ring in the D_{2d} complex. For such inclusion of Li(I) ion solvent molecules and a counter-anion must be removed before the complexation. These are costly energetic processes. On the other hand, the Li(I) cation still interacts with counter-anion in the C_{4v} complex. Although the complex of D_{2d} symmetry looked like ideal cation encapsulation with the crown ether, stabilization energy calculations showed that the C_{4v} complex was more stable than the D_{2d} complex. This strongly suggests that introduction of ligating sidearm into the 12-crown-4 ring, instead of counter-anion, can offer a more stable Li(I) complex.

We applied non-empirical "density functional calculation" to design Li(I) ion-specific armed aza-12-crown-4 derivatives (see Figure 5) *(16)* . "Armed crown ethers" are a new type of cation carriers and are characterized by a parent crown ring and a cation ligating sidearm *(17, 18)*. In this class of compounds, the donor group on the flexible sidearm provides further coordination with a guest metal cation trapped in the crown cavity. Such ionophores exhibit both the kinetically fast complexation properties of simple crown ethers and the three-dimensional binding characteristics of bicyclic cryptands. These features are similar to those of naturally occurring ionophores and offer effective binding and selective transport of target metal cations. The guest selectivity can be tuned by selection of the crown ether ring and the sidearm and several armed crown ethers were reported to show high guest selectivity and excellent transport efficiency.

The density functional method has recently been considered a promising alternative to the Hartree-Fock approach, but has rarely been employed in crown ether complex systems *(19)*. Figure 6 illustrates the optimized structure of Na(I) complex of N-methoxyethylaza-12-crown-4 (4) with predicted bond lengths. Na-O (crown ring) distances were estimated to range from 2.316 to 2.351 Å, which are longer by ca. 0.1 Å than that for sidearm O atom (2.242 Å). The Na-N distance was calculated to be

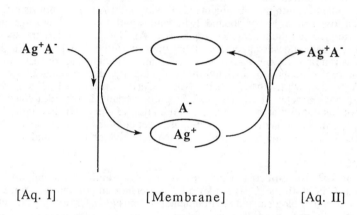

[Aq. I]   [Membrane]   [Aq. II]

Figure 3. An Ag$^+$ ion-selective liquid membrane.

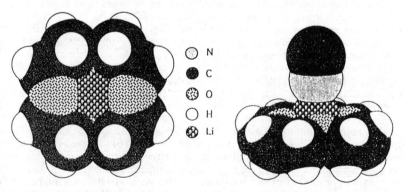

Figure 4. Space-filling drawings of 12-crown-4·Li$^+$ complexes with D$_{2d}$ symmetry (left) and C$_{4v}$ symmetry (right) (Adapted from ref. 15).

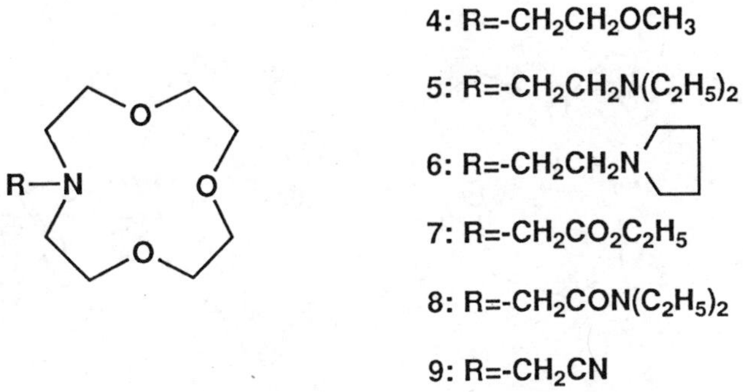

R—N
O
4: R=-CH2CH2OCH3
5: R=-CH2CH2N(C2H5)2
6: R=-CH2CH2N
7: R=-CH2CO2C2H5
8: R=-CH2CON(C2H5)2
9: R=-CH2CN

Figure 5. Armed crown ethers as Li^+ ion-specific carriers.

OMe
N
Na+
O
2.242(2.453)Å
2.320(2.459)Å
2.391(2.561)Å
2.351(2.455)Å
2.316(2.411)Å

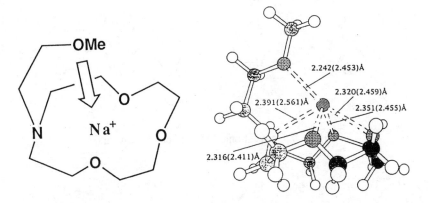

Figure 6. Optimized structure of Na^+ complex with ether-armed aza-12-crown-4 (**4**) (Adapted from ref. 16). The calculated bond distances are indicated, while the distances determined in the NaI complex crystal are shown in parentheses.

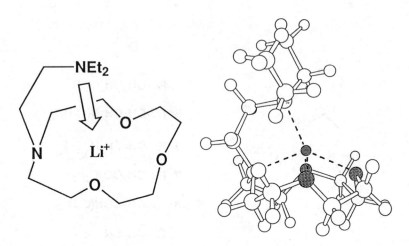

Figure 7. Optimized structure of Li+ complex with amine-armed aza-12-crown-4 (5) (Adapted from ref. 21).

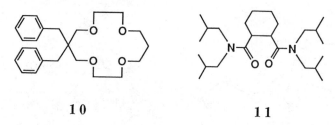

10   1 1

Figure 8. Conventional Li+ ion-specific carriers.

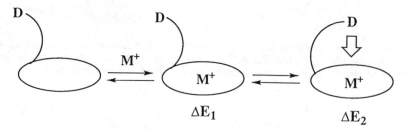

Figure 9. Possible binding modes of armed crown ether.

2.391 Å. Arnold et al. determined the crystal structure of the same crown ether 4-NaI complex *(20)*. Since their estimated bond lengths were comparable to those predicted, the density functional method is believed to be applicable for crown ether - alkali cation complex. We attempted to apply MM2 and MNDO calculations for several crown ether-alkali metal cation complexes *(21)*. Since the parameters for weak interactions between alkali metal cations and neutral donor atoms have not been optimized completely, these calculations sometimes gave results which differ from those in experiments.

Figure 7 displays the optimized structure of a Li(I) complex with N,N-diethylaminoethylaza-12-crown-4 (5) based on the density functional calculations in which the amino-nitrogen atom on the sidearm acts as a secondary binding site. We carried out similar calculations for Li complexes with various armed aza-12-crown-4 derivatives 4, 7 and 8 and obtained almost the same complex structures. Ether-, ester- and amide-functionalized sidearms were confirmed to act cooperatively with parent crown rings, but energy calculations revealed that the amine-functionalized sidearm offered larger stabilization than the other functionalized sidearms. We compared cation-binding properties of a series of armed aza-12-crown-4 derivatives with those expected by computer calculations. As predicted above by the calculations, amine-armed crown ethers 5 and 6 were confirmed to form stable Li(I) cation complexes and to transport Li(I) ion specifically across a CH_2Cl_2 liquid membrane. The 14-crown-4 compound 10 and podand 11, which are commercially available Li(I) ion-specific carriers (Figure 8), were also examined under the same conditions. They mediated selective transport of Li(I) ion but their transport rates were much less than those observed for our amine-armed crown ethers 5 and 6. 13 C NMR and 7 Li NMR binding studies demonstrated that amine-functionalized sidearms specifically coordinated with Li(I) ion which was trapped in the aza-12-crown-4 cavity.

We further analyzed the cation binding abilities of armed crown ethers in terms of stabilization energy. The structures of free armed crown ethers and their two different kinds of complexes were optimized and their stabilization energies ΔE_1 and ΔE_2 were estimated (Figure 9). ΔE_1 is obtained by complex formation in which the crown ring accommodates the guest cation, but sidearm does not interact with it. ΔE_2 is estimated for the three-dimensional complex in which the guest cation is cooperatively coordinated with donor atoms of both the crown ether ring and the sidearm. Thus, the energy difference $\Delta E_2 - \Delta E_1$ can be considered as a measure of the effectiveness of sidearm coordination. These values for amine-armed crown ether 5 were calculated as -13.7 kcal/mol and -10.1 kcal/mol for Li(I) and Na(I) complexes, respectively. These values mean that the amine moiety on the sidearm effectively recognized the two metal cations. Moreover by 3.6 kcal/mol, the amine-functionalized sidearm greatly favored Li(I) cation. In contrast, the ether group of the armed crown ether 4 binds these cations with similar stabilization since the $\Delta E_2 - \Delta E_1$ values are -14.0 kcal / mol for the Li(I) complex and -13.2 kcal / mol for the Na(I) complex. These calculation results compared well with the observed cation binding properties, indicating that the density functional method gave results consistent with the experimental observations.

Thus effective combinations of carrier chemistry with computer chemistry have established a useful guideline for the design of new, metal cation-specific carriers. We have many options in the selection of carrier structures from a variety of acyclic podands, cyclic crown ethers, bicyclic cryptands and related derivatives. When we adjust a carrier structure for a target guest, we can consult the computer and decide on the proper carrier structure in a non-empirical fashion. Although there are still many problems to be resolved, computer chemistry can be a good partner with carrier chemistry.

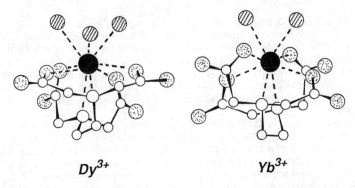

Dy^{3+} Yb^{3+}

Figure 10. Crystal structures of lanthanide-EDTA⁻ complexes (Adapted with permission from ref. 24. Copyright 1979 International Union of Crystallography)

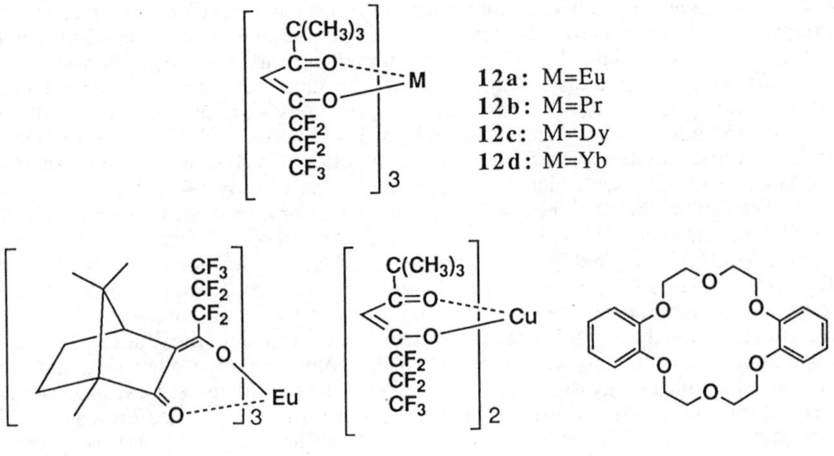

$$\begin{bmatrix} \underset{\displaystyle \underset{\displaystyle \underset{\displaystyle CF_3}{CF_2}}{CF_2}}{\overset{\displaystyle \overset{\displaystyle C(CH_3)_3}{C=O}}{\underset{\displaystyle C-O}{}}}\cdots M \end{bmatrix}_3 \quad \begin{matrix} 12a: M=Eu \\ 12b: M=Pr \\ 12c: M=Dy \\ 12d: M=Yb \end{matrix}$$

13   14   15

Figure 11. Lanthanide complexes as anion carriers.

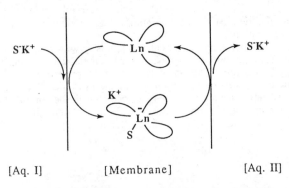

[Aq. I]   [Membrane]   [Aq. II]

Figure 12. Anion transport mediated by lanthanide tris(β-diketonate).

Rare Earth Metal Complexes as a New Membrane Carrier

Rare earth metal cations have been viewed as one of the most interesting guest cations in liquid membrane science and technology. Since they have unique chemical and physical properties, their detection and separation are of great importance in both pure and applied science *(22)*. Rare earth metal cations are also used to label organic molecules in the areas of clinical chemistry and molecular biology *(23)* and are particularly widely employed in immunoassay and nucleic acid hybridization. Coordination chemistry of rare earth metal cations differs considerably from that of transition metal cations, alkali metal cations or other metal cations. Figure 10 illustrates crystal structures of EDTA complexes with rare earth cations *(24)*. These complexes are rather stable and monoanionic, since the central trivalent cations are coordinated with four carboxylate anions of the EDTA ligand. The complexes also have two or three water molecules, so that highly coordinated complexes are formed. It is suggested that rare earth complexes such as tris(β-diketonate) complexes can coordinate with substrates without ligand exchange.

We examined the carrier activities of various lanthanide tris(β-diketonates) (Figure 11) in the liquid membrane system *(25)*. These complexes have several features of interest for required for membrane carriers, especially for anionic guests: (a) They have the potential to coordinate with the anionic group of one or more substrates; (b) Fluorinated ligands enhance the anion coordination ability of the central metal cation as well as the lipophilicity of the complex; (c) These are stable complexes and carry anionic guests without ligand exchange or decomposition under the transport conditions; (d) Complexes having chiral β-diketone ligands are available and may mediate enantio-selective transport; and (e) The ion radius of the trivalent rare earth metal cation changes with increase in atomic number, indicating that tuning of guest selectivity is possible by a proper selection of the central metal cation. We were the first to demonstrate that these complexes exhibit excellent transport abilities toward several anionic guests *(26)*.

Anion Carriers Derived from Rare Earth Complexes. The binding and transport of inorganic and organic anions play an important role in many biological and artificial processes and considerable effort has been devoted to the development of anion transporting agents *(27)*. Several surfactants, lipophilic metal complexes, rigid receptors and macrocycles with positively charged subunits have been reported as anion carriers. Although they mediate transport of inorganic and organic anions, the number of anion carriers remains quite limited *(28-31)*. The trivalent rare earth metal cations such as europium and praseodymium have proven useful as aqueous NMR shift reagents for examining the structure of small biological molecules. These bind phosphate, carboxylate and other anions, and induce characteristic NMR spectral changes. We applied lipophilic lanthanide tris(β-diketonates) as membrane carriers and found that they bind organic anions efficiently (Figure 11).

Anion bindings of the lanthanide tris(β-diketonates) were typically characterized using 13 C NMR titration experiments. When tetrabutylammonium acetate was employed as a guest, carbon signals of the acetate anion completely disappeared in the presence of a small amount of europium complex **12a**, indicating that the guest anion coordinates with the europium cation. Carbon signals of the butyl group in the counter-cation shifted greatly, but were still observable. Their spectral changes suggested 1:1 complexation between the europium tris(β-diketonate) **12a** and acetate anion. Since copper bis(β-diketonate) **14** induced negligible spectral changes, lanthanide tris(β-diketonates) are shown to have unique anion binding functions.

Figure 12 illustrates an anion transport system with a lanthanide tris(β-diketonate) as the carrier. When the lipophilic lanthanide complex is present in Membrane, a highly coordinated complex is formed with the anion guest at the interface between Aq. I and Membrane, and K(I) cation is extracted into Membrane as the counter-cation. The resulting ternary complex moves across Membrane. At the interface between Membrane and Aq. II, the guest anion is released into Aq. II together with its counter-cation. Crown ether carrier mediates anion transport in a different fashion (see Figure 3). In that case, the carrier binds K(I) cation rather than the anionic guest and transports both species in the same direction. These two types of carriers have different binding modes for anionic guests and exhibit different recognition and transport properties. When lanthanide complex 12 or crown ether 15 was employed as the carrier, a corresponding amount of K(I) cation moved from Aq. I into Aq. II. Thus, the transport mechanisms described above operate in each case. Lanthanide complex carriers 12a-12d effectively transported benzyloxycarbonyl-amino acid (Z-amino acid) anions with K(I) cation and their transport profiles were interestingly different from those of crown ether carrier 15. They exhibited larger transport rates for Z-Val and Z-Leu anions than for Z-Phe anion, while the crown ether carrier favored Z-Phe anion. The nature of the central lanthanide cation had a marked influence on the transport rate, but only slightly changed the transport selectivity. Since rare earth complexes having non-fluorinated ligands often produced insoluble materials under the transport conditions, a fluorinated ligand should be incorporated into the carrier of this type. Chiral europium tris(β-diketonate) 13 was also examined. This effectively bound several Z-amino acid salts and transported them, even though considerable steric hindrance between three bulky ligands and guest anion was anticipated. Since enantioselective transport of Z-amino acid salts was not realized, anion coordination of the lanthanide metal cation was not rigid enough for precise chiral discrimination (32, 33).

Further Applications of Rare Earth Complexes in Liquid Membranes. Rare earth complexes have a further possibility as a unique carrier in liquid membrane processes: Zwitterionic amino acids are one of the most interesting guest candidates for liquid membrane separation. Although a variety of synthetic carriers have been reported to transport amino acid derivatives, most of them bound ammonium or carboxylate moieties of the guests (1, 2). Unsubstituted amino acids exist in a zwitterionic form under the biologically neutral conditions. The ammonium and carboxylate ions of the guest considerably diminish the abilities to interact with a carrier. Since desolvation of the double ion is a costly energetic process, only a limited number of carriers extract zwitterionic amino acids and mediated their transport (34-37). Lanthanide tris(β-diketonates) such as 12a-12d may bind zwitterionic amino acid. First coordination with the carboxylate anion produces a highly coordinated complex with a negative charge. The resulting anionic species can interact with the ammonium moiety of the bound amino acid through electrostatic interaction, so that the parent lanthanide tris(β-diketonates) binds the zwitterionic amino acid via two-point binding. Both functional groups of the amino acid are positionally fixed on the lanthanide complex and chiral discrimination is sometime possible (38).

Chromogenic rare earth complexes can be used as a color indicator in the sensing of anionic or zwitterionic guests. Some rare earth porphyrin complexes have already been prepared (39). Since the ion radius of trivalent rare earth cations is larger than cavity size of the porphyrin ring, the metal cation is located above the porphyrin plane and coordinated with one other ligand to be electrically neutral. There is a space for guest binding without ligand exchange and this type of porphyrin complex is thought to bind guest species in a similar fashion to lanthanide tris(β-diketonates) and to change their spectral properties.

Concluding Remarks

Synthesis of a new, specific carrier is a particularly exciting job, because the carrier can remarkably modify the liquid membrane transport phenomenon. In addition to the traditional "trial-and-error" approach, a new, non-empirical methodology in designing synthetic carriers is emerging. As described here, the computational method offers great promise in carrier design. Application of "exotic" rare earth has complexes led to new and interesting transport phenomena. Although we have limited ourselves to dealing with organic carriers, the rare earth complex carriers exhibited uncommon and promising functions. The results clearly demonstrate that new concepts and methodologies developed in carrier chemistry are leading to more elegant and practical liquid membrane separations. Liquid membrane science and technology are coming of age.

Acknowledgements

The author expresses his thanks to Professors Richard A. Bartsch of Texas Tech University and J. Douglas Way of Colorado School of Mines for their encouragement. He is also grateful to Professor Kenzi Hori of Kyushu University, Professors Jun-ichi Uenishi and Osamu Yonemitsu of Okayama University of Science and Professor Hitoshi Tamiaki of Ritsumeikan University for helpful discussions. Permission to use Figure 10 which was granted by the International Union of Crystallography is appreciated. This research was supported in part by Grants-in-Aid for Scientific Research on Priority Areas, "New Development of Rare Earth Complexes", (Nos. 06241253 and 0723076) from the Ministry of Education, Science, Sports and Culture, Japan.

Literature Cited

1. Tsukube, H. In *Liquid Membranes: Chemical Applications*; Araki, T.; Tsukube, H., Eds.; CRC Press: Boca Raton, FL., 1990; pp 51-75.
2. Tsukube, H. In *Cation Binding by Macrocycles*; Inoue, Y.; Gokel, G.W., Eds.; Marcel Dekker: New York, 1990; pp 497-522.
3. Tsukube, H. In *Crown Ethers and Analogous Compounds*; Hiraoka, M., Ed.; Elsevier: Amsterdam, 1992; pp 100-197.
4. Wipff, G. *J. Coord. Chem.* **1992**, *27*, 7.
5. Grootenhuis, P. P. J.; Kollman, P. A. *J. Am. Chem. Soc.* **1989**, *111*, 4046.
6. Lifson, S.; Felder, C. E.; Shanzer, A.; Libman, J. In *Synthesis of Macrocycles*; Izatt, R. M.; Christensen, J. J., Eds.; John Wiley & Sons: New York, 1987; pp 241-307.
7. Still W. C.; Erickson, S.; Wang, X.; Li, G.; Armstrong, A.; Hong, J.-J.; Namgoong, S. K.; Liu, R. In *Molecular Recognition: Chemical and Biochemical Problems II.*; Roberts, S. M., Ed.; Royal Society of Chemistry: Cambridge, 1992; pp 171-182.
8. Tsukube, H.; Uenishi, J.; Kojima, N.; Yonemitsu, O. *Tetrahedron Lett.* **1995**, *36*, 2257.
9. Parker, D. *Chem. Soc. Rev.* **1990**, 271.
10. Kang, H. C.; Hanson, A. W.; Eaton, B.; Boekelheide, V. *J. Am. Chem. Soc.* **1985**, *107*, 1979.
11. Groot, B.; Giesbrecht, G. R.; Loeb, S. J.; Shimidzu, G. K. H. *Inorg.Chem.* **1991**, *30*, 177.
12. Vögtle, F.; Weber, E. In *Crown Ethers and Analogs*, Patai, S.; Rappoport, Z., Eds.; John Wiley & Sons: Chichester, 1989; pp 207-304.
13. Tsukube, H. In *Liquid Membranes: Chemical Applications*; Araki, T.; Tsukube, H., Eds., CRC Press: Boca Raton, FL., 1990; pp 27-50.
14. Kimura, K.; Shono, T. In *Cation Binding by Macrocycles*; Inoue, Y.; Gokel, G. W., Eds.; Marcel Dekker: New York, 1990; pp 429-463.
15. Hori, K.; Haruna, Y.; Kamimura, A.; Tsukube, H.; Inoue, T. *Tetrahedron* **1993**, *49*, 3959.
16. Tsukube, H.; Inoue, T.; Hori, K. *J. Org. Chem.* **1994**, *59*, 8047.
17. Gokel, G. W. *Chem. Soc. Rev.* **1992**, 39.
18. Tsukube, H. *Talanta* **1993**, *40*, 1313.
19. Andzelm, J.; Wimmer, E. *J. Chem. Phys.* **1992**, *96*, 1280.
20. Arnold, K. A.; Mallen, J.; Trafton, J. E.; White, B. D.; Fronczek, F. R.; Gehrig, L. M.; Gandour, R. D.; Gokel, G. W. *J. Org. Chem.* **1988**, *53*, 5652.

21. Tsukube, H.; Hori, K.; Inoue, T. *Tetrahedron Lett.* **1993**, *34*, 6749.
22. Adachi, G.; Hirashima, Y. In *Cation Binding by Macrocycles*; Inoue, Y.; Gokel, G. W., Eds.; Marcel Dekker: New York, 1990, pp 701-741.
23. Saha, A. K.; Kross, K.; Kloszewski, E. D.; Upson, D. A.; Toner, J. L.; Snow, R. A.; Black, C. D. V.; Desai, V. C. *J. Am. Chem. Soc.* **1993**, *115*, 11032.
24. Nassimbeni, L. R.; Wright, M. R. W.; Niekerk, J. C.; McCallum, P. A. *Acta Cryst.* **1979**, *B35*, 1341.
25. Tsukube, H.; Shiba, H.; Uenishi, J. *J. Chem. Soc., Dalton* **1995**, 181.
26. Willner, I.; Eichen, Y.; Susan, S.; Shoham, B. *New. J. Chem.* **1991**, *15*, 879. Tsukube, H.; Uenishi, J.; Higaki, H.; Kikkawa, K. *Chem. Lett.* **1992**, 2307.
27. Katz, H. E. In *Inclusion Compounds*; Atwood, J. L.; Davies, J. E. D.; MacNicol, D. D., Eds.; Oxford Science Publications: Oxford, 1991; pp 391-405.
28. Behr, J. P.; Lehn, J. M. *J. Am. Chem. Soc.* **1973**, *95*, 6108.
29. Tsukube, H. *Angew. Chem. Int. Ed. Engl.* **1982**, *21*, 304.'
30. Galan, A.; Andreu, D.; Echavarren, A. M.; Parados, P.; Mendoza, J. *J. Am. Chem. Soc.* **1992**, *114*, 1511.
31. Furuta, H.; Cyr. M. J.; Sessler, J. L. *J. Am. Chem. Soc.* **1991**, *113*, 6677.
32. Zinic, M.; Frkanec, L.; Skaric, V.; Trafton, J.; Gokel, G. W. *J. Chem. Soc., Chem. Commun.* **1990**, 1726.
33. Konishi, K.; Yahara, K.; Toshishige, H.; Aida, T.; Inoue, S. *J. Am. Chem. Soc.* **1994**, *116*, 1337.
34. Rebek, J.; Askew, B.; Nemeth, D.; Parris, K. *J. Am. Chem. Soc.* **1987**, *109*, 2432.
35. Aoyama, Y.; Asakawa, M.; Yamagishi, A.; Toi, H.; Ogoshi, H. *J. Am. Chem. Soc.* **1990**, *112*, 3145.
36. Reetz, M.; Huff, J.; Rudolph, J.; Töllner, K.; Deege, A.; Goddard, R. *J. Am. Chem. Soc.* **1994**, *116*, 11588.
37. Mohler, L. K.; Czarnik, A. W. *J. Am. Chem. Soc.* **1993**, *115*, 7037.
38. Tsukube, H.; Uenishi, J.; Kanatani, T.; Itoh, H.; Yonemitsu, O. *J. Chem. Soc., Chem. Comuun.* in press.
39. Wong, C. P.; Venteicher, R. F.; Horrocks, W. D. *J. Am. Chem. Soc.* **1974**, *96*, 7149.

Chapter 11

New Multidentate Ligand Carriers for Macrocycle-Facilitated Metal Ion Transport Across Liquid Membranes

Richard A. Bartsch, Mark D. Eley, Marty D. Utterback, Joseph A. McDonough, and Youngchan Jang

Department of Chemistry and Biochemistry, Texas Tech University, Lubbock, TX 79409–1061

To probe the influence of structural variation within macrocyclic multidentate compounds upon their ability to function as carriers for macrocycle-facilitated, metal ion transport across liquid membranes, a variety of new crown and lariat ether componds have been evaluated in a bulk liquid membrane system. For crown ethers substituted with one or more linear alkyl groups, the effect of varying the number and attachment sites for the lipophilic groups and replacement of one or more ring oxygen atoms with sulfur or nitrogen atoms has been investigated. For lariat ethers with dibenzo-16-crown-5 rings, the influence of amide- and thioamide-containing side arms has been assessed.

In the search for improved metal ion separation schemes, considerable attention has been focused upon the use of liquid membranes (1,2). In a liquid membrane system, a liquid or quasi-liquid phase separates two other liquid phases in which the membrane is immiscible. In the most common arrangement, a hydrophobic liquid phase, such as chloroform or toluene, separates two aqueous phases. If chemical species have some solubility in the membrane, they may pass from one aqueous phase through the membrane into the second aqueous phase by simple diffusion. More frequently, a carrier molecule which resides in the membrane provides carrier-facilitated transport of metal ions across the membrane. Compared with simple diffusion, the carrier-facilitated transport is usually more efficient and selective.

Due to their ability to form complexes selectively with metal cations and thereby solubilize such cations in liquid media of low dielectric constant, macrocyclic multidentate ligands have been widely investigated as carriers in metal ion transport processes for the past two decades (3-9). Such macrocycles include crown ethers, lariat ethers (crown ethers with a side arm which contains potential binding sites) (10), calixarenes, and some cryptands. Studies of structural variations within a given

0097–6156/96/0642–0155$15.00/0

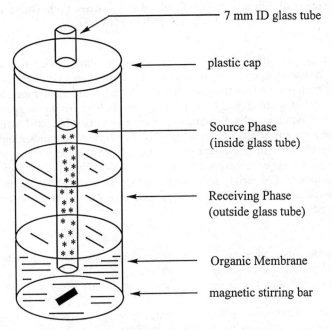

7 mm ID glass tube

plastic cap

Source Phase
(inside glass tube)

Receiving Phase
(outside glass tube)

Organic Membrane

magnetic stirring bar

Figure 1. Liquid Membrane Transport Cell.

type of macrocycle have been utilized to enhance the understanding of factors which produce efficient and selective metal ion transport. Such knowledge may then be employed in the design of new macrocyclic ligands for which efficient and highly selective transport of targeted metal ion species is envisioned.

In this study, several types of new crown and lariat ether compounds have been evaluated as metal ion carriers in a bulk chloroform membrane system. To probe the effect of varying the number and attachment sites of lipophilic groups, one series of new crown ether carrier molecules includes 15-crown-5 and 18-crown-6 compounds with one, two or four linear alkyl groups attached. In a second series, one or two ring oxygen atoms in selected members of the first series are replaced with nitrogen or sulfur atoms. To probe the influence of side arm group variation in lariat ether carriers, dibenzo-16-crown-5 compounds with amide- and thioamide-containing side arms are included in a third series.

Technique for Measurement of Metal Perchlorate Transport Across Bulk Chloroform Membranes.

A technique reported by others (*11*) for screening the effect of ionophore structure upon transport of metal salts across bulk chloroform membranes was utilized. The transport cell is illustrated in Figure 1. In this "tube-within-a-shell" transport cell, the metal salt in the upper aqueous source phase inside the glass tube passes through the lower chloroform phase (the organic membrane) containing the ionophore on its way to the upper aqueous receiving phase which surrounds the inner glass tube. The transport potential is provided by the metal salt concentration gradient between the source and receiving phases.

In the present system, the aqueous source phase was 1.0 mL of a 0.10 M solution of a metal perchlorate. The liquid membrane was 4.0 mL of a 1.0×10^{-3} M solution of the carrier molecule in chloroform and the receiving phase was 6.5 mL of deionized water. Gentle stirring of the organic phase was provided by a small magnetic stirring bar and an external 120 rpm constant speed stirring motor. Transport experiments were conducted in triplicate for 24 hours at 25 ± 1 °C. After this period, the metal ion concentrations in the aqueous receiving phases were determined by ion chromatography for the alkali metal perchlorates or atomic absorption for the other metal perchlorate species. Reported transport rates are the average for the triplicate transport experiments and the uncertainty is the standard deviation from the average. Selectivities are calculated from the transport rates determined for the two individual metal salt species.

Transport of Alkali Metal Perchlorates by Lipophilic 18-Crown-6 and 15-Crown-5 Derivatives.

Although 18-crown-6 (**1**) (Figure 2) efficiently complexes metal salts in homogenous solution, this ionophore is unsuitable for use as a metal salt carrier in liquid membrane transport due to its water solubility (*12*). Loss of the carrier from the membrane phase into the aqueous source and receiving phases is significant. On the other hand, the more lipophilic analogues dibenzo-18-crown-6 and particularly dicylohexano-18-crown-6 (**2**) have been widely utilized as metal salt carriers in liquid membrane transport processes. Compared with 18-crown-6, dibenzo-18-crown-6 is a weaker metal ion complexing agent due to its four aryl alkyl ether oxygens and

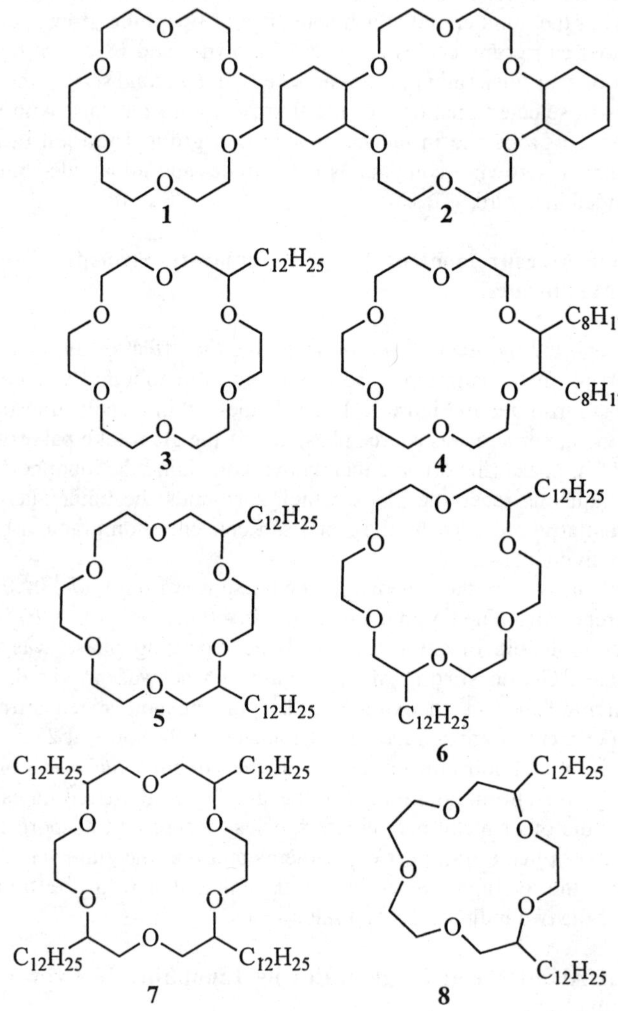

Figure 2. Lipophilic Crown Ether Carriers.

dicyclohexano-18-crown-6 suffers from conformational restrictions caused by fusing of the cyclohexane and macrocyclic polyether ring systems. Lamb *et al.* (*11*) have probed the effect of adding alkyl groups onto the benzo and cyclohexano rings of these macrocyclic polyethers upon the transport of alkali and alkaline earth nitrates across bulk chloroform membranes. The presence of methy, *t*-butyl, or tetradecyl groups on each benzo group of dibenzo-18-crown-6 or of methyl or decyl groups on each cyclohexano ring of dicyclohexano-18-crown-6 did not significantly influence the transport rates.

In this work, we have investigated the effect of attaching one, two or four lipophilic, linear alkyl groups directly to the macrocyclic ring of 18-crown-6 in ionophores **3-7** (Figure 2). The objective was to maintain the flexibility of the macrocyclic polyether ring while enhancing the lipophilicity of the crown ether compound. In addition to these lipophilic 18-crown-6 derivatives, the lipophilic 15-crown-5 compound **8** was also studied.

Transport rates of the five individual alkali metal perchlorate species across bulk chloroform membranes by dicyclohexano-18-crown-6 (**2**) and lipophilic crown ethers **3-8** are recorded in Table 1. In all cases, transport of lithium perchlorate was undetectable. The K^+/Na^+ and K^+/Rb^+ transport selectivities calculated from the single species transport rates are also presented in Table 1.

Table 1. Transport of Alkali Metal Perchlorates by Lipophilic Crown Ethers 2-8

Cmpd.	Transport Rate (moles/24 hours) X 10^6				Selectivity	
	Na^+	K^+	Rb^+	Cs^+	K^+/Na^+	K^+/Rb^+
2	2.1 ± 0.1	5.6 ± 0.6	1.9 ± 0.1	0.59 ± 0.03	2.7	2.9
3	2.3 ± 0.2	50.5 ± 0.8	7.0 ± 0.6	2.3 ± 0.2	22	7.2
4	3.4 ± 0.5	27.7 ± 0.8	3.7 ± 0.5	0.21 ± 0.01	8.1	7.5
5	3.8 ± 0.3	19.9 ± 0.3	4.4 ± 0.4	0.12 ± 0.01	5.9	4.5
6	4.7 ± 0.8	15.4 ± 0.3	4.6 ± 0.6	0.15 ± 0.01	3.3	3.3
7	0.66 ± 0.10	2.7 ± 0.1	0.52 ± 0.1	0.20 ± 0.01	4.1	2.0
8	0.84 ± 0.05	0.55 ± 0.08	0.27 ± 0.08	BDL[a]-	0.65	2.0

[a]BDL = Below detection limit.

The transport of alkali metal perchlorates across the bulk chloroform membrane by dicyclohexano-18-crown-6 (**2**) varies with the nature of the alkali metal cation in the order: $K^+>Na^+>Rb^+>Cs^+>Li^+$. For the 18-crown-6 ring size, strongest complexation of K^+ would be anticipated and the transport shows modest K^+/Na^+ and K^+/Rb^+ selectivities. For the transport of alkali metal nitrates across a bulk chloroform membrane by **2**, an ordering of $K^+>Rb^+>Na^+>Cs^+>Li^+$ has been reported (*11*). This reversal in the ordering for transport of Na^+ and Rb^+ is attributed to an effect of the co-transported anion upon transport selectivity (*13*).

For the series of substituted 18-crown-6 compounds **3-7**, one linear alkyl group is attached to the 18-crown-6 ring in **3**, two linear alkyl groups are attached in different positions of the macrocyclic polyether ring in **4-6** and four linear alkyl groups are bonded to the 18-crown-6 ring in **7**. As can be seen from the data presented in Table 1, both the efficiency and selectivity of K^+ transport are influenced by variations in the number and attachment sites of the lipophilic linear alkyl groups. In terms of transport efficiency, the highest K^+ transport rate is obtained for crown ether **3** which has a single pendent, lipophilic alkyl group. The K^+ transport rate is diminished when two lipophilic alkyl groups are attached in ionophores **4-6**. For this series of disubstituted 18-crown-6 compounds, the transport rate decreases as the spacing between the attachment sites for the two lipophilic groups increases from zero atoms in **4** to four atoms in **5** to seven atoms in **6**. The K^+ transport rate is the lowest for crown ether **7** which has four symmetrically placed lipophilic alkyl groups. Thus the transport of K^+ is facilitated when the lipophilicity is incorporated at just one point or on one side of the carrier molecule. Presumably this arrangement provides better metal salt extraction by the lipophilic crown ether at the source phase-membrane phase interface. It should be noted that compared with dicyclohexano-18-crown-6 (**2**), the four lipophilic 18-crown-6 compounds **3-6** are much more efficient in transporting potassium perchlorate across bulk chloroform membranes.

In addition to providing a nine-fold enhancement in the rate of K^+ transport, dodecyl-18-crown-6 (**3**) also exhibits K^+/Na^+ and K^+/Rb^+ selectivities of 22 and 7.2, respectively, which far surpass the selectivities of >3 which are found for dicyclohexano-18-crown-6 (**2**). For the series of substituted 18-crown-6 compounds **3-7**, higher K^+ transport selectivities are noted as the lipophility is confined to just one point or one side in the carrier molecule.

The dialkylated 15-crown-5 molecule **8** exhibits a transport selectivity order of $Na^+>K^+>Rb^+>Li^+,Cs^+$. This ordering and the rather low Na^+/K^+ selectivity are anticipated for the 15-crown-5 ring size.

Transport by Metal Perchlorates by Lipophilic Diaza-, Monothia- and Dithia-15-crown-5 and -18-Crown-6 Compounds.

According Pearson's Hard and Soft Acid and Base Theory, oxygen is a hard donor atom, sulfur is a soft donor atom and nitrogen is intermediate (*14,15*). To probe the influence of replacing one or two oxygen atoms in lipophilic crown ethers **5**, **6** and **8** upon metal perchlorate transport, the lipophilic macrocyclic carriers **9-16** (Figure 3) were investigated. Macrocycles **9** and **10** are lipophilic diaza-15-crown-5 compounds and **11**, **14** and **15** are diaza-18-crown-6 derivatives. Macrocyle **16** is a lipophilic monothia-15-crown-5 compound and macrocycles **12** and **13** are dithia-15-crown-5 and -18-crown-6 compounds, respectively.

Transport of Alkali Metal Perchlorates by Lipophilic Diazacrown Ethers. Transport rates for alkali metal perchlorates across bulk chloroform membranes by the lipophilic diazacrown compounds **9-11**, **14** and **15** are presented in Table 2. In all cases, transport of lithium and cesium perchlorates was undetectable.

Table 2. Transport of Alkali Metal Perchlorates by Lipophilic Diazacrown Ethers 9-11, 14 and 15

Cmpd.	Transport Rate (moles/24 hours) X 10^6		
	Na^+	K^+	Rb^+
9	0.42 ± 0.08	0.07 ± 0.01	BDL[a]
10	0.55 ± 0.05	0.11 ± 0.01	BDL
11	0.12 ± 0.01	0.41 ± 0.01	0.31 ± 0.05
14	0.60 ± 0.04	0.84 ± 0.04	0.72 ± 0.01
15	0.33 ± 0.02	2.09 ± 0.04	0.72 ± 0.01

[a]BDL = Below detection limit.

Comparison of the data for the lipophilic diazacrown ethers in Table 2 with that for the analogous crown ether compounds in Table 1 shows that replacement of two oxygen atoms in the macrocyclic ring with NH or NC_2H_5 units drastically reduces the alkali metal perchlorate transport. Presumably this results from diminished complexation of the hard alkali metal cations by the macrocyclic ligands when two hard oxygen donor sites are replaced by two nitrogen atoms.

Transport experiments were also conducted with metal perchlorates for which the metal ions were softer (Ag^+, Cd^{2+}, Co^{2+}, Hg^{2+}, Pb^{2+} and Zn^{2+}). For none of these metal perchlorates was transport detectable.

Transport of Silver Perchlorate by Lipophilic Monothia- and Dithiacrown Ethers. For lipophilic monothia-15-crown-5 **16**, dithia-15-crown-5 **12** and dithia-18-crown-6 **13**, transport of Na^+, K^+, Cd^{2+}, Hg^{2+}, Pb^{2+} and Zn^{2+} as the perchlorates was undetectable. However, all three macrocycles exhibited efficient transport of silver perchlorate. The Ag^+ transport rates for **16, 12** and **13** were 7.92 ± 0.05, 7.52 ± 0.15 and 6.87 ± 0.05 X 10^{-6} mole/24 hours, respectivley. Thus the soft Ag^+ cation was transported with almost equal efficiency by mono- and dithiacrown ethers which suggests that only one soft sulfur donor atom interacts with the soft Ag^+.

Transport of Metal Perchlorates by Lariat Ethers Based on Dibenzo-16-crown-5 Compounds with Amide- and Thioamide-Containing Side Arms.

Attachment of a side arm which contains potential binding sites to the crown ether framework gives a lariat ether (*10*). Simultaneous coordination of a metal ion species by both the polyether ring oxygen atoms and heteroatoms in the side arm enhances complexation (*16*).

We have developed efficient syntheses of *sym*-(hydroxy)dibenzo-16-crown-5 and *sym*-(alkyl)(hydroxy)dibenzo-16-crown-5 compounds (*17,18*). These lariat ether alcohols, in which the alcohol function and a hydrogen atom or alkyl group are attached to the central carbon of the three-carbon bridge serve as platforms for attachment of various potential coordinating groups to provide a variety of other lariat ether compounds. For example, dibenzo-16-crown-5 lariat ethers with pendent amide (*19*), ester (*20*) and thioamide (*21*) groups have been prepared. To further probe the effect of structural variation within lariat ethers, studies of metal perchlorate transport across bulk chloroform membranes by dibenzo-16-crown-5 lariat ethers with oxyacetamide and oxythioacetamide side arms were undertaken.

Transport of Alkali Metal Perchlorates by Lariat Ether Amides. The structures for two series of *sym*-(R)dibenzo-16-crown-5-oxacetamide compounds are shown in Figure 4. The parent for the first series is *sym*-dibenzo-16-crown-5-oxyacetamide (**17**) in which the amide nitrogen bears two hydrogen atoms. In **18**, these two hydrogens have been replaced with ethyl groups. Then in **19-22** the N,N-dialkyl substituents are sequentially replaced with propyl, butyl, pentyl and hexyl groups. The parent for the second series is *sym*-(propyl)dibenzo-16-crown-5-oxyacetamide (**23**) in which both propyl group and oxyacetamide side arms are attached to the central carbon of the three-carbon bridge in the dibenzo-16-crown-5 ring. In **24-28**, N,N-dialkyl groups of ethyl, propyl, butyl, pentyl and hexyl have been introduced. Thus in both series of lariat ether amides, the lipophilicity is systematically varied.

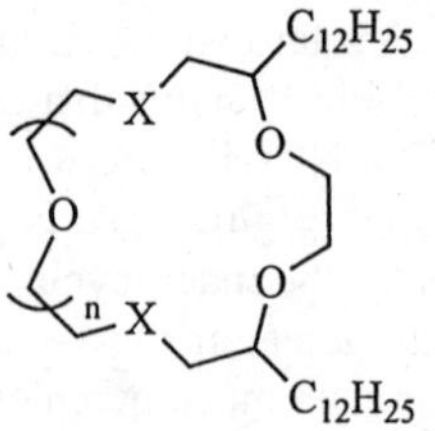

n	X
9	1	NH
10	1	NC$_2$H$_5$
11	2	NH
12	1	S
13	2	S

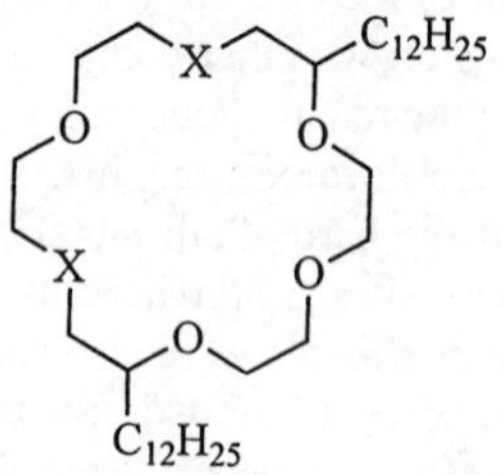

	X
14	NH
15	NC$_2$H$_5$

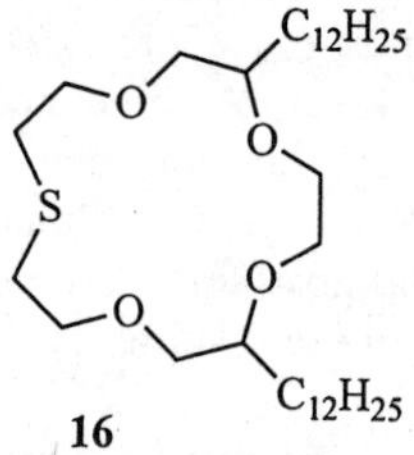

16

Figure 3. Lipophilic Diaza-, Monothia- and Dithiacrown Ether Carriers.

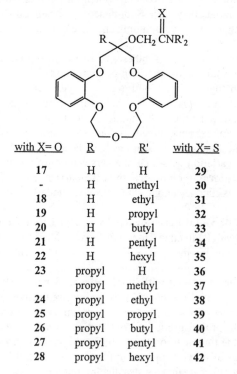

with X= O	R	R'	with X= S
17	H	H	29
-	H	methyl	30
18	H	ethyl	31
19	H	propyl	32
20	H	butyl	33
21	H	pentyl	34
22	H	hexyl	35
23	propyl	H	36
-	propyl	methyl	37
24	propyl	ethyl	38
25	propyl	propyl	39
26	propyl	butyl	40
27	propyl	pentyl	41
28	propyl	hexyl	42

Figure 4. Lariat Ether Amide and Thioamide Carriers.

Transport rates for the five alkali metal perchlorates across bulk chloroform membranes by lariat ether amides **17-28** are presented in Table 3. In all cases, transport of lithium, rubidium and cesium perchlorates was undetectable. The Na^+/K^+ transport selectivities calculated from the single species transport rates are also given in Table 3.

Table 3. Transport of Alkali Metal Perchlorates by Lariat Ether Amides 17-28

	Transport Rate (mole/24 hours) X 10^6		Na^+/K^+
Compound	Na^+	K^+	Selectivity
17	0.03 ± 0.01	0.13 ± 0.01	0.2
18	0.29 ± 0.02	0.31 ± 0.03	0.9
19	0.44 ± 0.02	0.19 ± 0.02	2.3
20	0.53 ± 0.05	0.15 ± 0.01	3.5
21	0.56 ± 0.01	0.27 ± 0.02	2.1
22	0.38 ± 0.01	0.16 ± 0.01	2.4
23	0.03 ± 0.01	0.10 ± 0.02	0.3
24	2.54 ± 0.09	0.06 ± 0.01	42
25	3.26 ± 0.62	0.06 ± 0.01	54
26	4.04 ± 0.18	0.07 ± 0.01	58
27	12.2 ± 1.1	0.29 ± 0.01	42
28	4.72 ± 0.33	0.10 ± 0.01	47

Based upon the relative sizes of the macrocyclic polyether ring and the alkali metal cations, dibenzo-16-crown-5 compounds are anticipated to be somewhat selective for complexation of Na^+. For lariat ether amide series of **17-22** which is based upon *sym*-dibenzo-16-crown-5-oxyacetamide (**17**), transport levels for Na^+ and K^+ are uniformly low with somewhat faster transport of Na^+ than K^+ when the length of the N,N-dialkyl groups is increased. When a geminal propyl group is incorporated to provide the second series of lariat ether amides, the rate of K^+ transport is low and remains virtually constant throughout the series **23-28**. On the other hand, the rate of Na^+ transport is strongly influenced by the nature of the atoms or groups attached to the nitrogen of the amide group in the coordinating side arm. Thus the Na^+ transport rate jumps from a very low level with *sym*-(propyl)dibenzo-16-crown-5-oxy-acetamide (**23**) to much higher levels with the N,N-dialkyl lariat ether amides **24-28**. The Na^+ transport rate shows a modest increase as the N,N-dialkyl groups are lengthened from ethyl to propyl to butyl to hexyl groups in compounds **24-26** and **28**, respectively. For lariat ether amide **27** which is the N,N-di(pentyl) member of the series, both the Na^+ and K^+ transport rates are higher than those for the N,N-di(butyl) and N,N-di(hexyl) analogues, **25** and **26**, respectively, by a factor of 3-4. This anamolous result was verified in three separate sets of transport experiments (each comprised of triplicate transport runs) conducted with the N,N-di(pentyl) lariat ether amide **27**. Since the effect is reproducible, it must reflect some optimal balance in the lipophilicity of the lariat ether carrier.

The difference in Na^+/K^+ transport selectivities for the N,N-di(propyl) through N,N-di(hexyl) members of the two series of lariat ether amides is striking. Thus the selectivity of 2-3 which is calculated for the N,N-dialkyl *sym*-dibenzo-16-crown-5-oxyacetamides increases to the range of 40-60 for the N,N-dialkyl *sym*-(propyl)dibenzo-16-crown-5-oxyacetamide analogues. Conformational differences of the amide-containing side arms in the two series is proposed as the causitive factor.

In the former series, the side arm is conformationally flexible and would require considerable restriction in its mobility for coordination with a metal ion which is bound within the crown ether cavity. On the other hand, the geminal propyl group in the second series, orients the amide-containing side arm over the crown ether cavity (22). Such pre-organization of the binding site is known to enhance selectivity in metal ion complexation (23).

Transport of Silver Perchlorate by Lariat Ether Thioamides. Amides may be readily transformed into the corresponding thioamide compounds by reaction with Lawesson's reagent (24). The structures for 12 dibenzo-16-crown-5 compounds with oxythioacetamide side arms **29-42** are shown in Figure 4. The series of compounds **29-35** is based on *sym*-dibenzo-16-crown-5-oxythioacetamide (**29**). In this series, the structural variation is systematic replacement of the two hydrogen atoms attached to the amide nitrogen in **29** with alkyl groups ranging from methyl to hexyl groups. The second series of lariat ether thioamides **36-42** is based on *sym*-(propyl)dibenzo-16-crown-5-oxythioacetamide (**36**) and includes replacement of the two hydrogen atoms on the amide nitrogen of **36** with methyl, ethyl, propyl, butyl, pentyl and hexyl groups.

Transport rates for potassium and silver perchlorates across bulk chloroform membranes by lariat ether thioamides **29-42** are reported in Table 4. Transport of sodium, cadmium and zinc perchlorates was undetectable.

Table 4. Transport of Potassium and Silver Perchlorates by Lariat Ether Thioamides 29-42

Compound	Transport Rate (mole/24 hours) X 10^6	
	K^+	Ag^+
29	Ppt[a]	Ppt
30	BDL[b]	13.7 ± 0.8
31	0.20 ± 0.01	13.1 ± 0.7
32	0.21 ± 0.01	14.3 ± 1.1
33	0.20 ± 0.02	14.2 ± 0.6
34	0.20 ± 0.01	20.1 ± 1.3
35	0.20 ± 0.01	18.9 ± 1.0
36	BDL	8.86 ± 0.67
37	BDL	7.44 ± 0.21
38	BDL	9.67 ± 0.49
39	BDL	7.89 ± 0.18
40	BDL	8.10 ± 0.22
41	0.23 ± 0.03	4.23 ± 0.51
42	0.19 ± 0.03	4.31 ± 0.60

[a]Ppt = Precipitate formed.
[b]BDL = Below detection limit.

For the carriers **29-42**, the transport of potassium perchlorate was uniformly low or undetectable. On the other hand, the lariat ether thioamides exhibit a propensity for transport of Ag^+ across bulk chloroform membranes. Thus substitution of the amide oxygen in a dibenzo-16-crown-oxyacetamide by sulfur to produce the corresponding thioamide transforms an effective Na^+ carrier into an efficient Ag^+

carrier. The introduction of a soft thioamide sulfur coordinating site produces good lariat ether carriers for Ag^+.

As described in the previous section, the introduction of a geminal propyl group into a N,N-dialkyl dibenzo-16-crown-5 lariat ether amide markedly enhanced the propensity for Na^+ transport. In contrast for the corresponding lariat ether thioamides the presence of a geminal propyl group in **36-42** somewhat diminishes the Ag^+ transport rate compared with those for the corresponding compounds **30-35**, which do not have an alkyl group attached to the central carbon of the three-carbon bridge. Thus pre-organization of the binding site in the lariat ether thioamides is detrimental to Ag^+ transport, which suggests predominant interaction of the soft Ag^+ with the soft sulfur atom of the thioamide group in the side arm.

Conclusions

Compared with the frequently utilized lipophilic crown ether dicyclohexano-18-crown-6 (**2**), attachment of one or two linear alkyl groups to the framework of 18-crown-6 provides carrier molecules **3-6** which exhibit enhanced efficiency for transport of K^+ across bulk chloroform membranes. The K^+/Na^+ and K^+/Rb^+ selectivities are also much higher with **3-5**. The K^+ transport efficiency and selectivity are enhanced when the lipophilicity is restricted to one side of the carrier molecule and the highest rate of transport and selectivity are obtained with dodecyl-18-crown-6.

Replacement of one or two oxygen donor atoms in a lipophilic 15-crown-5 or 18-crown-6 compound with nitrogen or sulfur atoms strongly influences the type of metal ion species which can be transported through the bulk chloroform membranes. Incorporation of one or two soft sulfur donor atoms provides carriers **12, 13** and **16** which transport soft Ag^+, but not the hard alkali metal cations.

For alkali metal cation transport across bulk chloroform membranes by lariat ether amides based upon dibenzo-16-crown-5 rings, preorganization of the coordinating side arm over the polyether cavity enhances both the efficiency and selectivity of Na^+ transport. For the corresponding lariat ether thioamides, efficient transport of Ag^+, but not the alkali metal cations, is observed. In this case, preorganization of the side arm is detrimental which indicates predominant coordination of the soft Ag^+ with the soft sulfur atom of the thioamide group.

Through such studies of the effect of structural variation within the metal ion carriers, new carrier species may be designed and synthesized which will have high efficiency and selectivity for the transport of targeted metal ion species in liquid membrane systems.

Acknowledgement

This research was supported by the Division of Chemical Sciences of the Office of Basic Energy Sciences of the U.S. Department of Energy (Grant DE-FG03-94ER14416).

References

1. *Liquid Membranes. Theory and Applications*, ACS Symposium Series 347; R. D. Noble, J. D. Way, Eds.; American Chemical Society: Washington, DC, 1987.

2. *Liquid Membranes: Chemical Applications*; T. Araki, H. Tsukube, Eds.; CRC Press: Boca Raton, FL, 1990.
3. Lamb, J. D.; Izatt, R. M.; Christensen, J. J. *Progress in Macrocyclic Chemistry*; R. M. Izatt, J. J. Christensen, Eds.; Wiley: New York, NY, 1981, Vol. 2; Chapter 2, pp 41-90.
4. McBride, D. W., Jr.; Izatt, R. M.; Lamb, J. D.; Christensen, J. J. *Inclusion Compounds. Physical Properties and Applications*; J. L. Atwood, J. E. D. Davies, D. D. MacNicol, Eds.; Academic Press: New York, NY, 1989, Vol. 3; Chapter 16, pp 571-628.
5. Okahara, M.; Nakatsuji, Y. *Top. Curr. Chem.* **1985**, *128*, 37.
6. Bartsch, R. A.; Charewicz, W. A.; Kang, S. I.; Walkowiak, W. *Liquid Membranes. Theory and Applications*; R. D. Noble, J. D. Way, Eds; American Chemical Society: Washington, DC, 1987; Chapter 6, pp 86-97.
7. Bruening, R. L.; Izatt, R. M.; Bradshaw, J. S. *Cation Binding by Macrocycles. Complexation of Cationic Species by Crown Ethers*; Y. Inoue, G. W. Gokel, Eds.; Marcel Dekker: New York, NY, 1990; Chapter 2, pp 111-132.
8. Tsukube, H. *Liquid Membranes: Chemical Applications*; T. Araki, H. Tsukube, Eds.; CRC Press: Boca Raton, FL, 1990; Chapter 4, pp 51-102.
9. Hiratani, K.; Yamaguchi, T. *Liquid Membranes: Chemical Applications*; T. Araki and H. Tsukube, Eds.; CRC Press: Boca Raton, FL, 1990, Chapter 6, pp 103-120.
10. Gokel, G. W.; Dishong, D. M.; Diamond, C. J. *J. Chem. Soc., Chem. Commun.* **1980**, 1053.
11. Lamb, J. D.; Izatt, R. M.; Garrick, D. G.; Bradshaw, J. S.; Christensen, J. J. *J. Membrane Sci.* **1981**, *9*, 83.
12. Hiraoka, M. *Crown Compounds. Their Characteristics and Applications*; Elsevier: New York, NY, 1982; p 32.
13. Olsher, U.; Hankins, M. G.; Kim, Y. D.; Bartsch, R. A. *J. Am. Chem. Soc.* **1993**, *115*, 3370.
14. Pearson, R. G. *J. Am. Chem. Soc.* **1963**, *85*, 3533.
15. Pearson, R. G. *J. Chem. Educ.* **1968**, *45*, 581, 643.
16. Gokel, G. W.; Trafton, J. E. *Cation Binding by Macrocycles. Complexation of Cationic Species by Crown Ethers*; Y. Inoue, G. W. Gokel, Eds.; Marcel Dekker; New York, NY, 1990; Chapter 6, pp 253-310.
17. Heo, G. S.; Bartsch, R. A.; Schlobohm, L. L.; Lee, J. G. *J. Org. Chem.* **1981**, *46*, 3574.
18. Bartsch, R. A.; Liu, Y.; Kang, S. I.; Son, B.; Heo, G. S.; Hipes, P. G.; Bills, L. J. *J. Org. Chem.* **1983**, *48*, 4864.
19. Kasprzyk, S. P.; Bartsch, R. A. *J. Heterocycl. Chem.* **1993**, *30*, 119.
20. Ohki, A.; Lu, J. P.; Hallman, J. L.; Huang, X.; Bartsch, R. A. *Anal. Chem.* **1995**, *67*, 2405.
21. Jang, Y.; Bartsch, R. A. *J. Heterocycl. Chem.* **1995**, *32*, 1441.
22. Ohki, A; Lu, J.-P.; Bartsch, R. A. *Anal. Chem.* **1994**, *66*, 651.
23. Cram, D. J. *Angew. Chem., Int. Ed. Engl.* **1986**, *25*, 1039.
24. Yde, B.; Yousif, N. M.; Pedersen, U.; Thomsen, I.; Lawesson, S.-O. *Tetrahedron* **1984**, *40*, 2047.

Chapter 12

Highly Selective Transport of Heavy Metal Ions by Novel Amide Compounds

Kazuhisa Hiratani and Kazuyuki Kasuga

National Institute of Materials and Chemical Research, 1–1 Higashi, Tsukuba, Ibaraki 305, Japan

Several types of acyclic amides have been synthesized and evaluated as carriers for metal ion transport across liquid membranes. The abilities of oligoamides containing 8-quinolyl groups to transport heavy metal ions, such as copper(II), mercury(II), silver(I), etc., across bulk chloroform membranes are assessed. Among the different types of amides, malonamide and glutaramide derivatives are found to exhibit high selectivity for copper(II) transport and different methionine derivatives for mercury(II) and for silver(I) transport. The relationship between the structure of the ionophore and its transport behavior is discussed.

Separation, concentration, and recovery of both valued and toxic metal ions are becoming increasingly important due to dwindling mineral resources and the need to protect the environment, respectively (*1,2*). Therefore the development of efficient and selective metal ion complexing agents which may be utilized in such operations is of great interest. In this paper, we describe the preparation of several types of acyclic amide ligands and their evaluation in heavy metal ion separations.

Within the general field of host-guest chemistry, our attention has been focused upon acyclic compounds for many years (*3-6*). We have synthesized acyclic ligands and investigated their extractability, transport ability, response, etc. for metal ions for the following reasons. First, acyclic ligand can be prepared more easily than macrocyclic analogues. Special techniques, such as high dilution, usually are not necessary and systematic structural variations may be readily incorporated into the potential ionophores. Second, the acyclic host may adopt a pseudocyclic arrangement in metal ion complexation, as illustrated in Figure 1. In addition to rigidity and chain structure, the size and shape of the pseudocavity may be influenced by pi-stacking and hydrogen bonding interactions of functional end groups.

In this study, the target molecules are acyclic oligoamides with the generalized structure shown in Figure 2. Within this structure, constituents R, X, Y, and Z, as well as the chain length, may be systematically varied. In addition, incorporation of appropriate substituents into X and Y may be utilized to enhance the propensity for formation of a pseudocyclic cavity. Effectiveness of these new ionophores as metal ion carriers can be evaluated by proton-driven cation transport from a weakly acidic aqueous solution through a liquid membrane into a more acidic aqueous solution (see Figure 3).

0097–6156/96/0642–0167$15.00/0

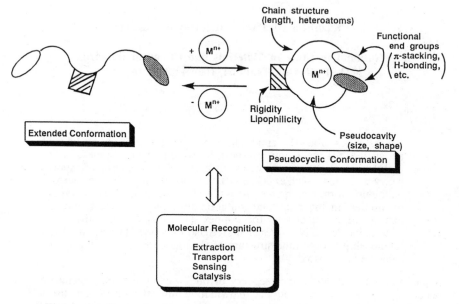

Figure 1. Dynamic Conformational Change of Acyclic Ligands Induced by the Uptake of Cations.

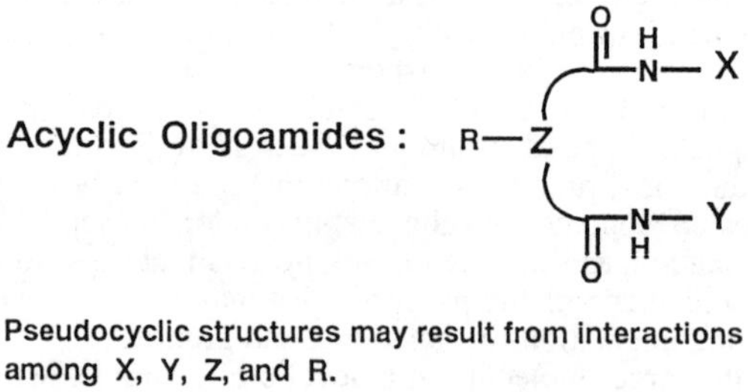

Acyclic Oligoamides :

$$R - Z \begin{cases} \overset{O}{\underset{}{\parallel}} \overset{H}{\underset{}{N}} - X \\ \underset{O}{\overset{}{\parallel}} \underset{H}{\overset{}{N}} - Y \end{cases}$$

Pseudocyclic structures may result from interactions among X, Y, Z, and R.

Figure 2. Design of Acyclic Oligoamides.

N,N'-Bis(8-quinolylcarbamoyl) Derivatives.

We have previously reported that malonamides with two 8-quinolyl groups, such as **1** (Scheme 1), provide very selective extraction of Cu(II) from weakly acidic aqueous solutions (7, 8). These malonamides form 1:1 complexes with Cu(II) (Scheme 1) both in solution and the solid state. Analogues of **1** have been prepared according to the procedure which depicted in Scheme 2. Extractability of Cu(II) from aqueous solutions at pH 6.2 into chloroform decreases in the order: **1** >> **2** > **3** > **4** (9). Thus lengthening the chain between the amide groups in the acyclic ligand diminishes its extraction ability toward Cu(II).

The ability of ligands **1-4** to function as carriers for transport of heavy metal ions across bulk liquid membranes has also been assessed (9,10). For the metal ion transport studies, the U-tube type glass cell shown in Figure 4 (11) was utilized. Initial compositions of the aqueous source and receiving phases and the chloroform membrane phase are also given in Figure 4. Both single species and competitive transport of the heavy metal ions Cu(II), Ni(II), Co(II), and Zn(II) were investigated. Table 1 summarizes the results for the single species transport of Cu(II). Although it exhibits only low extractability, glutaramide derivative **3** is the most efficient carrier for transport Cu(II). After two days, more than 60% of the Cu(II) initially present in the aqueous source phase has been transported into the receiving phase by glutaramide **3**. This means that Cu(II) is being transported against its concentration gradient during the latter stages of the experiment. On the other hand, malonamide **1**

Table 1. Amount of Cu(II) Transported at 25 °C by Acyclic Diamides 1-4 after 48 Hours.[a]

Acyclic diamide	Cu(II) transported into the receiving phase		Cu(II) remaining in the source phase	
	%	mmol	%	mmol
1	0.2	0.3	94	141
2	29	44	60	90
3	63	95	35	53
4	6	51	93	140

[a]Initial conditions: Source phase, 15 ml of a 10 mM aqueous solution of Cu(OAc)$_2$ at pH = 6.2; Membrane phase, 0.30 mmol of carrier in 30 ml of chloroform; Receiving phase, 15 ml of 0.050 M sulfuric acid.

barely transports Cu(II). This is attributed to the strong complexation of Cu(II) by **1** in the membrane which renders release of the metal ion into the aqueous receiving phase difficult. As Cu(II) carriers, both succinamide **2** and adipamide **4** are inferior to **3**. Thus an optimal chain length in carriers **1-4** is clearly evident. For systems in which Cu(II) and the competing ions Co(II), (Ni) and Zn(II) were present in the aqueous source phase, glutaramide **3** only transported Cu(II) into the receiving phase even when the initial concentrations of the competing ions were ten times that of Cu(II) (Table 2). Thus, it has been shown that glutaramide **3** can transport Cu(II) from a weakly acidic aqueous solution across chloroform membrane and into a more acidic aqueous solution with high selectivity.

Table 2. Competitive Metal Ion Transport by Glutaramide 3.[a]

Percent (μmol) of cations transported after 48 hours at 25 °C			
Cu(II)	Ni(II)	Co(II)	Zn(II)
71 (11)	0	0	0

[a]Initial conditions: Source phase, 15 ml of an aqueous solution 1.0 mM in Cu(OAc)$_2$ and 10.0 mM each in Co(OAc)$_2$, Ni(OAc)$_2$, and Zn(OAc)$_2$ at pH = 6.2; Membrane phase, 0.30 mmol of **3** in 30 ml of chloroform; Receiving phase, 15 ml of 0.050 M sulfuric acid.

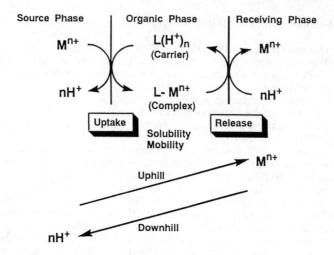

Figure 3. Proton-Driven Transport of Metal Ions Through a Liquid Membrane.

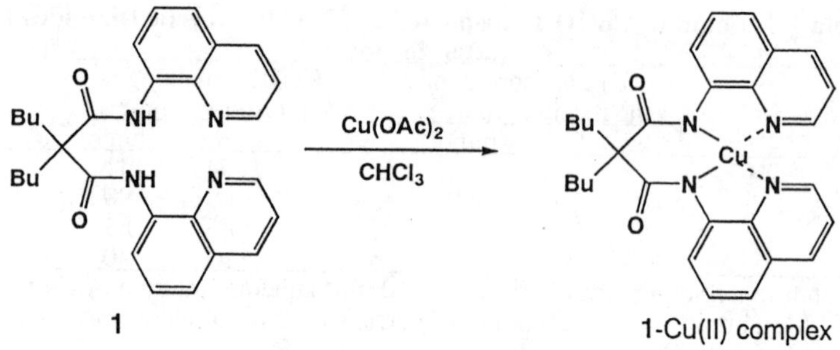

Scheme 1. Complexation of malonamide **1** with Cu(II)

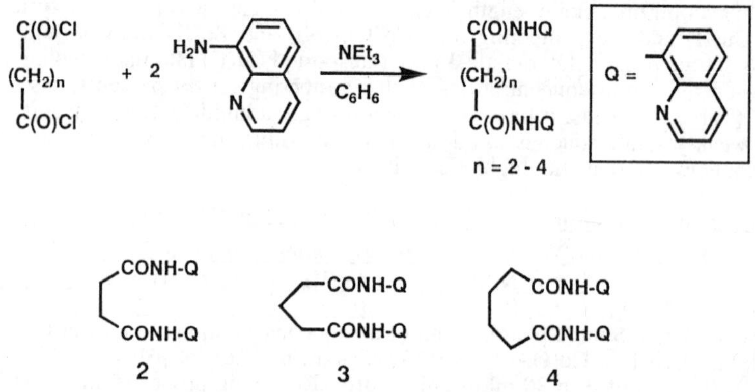

Scheme 2. Preparation of N,N'-bis(8-quinolyl)diamide derivatives

Unsymmetric Malonamide Derivatives as Heavy Metal Ion Carriers.

In an effort to produce even more selective and efficient carriers than the glutaramide derivative **3** described above, we designed additional potential ionophores with a malonamide unit in the structure. Based on our previous findings (*9,10*), we considered two points to be important. The new compound should have better extractability than glutaramide derivative **3**, but provide a metal ion complex of lower stability than that of malonamide derivative **1**. Several malonamide derivatives containing quinolyl and/or pyridyl groups were synthesized by the method shown in Scheme 3. For the series of malonamide derivatives **1** and **6-9**, the chain length and the terminal group have been systematically varied. The series contains two unsymmetric malonamides **6** and **8**.

First, the transport of Cu(II) with these malonamides as carriers was investigated under the same conditions as those shown in Table 1. Surprisingly, the unsymmetric malonamide **6** was found to transport Cu(II) most efficiently (Table 3). After 2 days, only 4 % of the Cu(II) remained in the source phase. As shown in Table 3, the

Table 3. Amount of Cu(II) Transported at 25 °C by Glutaramide Derivatives 1 and 6-9 after 48 Hours.[a]

Carrier	Cu(II) transported into the receiving phase, %	Cu(II) remaining in the source phase, %
1	0.2	94
6	93	4
7	56	43
8	15	85
9	0	100

[a]For initial conditions, see Table 1.

transport ability decreases in the order: **6 > 7 > 8 > 1 > 9**. Thus, the transport ability for Cu(II) was found to be strongly influenced by the identity of the end groups for the malonamide carrier (*12,13*).

Figure 5 shows the time dependence of Cu(II) transport for **3**, 8-hydroxy-7-undecylquinoline (**5**), and **6**. In these experiments, the initial conditions differ from those under which the data presented in Table 3 were obtained and are given in the figure caption. Under these conditions, the ability of the 8-quinolyl derivative **5** to transport Cu(II) into the aqueous receiving phase is found to be identical with that of the glutaramide derivative **3**. However, it should be noted that the concentration of **5** in the membrane phase was twice that of **3**. Figure 5 clearly illustrates that the unsymmetric malonamide carrier **6** transports Cu(II) much more efficiently than do carriers **3** and **5**. Competitive transport experiments also show high selectivity for transport of Cu(II) by malonamide carrier **6** in the presence of Co(II), Ni(II), and Zn(II). The symmetry of the lines in Figure 5 for the percentages of Cu(II) in the source and the receiving phases for carriers **3** and **6** means that transport of the metal ion from the aqueous source phase into the membrane phase is the rate-determining step.

The pH-dependence for Cu(II) extraction by malonamide derivatives **1, 3, 6, 7,** and **9** was also examined. The results which are presented in Figure 6 show that the order of extractability is: **1 > 6 > 7 > 3 > 9**. Thus acyclic diamide **6** which gave the greatest transport rate for Cu(II) does not give the highest extraction. This indicates that a balance between extractability and the ease of metal ion release into the aqueous receiving phase controls the transport efficiency.

Figure 7 shows the results for transport experiments conducted under different conditions than those described above. In these experiments, a much higher concentration of Cu(II) is present in the weakly acidic aqueous source phase (pH = 5.5). The initial ratio of Cu(II) in the source phase to carriers **3** or **6** in the chloroform membrane was 30. Under these conditions, the unsymmetric malonamide

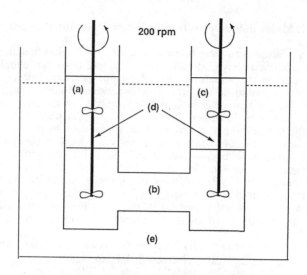

Initial transport conditions (25 °C):

(Source Phase)	(Organic Phase)	(Receiving Phase)
(I) 10 mM solution of each metal ion (Cu^{2+}, Ni^{2+}, Co^{2+}, Zn^{2+}), pH 6.2, 15 ml	Carrier (0.3 mmol) in 30 ml of $CHCl_3$	0.05 M H_2SO_4, 15 ml
(II) 10mM Cu^{2+}, pH 6.2, 15 ml		

Figure 4. Initial Conditions and Apparatus for Measuring Metal Ion Transport. (a) source phase, (b) membrane phase, (c) receiving phase, (d) glass stirrer, and, (e) constant temperature bath at 25 °C.

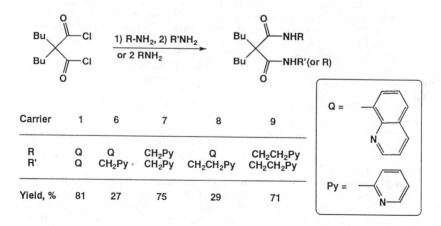

Carrier	1	6	7	8	9
R	Q	Q	CH_2Py	Q	CH_2CH_2Py
R'	Q	CH_2Py	CH_2Py	CH_2CH_2Py	CH_2CH_2Py
Yield, %	81	27	75	29	71

Scheme 3. Preparation of malonamide derivatives with various end groups

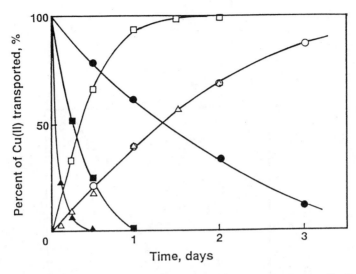

Figure 5. Time Dependence of Cu(II) Transport by Carrier **3**, 8-Hydroxy-7-undecylquinoline (**5**), and **6**. Initial conditions: 25 °C; source phase, 15 ml of a 1.0 mM aqueous solution of Cu(II) at pH = 6.2; membrane phase, 0.10 mmol of **3**, 0.20 mmol of **5**, or 0.10 mmol of **6** in 30 ml of chloroform; receiving phase, 15 ml of 0.050 M sulfuric acid. (S = source phase and R = receiving phase.) **3** = ○(R), ●(S); **5** = △(R), ▲(S); **6** = □(R), ■(S).

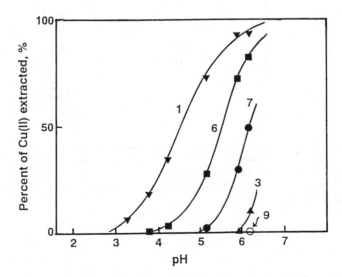

Figure 6. pH-Dependence for Cu(II) Extraction by **1**, **3**, **6**, **7**, and **9**.

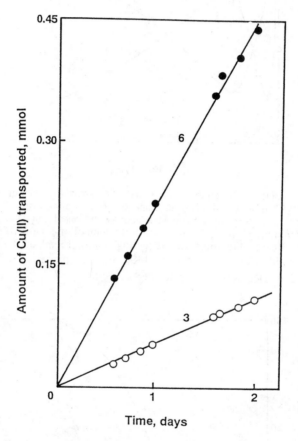

Figure 7. Amount of Cu(II) Transported by Carriers **3** and **6** Versus Time. Initial conditions: 25 °C; source phase, 15 ml of a 0.20 M aqueous solution of Cu(II) (3.0 mmol) at pH = 5.5; membrane phase, 0.10 mmol of carrier in 30 ml of chloroform; receiving phase, 15 ml of 0.050 M sulfuric acid.

6 transported Cu(II) much more efficiently than the symmetric glutaramide **3**. Under the same conditions, neither carrier **3** nor **6** transports Co(II), Ni(II), or Zn(II)(*8*).

The mechanism proposed for Cu(II) transport through the liquid membrane by carrier **6** is shown in Figure 8. Cu(II) transport from the weakly acidic aqueous source phase through the chloroform membrane and into the more acidic aqueous receiving phase is coupled with back transport of protons. Thus a pH gradient is utilized to drive the Cu(II) transport.

Amino Acid Derivatives as Heavy Metal Ion Carriers.

As described above, acyclic diamide compounds with appropriate structures are effective metal ion carriers in liquid membrane systems. Our study was extended next to amino acid derivatives because they may also have multiple amide groups but with a reversed sequence from that found in the malonamide derivatives (see Figure 9). It is interesting to note that the *N*-terminal oligopeptide moiety in human serum albumin (Figure 9) plays an important role in the storage and transport of Cu(II) in the body (*14*). We decided to mimic the terminal portion of this physiologically important biomolecule by use of a sequence of amide groups and with an aromatic 8-quinolyl group in place of the imidazole moiety.

Triamide derivatives derived from glutamic acid and diamide derivatives from phenylalanine were prepared as outlined in Scheme 4. Figure 10 presents the results for Cu(II) transport using glutaramide **3**, glutamic acid derivative **10** (*15*), and phenylalanine derivative **11** (*16*). For these three carriers, the phenylalanine derivative **11** gives the best Cu(II) transport. It is noted that the transport ability of **11** for Cu(II) is almost the same as that of the unsymmetric malonamide **6** in Table 3.

Next sulfur-containing amino acid derivatives were synthesized from methionine by the method shown in Scheme 5 (*17*). Such molecules might be anticipated to function as Hg(II) or Ag(I) ionophores because these soft heavy metal ions should be coordinated by the sulfur atom in the ligand. Parts "a" and "b" of the methionine derivatives (Figure 11) can be systematically varied by changing the R and R' groups, respectively. The series of R groups included methyl, *t*-butyl and *p*-tolyl and the R' group was chosen from 8-quinolyl, 2-methoxyphenyl, 2-methylfuranyl and 2-methyltetrahydrofuranyl. For carrier **12** in which R = methyl and R' = 8-quinolyl transport of Cu(II) has been found to be much more efficient than Hg(II) or Ag(I) (*18*). When the R group is changed from methyl to *t*-butyl in carriers **13-16**, the metal ion transport selectivity changes markedly (Table 4). For carriers **12** and **13**, the R'

Table 4. Percentage of Metal Ions Transported at 25 °C by Methionine Derivative Carriers 13-18 after 24 Hours (48 Hours).[a]

Carrier	Cu(II) Source/Receiving		Hg(II) Source/Receiving		Ag(I) Source/Receiving
13	(70)/	(30)	0 (0)/	100(100)	84(78)/10(20)
14	(64)/	(1)	90(66)/	8(19)	82(75)/ 8(18)
15	(97)/	(0)	70(55)/	18(42)	50(25)/43(71)
16	(93)/	(0)	70(53)/	27(47)	60(50)/23(44)
17	80(69)/16(27)		1 (0)/	92(95)	not measured
18	90(90)/ 1 (1)		44(35)/	0 (3)	35(11)/51(78)

[a]Initial conditions: Source phase, 15 ml of an aqueous solution 10 mM in Cu(II), Hg(II) or Ag(I) at pH = 6.2; Membrane phase, 0.30 mmol of carrier in 30 ml of chloroform; Receiving phase, 15 ml of 0.05 M sulfuric acid.

group is the same, but the R group is changed from methyl in the former to *t*-butyl in the latter. This structural variation drastically decreases the Cu(II) transport ability, presumably because the *t*-butyl group in **13** blocks coordination of Cu(II) with the amide group in part "a" of the carrier. Methionine derivative **13** transports Hg(II) with high efficiency and selectivity over Cu(II) and Ag(I). Carriers **14-16** combine R = *t*-butyl with R' groups containing oxygen atoms as potential metal ion coordination

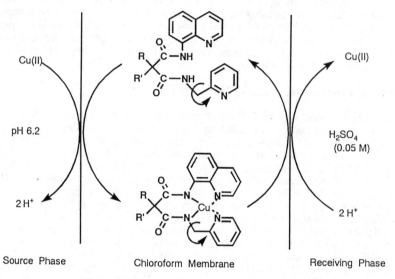

Cu(II)
Cu(II)
pH 6.2
H_2SO_4
(0.05 M)
2 H+
2 H+
Source Phase
Chloroform Membrane
Receiving Phase

Figure 8. Schematic of Carrier-Mediated Uphill Transport of Cu(II) by Carrier **6**.

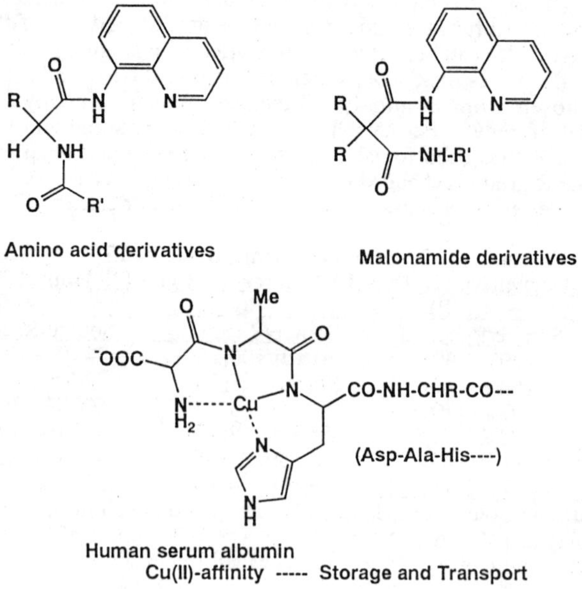

Amino acid derivatives
Malonamide derivatives
CO-NH-CHR-CO---
(Asp-Ala-His----)
Human serum albumin
Cu(II)-affinity ----- Storage and Transport

Figure 9. Amide Derivatives with Metal Ion Affinity.

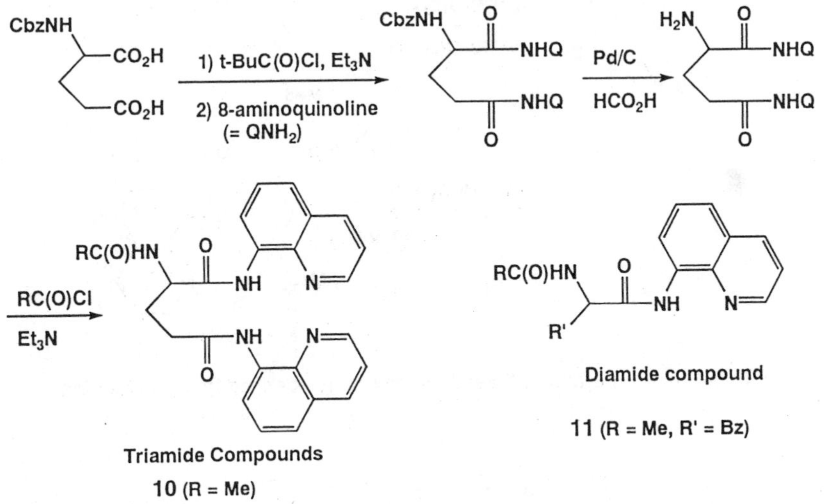

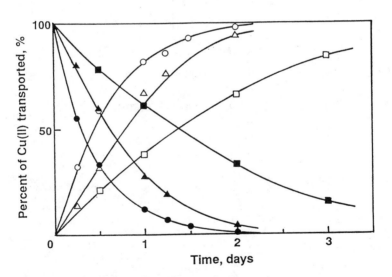

Scheme 4.   Preparation of oligoamide derivatives derived from amino acids

Figure 10. Time Dependence of Cu(II) Transport by Carriers **3**, **10**, and **11**. Initial conditions: 25 °C; source phase, 15 ml of 10 mM aqueous solution of Cu(II) at pH = 6.2; membrane phase, 0.30 mmol of carrier in 30 ml of chloroform; receiving phase, 15 ml of 0.050 M sulfuric acid. (S = source phase and R =receiving phase.) **3** = □ (R), ■(S); **10** = △(R), ▲(S); **11** = ○(R), ●(S). Reproduced with permission from reference 16.

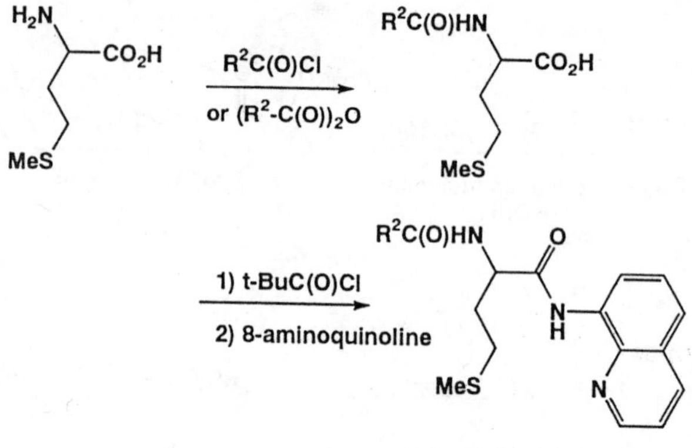

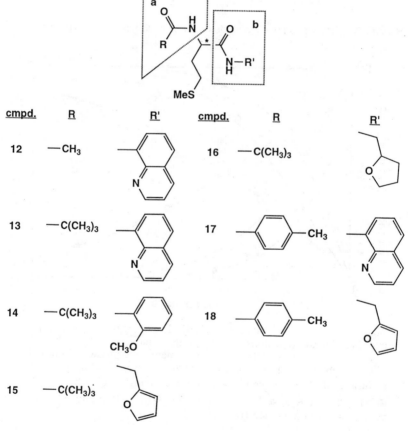

Scheme 5. Preparation of diamide compounds derived from methionine

Figure 11. Methionine Derivatives with Various R and R' Substituents.

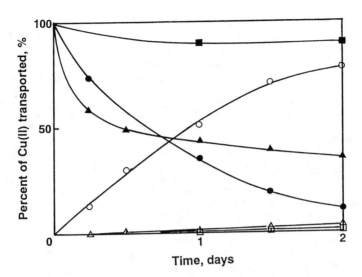

Figure 12. Time Dependence of Metal Ion Transport by Carrier **18**. For initial conditions, see Table 4. (S = source phase and R = receiving phase.) Cu(II) = □(R), ■(S); Hg(II) = Δ(R), ▲(S); Ag(I) = ○(R), ●(S).

sites. This structural variation markedly diminishes the efficiency and selectivity for Hg(II) transport which is observed with **13**. It is interesting to note that Cu(II) transport is almost completely suppressed for carriers **14-16**. When R = *p*-tolyl and R' = 8-quinolyl in carrier **17**, highly efficient and selective transport of Hg(II) is observed. Thus replacement of the *t*-butyl group in **13** with *p*-tolyl in **17** retains the high transport efficiency and selectivity towards Hg(II). Apparently a *p*-tolyl group is also sufficiently bulky to diminish coordination of Cu(II) by the amide group in part "a" of the carrier. Carrier **18** in which R = *p*-tolyl and R' = 2-methylfuranyl exhibits highly selective transport of Ag(I) over Cu(II) and Hg(II). This is clearly illustrated in Figure 12 which shows plots of the amount of metal ion transported by carrier **18** versus time for the single species transport of Cu(II), Hg(II), and Ag(I) ions. Thus by varying the structure of the methionine derivative carrier, very good transport selectivity for Cu(II) (with **12**), for Hg(II) (with **13** and **17**) and for Ag(I) (with **18**) may be obtained.

Conclusions.

Oligoamide compounds are effective carriers for heavy metal ions and can be designed to exhibit high selectivity for a particular metal ion. In this paper we have demonstrated several examples of carriers which can transport target metal ions with high selectivity and efficiency. Glutaramide **3** with two quinolyl groups, unsymmetric malonamide **6**, and amino acid derivatives **10** and **11** transport Cu(II) with excellent selectivity from weakly acidic aqueous solutions containing Cu(II), Ni(II), Co(II), and Zn(II) through bulk chloroform membranes into more acidic aqueous receiving solutions. Methionine derivatives with high selectivity for Hg(II) or Ag(I) transport have also been identified. For the methionine derivative carriers, the selectivity is largely dependent on the N-substituent adjacent to the alpha-carbon.

References

1. Noble, R.D.; Way, J.D., Eds.; *Liquid Membranes*; ACS Symposium Series 347; American Chemical Society.: Washington, D.C., 1987.
2. Araki, T.; Tsukube, H., Eds.; *Liquid Membranes: Chemical Application*; CRC Press, Boca Raton, Fl., 1990.
3. Hiratani, K.; Yamaguchi, T. In *Liquid Membranes: Chemical Applications,*; Araki, T.; Tsukube, H., Eds.; CRC Press: Boca Raton, Fl., 1990; Chapter 6, pp 103 - 120.
4. Hiratani, K. *J. Synth. Org. Chem. Jpn.* **1987**, *45*, 1186.
5. Kasuga, K.; Hirose, T.; Hiratani, K. *J. Synth. Org. Chem. Jpn.* **1993**, *51*, 317.
6. Sugihara, H.; Hiratani, H. *J. Synth. Org. Chem. Jpn.* **1994**, *52*, 530
7. Hiratani, K.; Taguchi, K.; Ohashi, K.; Nakayama, H. *Chem. Lett.* **1989**, 2073.
8. Hirose, T.; Hiratani, K.; Kasuga, K.; Saito, K.; Koike, T.; Kimura, E.; Nagawa, Y.; Nakanishi, H. *J. Chem. Soc., Dalton Trans.,* **1992**, 2679.
9. Hiratani, K.; Kasuga, K.; Hirose, T.; Taguchi, K.; Fujiwara, K. *Bull. Chem. Soc. Jpn.* **1992**, *65* 2381.
10. Hiratani, K.; Taguchi, K. *Chem. Lett.* **1990**, 725.
11. Hiratani, K.; Nozawa, I.; Nakagawa, T.; Yamada, S. *J. Membr. Sci.* **1982**, *12*, 207.
12. Hiratani, K.; Hirose, T.; Fujiwara, K.; Saito, K. *Chem. Lett.* **1990**, 1921.
13. Hiratani, K.; Hirose, T.; Kasuga, K.; Saito, K. *J. Org, Chem.* **1992**, *57*, 7083.
14. Laussac, J.-P.; Sarkar, B. *J. Biol. Chem.* **1980**, *255*, 7563; *Biochemistry* **1984**, *23*, 2832.
15. Kasuga, K.; Hiratani, K. *Anal. Sci.* **1991**, *7*, 95.
16. Kasuga, K.; Hirose, T.; Takahashi, T.; Hiratani, K. *Bull. Chem. Soc. Jpn.* **1993**, *66*, 3768.
17. Kasuga, K.; Hirose, T.; Takahashi, T.; Hiratani, K. *Chem. Lett.* **1993**, 2183.
18. Kasuga, K; Kubota, T. ; Akabori, S.; Hiratani, K., unpublished results.

Chapter 13

Transition Metal Cation Separations by Organophosphorus Compounds in Liquid Membrane Processes

W. Walkowiak[1] and J. Gega[2]

[1]Institute of Inorganic Chemistry and Metallurgy of Rare Elements,
Technical University of Wroclaw, Wybrzeze Wyspianskiego 27,
50–370 Wroclaw, Poland
[2]Department of General Chemistry, Technical University of Czestochowa,
Aleja Armii Krajowej 19, 42–200 Czestochowa, Poland

The use of di-(*p*-alkylphenyl)phosphoric acids containing butyl, hexyl, octyl and nonyl alkyl groups as carriers for separations of Co(II), Cu(II), Ni(II), and Zn(II) from aqueous sulfate solutions by bulk and emulsion liquid membrane processes has been explored. The organic phase was the di-(*p*-alkylphenyl)phosphoric acid in kerosene with the inclusion of Span 80 as an emulsifier for the emulsion liquid membrane systems. Both single metal ion species and competitive transport of the transition metal cations were investigated. For comparison, the transport of these metal cations by commercially available Cyanex 272™ and D2EHPA as carriers was studied also. To probe the mechanism of the liquid membrane transport processes, interfacial tension measurements were conducted. Multistage emulsion liquid membrane processes for the separation of the transition metal cation mixtures have been evaluated.

Solvent extraction of metal ions plays an important role in hydrometallurgy. Further development of solvent extraction is constrained by the limited number of commercially available extractants which include di(2-ethylhexyl)phosphoric acid (D2EHPA), bis(2,2,4-trimethylpentyl)phosphinic acid (Cyanex 272™), trioctylphosphine oxide (TOPO), tri-*n*-butylphosphate (TBP), some hydroxyoximes (extractants in the LIX or Acorga series) and derivatives of 8-hydroxyquinoline (extractants in the Kelex and LIX-26 series).

In the near future, resource exhaustion will become a serious problem worldwide. This will require the development of more efficient techniques for the recovery of metals from low-grade ores and secondary resources. The use of liquid membranes containing metal ion carriers offers as an alternative to solvent extraction

0097–6156/96/0642–0181$15.00/0
© 1996 American Chemical Society

for selective separation and concentration of metal ions from dilute aqueous solutions. Two type of liquid membranes are commonly employed. One is a thin layer of highly microporous polymer film which is impregnated with a solution of the carrier (*i.e.* a supported liquid membrane) and the second is the emulsion liquid membrane (*1,2*). Liquid membranes were invented in 1968 by Li (*3*) and now have many applications for the separation of metal ions from aqueous solutions (*4-9*). Among the experimental systems which have been utilized in the study of mass transport across liquid membranes are so-called bulk liquid membranes for which a beaker-in-a-beaker type of cell (*1,2*) may be employed. With such cells, facilitated transport of metal ions across a liquid membrane of known interfacial area and membrane thickness may be investigated.

We now report the results for single species and competitive transport of the transition metal cations Co(II), Cu(II), Ni(II), and Zn(II) through bulk and emulsion liquid membranes by a new series of di-(*p*-alkylphenyl)phosphoric acids carriers and by commercially available D2EHPA and Cyanex 272™. In addition, the results of interfacial tension measurements are presented.

Experimental.

Reagents. Structures of the organophosphorus carriers are shown in Figure 1. The first two compounds are the commercial reagents di(2-ethylhexyl)phosphoric acid (DEHPA, **1**) from Fluka AG and bis(2,2,4-trimethylpentyl)phosphinic acid (Cyanex 272™, **2**) from Cyanamid Company. The di-(*p*-alkylphenyl)phosphoric acids **3-6** were prepared in the Institute of Organic Technology of the Technical University of Wroclaw. The di-(*p*-alkylphenyl)phosphoric acids were synthesized as mixtures of mono- and diesters and the monoesters were separated with ethylene glycol (*10*). The purities of these organophosphorus compounds were assessed by potentiometric titration of an acetone solution of the carrier with a solution of potassium hydroxide in ethanol and by phosphorus content determination with phosphoromolibdinate blue (*11*). The results are presented in Table I. Only D2EHPA (**1**) was determined to be the pure acid. In the case of Cyanex 272™, the content of monoprotic acid was 92.0 % and the remainder was solvent. The di-(*p*-alkylphenyl)phosphoric acids **3-6** con-

Table I. Characteristics of the Organophosphorus Compounds

Compound	Molecular formula	Molecular weight $(g\ mol^{-1})$	Monoprotic acid content (%)	Phosphorus content (%) theory	found
1	$C_{16}H_{35}O_4P$	322.4	99.8	9.61	9.60
2	$C_{18}H_{39}O_2P$	318.5	92.0	9.72	8.96
3	$C_{20}H_{27}O_4P$	362.4	96.5	8.55	8.15
4	$C_{24}H_{35}O_4P$	418.5	95.2	7.40	7.10
5	$C_{28}H_{43}O_4P$	474.6	94.6	6.53	6.32
6	$C_{30}H_{47}O_4P$	502.7	92.3	6.16	5.85

tained small amounts (0.2-0.5 %) of the corresponding diprotic acids and some solvent.

A Polish commercial kerosene was utilized. As the emulsifier in the emulsion liquid membrane experiments, sorbitan monooleate (Span 80) from Fluka AG was employed. The Co(II), Cu(II), Ni(II) and Zn(II) sulfates and sulfuric acid, and toluene were reagent grade from POCH (Gliwice, Poland).

Interfacial Tension Measurements. The interfacial tension measurements were performed by the drop method at 20 °C using a Lauda TVT1 apparatus (*12*). This method determines the interfacial tension according to the equation:

$$\gamma = V_d \Delta dr^{-1} F \qquad (1)$$

where γ is the interfacial tension in N m^{-1}, V_d is the drop volume in m^3, Δ is the difference in densities of the aqueous and organic phases in kg m^{-3}, r is the capillary radius in m, and F is a correction factor which depends on the capillary radius and volume of the outflowing drop. Drops of the aqueous phase were formed in the organic phase (a toluene solution of the organophosphorus compounds at a concentration of 1.0×10^{-9} - 0.10 M).

Transport Across Bulk Liquid Membranes. The studies of transition metal cation transport across bulk liquid membranes utilized the beaker-in-a-beaker type of cell shown in Figure 2a. The aqueous source phase was placed in the external beaker and the aqueous receiving phase was placed in the internal beaker. The less dense organic (membrane) phase, which was a kerosene solution of the carrier, was placed on top of the aqueous phases. Volumes of the source, receiving, and organic phases were 96, 48, and 24 cm^3, respectively. The areas of the source phase/organic phase and organic phase/receiving phase interfaces were 13.5 and 10.2 cm^2, respectively. The transport experiments were conducted for 170 hours at room temperature (18-22 °C).

Each of the three phases was stirred separately at 100 rpm. A mechanical stirrer was used for the organic phase, while magnetic stirrers were employed to stir the aqueous phases. Such stirring did not disturb the interfaces and only slight motion of the interfaces was observed.

The source phase was an aqueous solution of Co(II), Cu(II), Ni(II), or Zn(II) sulfate. Concentrations of metal cations in the source phase were 0.005-0.010 M. The receiving phase was 1.0 M sulfuric acid. Samples (0.10 ml) of the aqueous phases were periodically removed for determination of the transition metal cation concentrations by atomic absorption spectroscopy. Concentrations of metal in the organic membrane phase were calculated by mass balance. The source and receiving phase volumes were maintained by addition of 0.10 ml of the appropriate initial aqueous solutions each time that samples were removed.

To provide a quantitative description of the metal ions transported across the source phase/organic phase and the organic phase/receiving phase interfaces, the fluxes, J (mol h^{-1}m^{-2}), were calculated for the source phase/organic phase interface:

$$J_{S/O} = V_S \,(c^0_{Me,S} - c^t_{Me,S}) \,/\, tS_{S/O} \qquad (2)$$

and for the organic phase/receiving phase interface:

$$J_{O/R} = V_R \, c^t_{Me,R} \,/\, tS_{O/R} \qquad (3)$$

where $c^t_{Me,S}$ and $C^0_{Me,S}$ are the molar concentrations of the metal cation in the source phase after the period of time t and in the initial source phase solution, respectively;

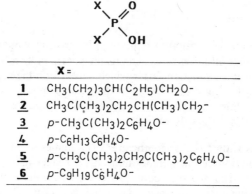

X =	
1	$CH_3(CH_2)_3CH(C_2H_5)CH_2O-$
2	$CH_3C(CH_3)_2CH_2CH(CH_3)CH_2-$
3	$p-CH_3C(CH_3)_2C_6H_4O-$
4	$p-C_6H_{13}C_6H_4O-$
5	$p-CH_3C(CH_3)_2CH_2C(CH_3)_2C_6H_4O-$
6	$p-C_9H_{19}C_6H_4O-$

Figure 1. Structures of the Organophosphorus Carriers.

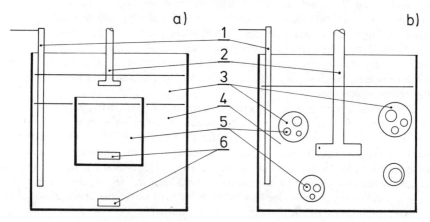

Figure 2. Cells for a) Bulk Liquid Membrane and b) Emulsion Liquid Membrane Transport. (1) pH electrode, (2) mechanical stirrer, (3) organic (membrane) phase, (4) aqueous source phase, (5) aqueous receiving phase, (6) magnetic stirrers.

$c^t_{Me,R}$ is the molar concentration of the metal cation in the receiving phase after the period of time t; V_S and V_R are the volumes of the source and receiving phases, respectively, in dm^3; $V_{S/O}$ and $V_{O/R}$ are the surface areas of the source phase/organic phase and organic phase/receiving phase interfaces, respectively in m^2; and t is the time for the transport process in hours.

Also the separation degree, E, was calculated:

$$E = n^t_{Me,S} / n^0_{Me,S} \qquad (4)$$

where $n^t_{Me,S}$ and $n^0_{Me,S \ are}$ are the moles of the metal cation in the source phase after the period of time t and in the initial source phase solution, respectively.

Transport Across Emulsion Liquid Membranes. The transport of transition metal cations across emulsion liquid membrane experiments were conducted in the cell shown in Figure 2b. The liquid membrane consisted of the carrier and an emulsifier dissolved in kerosene. The liquid membrane system was a water-in-oil-in-water type of emulsion which was obtained by stirring the aqueous receiving phase solution with the organic phase to form an emulsion which was then mixed with the aqueous source phase solution.

The emulsions were prepared in the following stages: a) A mixture of 10 ml of the aqueous receiving phase and 10 ml of the organic phase was stirred at 1500 rpm with a mechanical stirrer for 30 minutes. b) The resulting emulsion was stirred with 100 ml of the aqueous source phase at 400 rpm for 30 minutes. c) The source phase was separated from the emulsion. d) The emulsion was broken by centrifuging at 7000 rpm.

The following parameters were calculated to characterize the emulsion liquid membrane transport processes:

W, the enrichment ratio of the receiving phase,

$$W = c^t_{Me,R} / c^0_{Me,S} \qquad (5)$$

where $c^t_{Me,R}$ and $c^0_{Me,S}$ are molar concentrations of transition metal cations in the receiving phase after time t and in the source phase at the beginning of the transport process, respectively;

N_{car}, the utilization degree of the carrier,

$$N_{car} = n^{real}_{Me,R}/n^{theor}_{Me,R} \qquad (6)$$

where $n^{real}_{Me,R}$ is the moles of metal ions transported into the receiving phase during the transport experiment and $n^{theor}_{Me,R}$ is the moles of metal ions extracted into the organic phase during a one-step solvent extraction;

$S_{Me1/Me2,}$ the separation ratio,

$$S_{Me1/Me2} = (c^t_{Me1,R}/c^t_{Me2,R})/(c^0_{Me1,S}/c^0_{Me2,S}) \qquad (7)$$

where $c^t_{Me1,R}$ and $c^t_{Me2,R}$ are the concentrations of metal ions in the receiving phase after the period of time t for the "1" and "2" metal ion species, respectively, and $c^0_{Me1,S}$ and $c^0_{Me2,S}$ are the initial source phase concentrations for the "1" and "2" metal ion species, respectively.

Results and Discussion.

Surface Tension Measurements. Isotherms determined for the interfacial tensions of toluene solution of organophosphorus compounds **1-6** are presented in Figure 3. Much larger changes in interfacial tension with increasing concentrations of the organophorus compounds are noted for the di-(p-alkylphenyl)phosphoric acids **3-6** than for Cyanex 272™ (**2**) and D2EHPA (**1**). For the di-(p-alkylphenyl)phosphonic acids, the interfacial tension values decrease in the order **3** > **4** > **5** > **6**.

The interfacial activity of organophosphorus compounds **1-6** can be described by Szyszkowski's equation (*12*):

$$\gamma = \gamma_0 [1- B \ln\{(c_{car}/A) + 1\}] \qquad (8)$$

where γ and γ_0 are the interfacial tensions measured for carrier concentrations of C_{car} and 0, respectively, in N m^{-1} and A and B are constants.

Knowing the A and B values, the following parameters can be calculated:

Γ^{max}, the surface excess of saturated interface,

$$\Gamma^{max} = B\gamma_0/RT \qquad (9)$$

where R is the gas constant in J mol^{-1} °K and R is the temperature in °K.

A_{min}, the surface area for a single molecule of absorbed carrier,

$$A^{min} = 1/N_0\Gamma^{max} \qquad (10)$$

where N_0 is Avogadro's number.

ΔG^{ad}, the adsorption free energy, in kJ mol^{-1}

$$\Delta G^{ad} = -RT \ln A \qquad (11)$$

and c^π, the carrier concentration through which the reduction in interfacial tension takes place, where $\pi = \gamma_0 - \gamma = 0.01$ N m^{-1}.

Values of the calculated parameters are presented in Table II. The values of the excess of saturated interface (Γ^{max}) for **1** and **2** are very similar and are smaller than those for **3-6**. For **1** and **2**, the values of the surface area for a single molecule of adsorbed carrier (A^{min}) are similar, but larger than those for **3-6**. Also the values of c_π for **1** and **2** are much larger than those for **3-6**. For example, the reduction of surface tension to a value of 0.01 N m^{-1} for **6** takes place at a concentration of 1 X 10^{-5} M,

Table II. Absorption Parameters for Organophosphorus Compounds 3-6

Compound	Szyskowski's isotherm parameter $A \times 10^5$	$B \times 10^2$	$\Gamma^{max} \times 10^7$, mol m^{-2}	$A^{min} \times 10^{18}$, m^2 molecule^{-1}	ΔG^{ad}, kJ mol^{-1}	$c^{\pi} \times 10^4$, mol dm^{-3}
1	1.48	4.41	6.60	2.52	-27.1	100
2	33.8	5.92	6.87	2.42	-19.5	600
3	0.554	6.44	9.64	1.72	-29.5	7.00
4	1.19	8.00	12.0	1.38	-27.6	8.00
5	2.59	10.1	15.1	1.10	-25.7	6.00
6	0.00252	5.42	8.12	2.05	-42.6	0.10

while the same surface tension reduction of **2** occurs at a concentration of 1 X 10^{-2} M. These considerations, as well as the results presented in Figure 3 demonstrate that the di-(*p*-alkylphenyl)phosphoric acids **3-6** adsorb at the interface stronger than do the organophosphorus compounds **1** and **2**.

Transport of Transition Metal Cations Across Bulk Liquid Membranes. Kinetic curves for transport of Cu(II) across bulk kerosene membranes using organophosphorus compounds **1-6** as carriers are presented in Figure 4. The Cu(II) concentrations in the source, organic, and receiving phases are shown. The Cu(II) concentration in the source phase decrease smoothly with time and the Cu(II) concentration in the receiving phase smoothly increases. A quite different situation is noted for the Cu(II) concentration in the organic phase. The Cu(II) concentration in the membrane first increases. Then after reaching a maximum value, it exhibits a smooth decrease. Similar kinetic curves were observed in the single metal ion species transport of Co(II), Ni(II), and Zn(II). Strzelbicki *et al.* (*13,14*) have reported similar tendencies in other bulk liquid membrane transport processes.

The separation degree (E) values after 168 hours of bulk liquid membrane transport , as well as values for the fluxes across the two interfaces after 24 hours, are presented in Table III. From analysis of the E and $J_{S/O}$ values, the selectivity sequences for transport of the transition metal cations are estimated to be: Zn(II) > Cu(II) > Co(II) > Ni(II) for **1**, **5**, and **6**; Cu(II) > Zn(II) > Co(II) > Ni(II) for **2** and **3**; and Cu(II), Zn(II) > Co(II) > Ni(II) for **4**. For carrier **2**, no Ni(II) was detected in the aqueous receiving phase. Although there are only relatively small variations in the rate of transition metal transport from the organic phase into the receiving phase, all carriers exhibit the highest flux for Cu(II). By comparison of the metal ion fluxes across the source phase/organic phase and the organic phase/receiving phase interfaces, the selectivity sequence for the di-(*p*-alkylphenyl)phosphoric acids is **3** < **4** < **5** < **6**. This sequence corresponds to the increasing number of carbons in the *p*-substituents. Since the increase in flux with enhancement in the number of carbon atoms in the *p*-substituent is greater for the organic phase/receiving phase interface than the source phase/organic phase interface, the increase in overall transport is found to arise primarily from more efficient re-extraction of the transition metal cations at the organic phase/receiving phase interface.

Transport of Transition Metal Cations Across Emulsion Liquid Membranes. The next series of experiments involved emulsion liquid membranes and competitive transport of the four transition metal cations, Co(II), Cu(II), Ni(II),

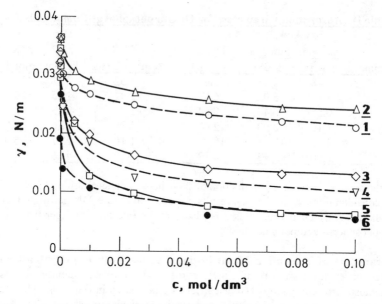

Figure 3. Interfacial Tension at the Toluene/Water Interface as a Function of the Concentration of Organophorphorus Compounds 1-6.

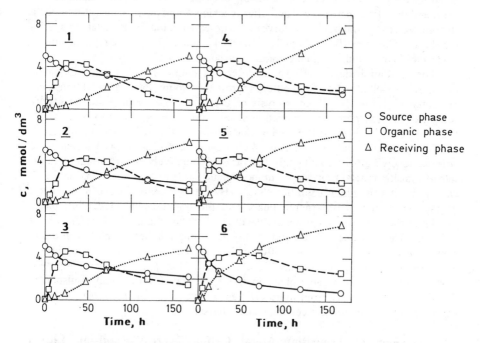

Figure 4. Kinetics of Cu(II) Transport Through Bulk Liquid Membranes by Carriers 1-6. Source phase, 0.01 M Cu(II), pH = 3.0; Organic phase, 0.01 M carrier in kerosene; Receiving phase, 1.0 M sulfuric acid.

**Table III. Transport of Single Metal Cation Species
Through a Bulk Kerosene Membrane[1]**

Carrier	Metal ion	Separation degree, $(E)^2$	$J_{S/O}$ X 10^6, mol h^{-1} m^{-2}	$J_{O/R}$ X 10^6, mol h^{-1} m^{-2}
1	Zn(II)	0.84	2.75	0.850
	Cu(II)	0.56	2.25	0.915
	Co(II)	0.16	1.74	0.841
	Ni(II)	0.050	1.09	0.814
2	Zn(II)	0.79	2.79	0.752
	Cu(II)	0.94	2.84	0.980
	Co(II)	0.041	1.54	1.32
	Ni(II)	0.0	0.0	0.0
3	Zn(II)	0.95	3.16	0.651
	Cu(II)	0.97	2.68	0.845
	Co(II)	0.23	1.76	0.815
	Ni(II)	0.18	1.78	0.769
4	Zn(II)	0.83	2.81	0.850
	Cu(II)	0.81	2.85	0.947
	Co(II)	0.067	2.05	0.951
	Ni(II)	0.021	1.95	0.935
5	Zn(II)	0.89	3.46	1.26
	Cu(II)	0.77	3.10	1.58
	Co(II)	0.40	2.78	1.05
	Ni(II)	0.19	2.48	1.14
6	Zn(II)	0.89	4.26	1.58
	Cu(II)	0.72	3.70	1.79
	Co(II)	0.37	2.98	1.15
	Ni(II)	0.065	2.72	1.24

[1]Source phase, 0.0050 M metal cation, pH = 3.0; Organic phase, 0.0050 M carrier in kerosene; Receiving phase, 1.0 M sulfuric acid.
[2]E is defined in the Experimental section.

and Zn(II). Enrichment ratios, W, of the metals in the receiving phase and the utilization degree of the carrier, N_{car}, values are presented in Table IV.

The enrichment ratio values rise as the carrier concentration is increased from 0.001 to 0.1 M. Selectivity sequences for competitive transition metal cation transport by 0.1 M concentrations of the carrier are: Zn(II) > Cu(II) > Co(II) > Ni(II) for **1**; Cu(II) > Zn(II) > Co(II) > Ni(II) for **2**; Cu(II), Zn(II) >> Co(II), Ni(II) for **3**; Zn(II) ≥ Cu(II) >> Co(II), Ni(II) for **4**; Zn(II) > Cu(II) > Ni(II) > Co(II) for **5**; and Zn(II) > Cu(II) > Ni(II) > Co(II) for **6**. These sequences are quite similar to those extrapolated from the single metal species cation transport experiments conducted with bulk liquid membranes. Carriers **3** and **4** allow Cu(II) and Zn(II) to be separated from Co(II) and Ni(II). In certain cases, particularly at a carrier concentration of 0.001 M, enrichment ratio values of less than 1 were observed which means that no enrichment was observed. The utilization degree values of 18-41 are high and demonstrate that a carrier molecule in the organic membrane is used repeatedly during the transport experiment to transfer metal ions from the source phase into the receiving phase. (For a one-step solvent extraction, the utilization degree is 1.0.)

Table IV. Metal Ion Enrichment in the Receiving Phase (W) and the Utilization Degree of the Organophosphorus Carrier (N_{car}) for the Emulsion Liquid Membrane Transport of Transition Metal Cations[1,2]

Carrier	pH	[Carrier], M	W				
			Zn(II)	Cu(II)	Co(II)	Ni(II)	N_{car}
1	3.0	0.001	0.940	0.960	0.720	0.090	37.1
		0.01	7.54	5.42	2.06	0.440	21.5
		0.1	13.4	9.82	3.65	0.745	5.60
2	4.0	0.001	0.450	0.960	0.190	0.0	22.0
		0.01	3.62	7.42	1.56	0.210	15.8
		0.1	9.03	17.9	3.78	0.540	6.30
3	2.0	0.001	0.740	0.600	0.20	0.010	17.7
		0.01	6.87	6.75	0.150	0.130	17.9
		0.1	12.8	12.9	0.310	0.280	5.29
4	2.0	0.001	0.75	0.640	0.150	0.110	34.1
		0.01	6.31	5.54	0.740	0.666	17.4
		0.1	10.6	9.10	1.28	1.05	4.55
5	3.0	0.001	0.94	0.620	0.280	0.460	40.6
		0.01	7.15	3.51	1.21	2.45	19.1
		0.1	9.74	6.42	3.09	4.86	3.31
6	4.0	0.001	0.74	0.440	0.050	0.130	18.5
		0.01	6.94	3.45	0.270	1.21	12.0
		0.1	8.94	4.32	0.280	1.43	3.14

[1]W and N_{car} are defined in the Experimental section.
[2]Source phase, 0.01 M in Co(II), Cu(II), Ni(II), and Zn(II); Organic phase, a solution of the carrier and 5 volume % of SPAN 80 in kerosene; Receiving phase, 1.0 M sulfuric acid.

Separation ratio values for competitive transport of pairs of transition metal cation species across emulsion liquid membranes with organophosphorus carriers **1-6** at concentrations in the organic phase of 0.001, 0.01 and 0.1 M are shown in Table V. The separation ratios depend mainly on the carrier concentration and are highest when the carrier concentration is the lowest. From comparison of the separation ratio values with 0.1 M carrier concentrations, the following conclusions may be drawn: a) For separation of Zn(II) from Co(II), Zn(II) from Ni(II), Cu(II) from Co(II), and Cu(II) from Ni(II), the best carrier is **3** and the worst is **5**. b) For none of the carriers is a reasonable separation of Zn(II) from Cu(II) obtained. c) For the separation of Co(II) from Ni(II), only carriers **1** and **2** are effective. Among the carriers **3-6**, only **6** possesses a reasonable selectivity for Ni(II) over Co(II).

Separation of Transition Metal Cations by Multistage Emulsion Liquid Membrane Processes. In a final series of experiments, multistage emulsion liquid membrane processes were studied. Flow sheets for these process are shown in Figure 5. In the first process (Figure 5a), Cyanex 272™ (**2**) was utilized as the carrier for separation of Co(II), Cu(II), Ni(II), and Zn(II) by competitive transport in four steps. The receiving phases from the four consecutive steps were enriched in Cu(II), Zn(II), Co(II), and Co(II), respectively. The effluent after the fourth step contained only

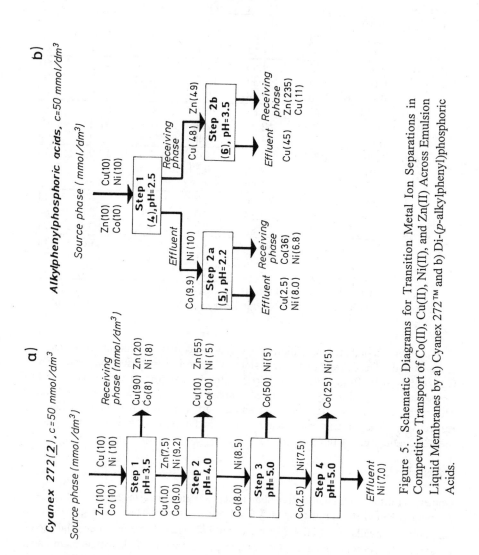

Figure 5. Schematic Diagrams for Transition Metal Ion Separations in Competitive Transport of Co(II), Cu(II), Ni(II), and Zn(II) Across Emulsion Liquid Membranes by a) Cyanex 272™ and b) Di-(*p*-alkylphenyl)phosphoric Acids.

CHEMICAL SEPARATIONS WITH LIQUID MEMBRANES

Table V. Separation Ratios for Pairs of Transition Metal Cations as a Function of the Carrier Structure and Concentration in Competitive Transport Across Emulsion Liquid Membranes [1]

Carrier	pH	[Carrier], M	$S_{Me1/Me2}$					
			Zn/Cu	Zn/Co	Zn/Ni	Cu/Co	Cu/Ni	Co/Ni
1	3.0	0.001	5.48	12.1	34.2	3.50	13.4	5.20
		0.01	1.40	3.68	17.5	2.62	12.6	4.79
		0.1	1.20	3.14	15.4	2.58	9.4	4.81
2	4.0	0.001	0.32	4.50	19.8	8.20	68.1	9.50
		0.01	0.49	2.34	18.9	4.84	38.9	8.10
		0.1	0.50	2.68	18.3	5.20	35.8	8.30
3	2.0	0.001	1.48	64.8	76.1	60.4	67.6	1.52
		0.01	1.02	52.6	62.2	51.8	61.2	1.18
		0.1	1.01	44.1	54.8	42.5	53.2	1.09
4	2.0	0.001	1.89	12.4	14.6	11.8	12.9	1.54
		0.01	1.14	8.50	9.53	7.46	8.36	1.12
		0.1	1.05	6.80	7.10	5.40	5.68	1.06
5	3.0	0.001	2.48	3.50	2.20	2.54	1.52	0.48
		0.01	1.53	3.14	2.00	2.05	1.31	0.64
		0.1	1.27	2.83	1.86	1.78	1.22	0.76
6	4.0	0.001	3.51	34.0	7.25	15.4	3.60	0.14
		0.01	2.01	27.7	5.82	13.7	2.88	0.21
		0.1	1.58	16.8	3.20	8.60	1.81	0.34

[1]Source phase, 0.01 M in each metal ion species; Organic phase, a solution of the carrier and 5 volume % of SPAN 80 in kerosene; Receiving phase, 2.0 M sulfuric acid.

Ni(II). In the second process (Figure 5b), the di-(p-alkylphenyl)phosphoric acids **4, 5,** and **6** were used as carriers in different steps. Carrier **4** was chosen for separation of Cu(II) and Zn(II) from Co(II) and Ni(II) in the first step instead of **3** because the former has lower solubility in an aqueous phase (J. Gega, private communication). The receiving phases after Steps 2a and 2b are enriched in Cu(II) and Zn(II), respectively. The effluent from Step 2a is enriched in Ni(II), while the effluent from Step 2b contains only Cu(II).

Conclusions.

Isotherms for the interfacial tension at an aqueous/organic interface show that the new di-(p-alkylphenyl)phosphoric acids **3-6** adsorb at the interface stronger than do commercially available DEHPA (**1**) or Cyanex 272™ (**2**). From the interfacial fluxes for single transition metal cation species transport across a bulk kerosene membrane, the following selectivity orders were derived: Zn(II) > Cu(II) > Co(II) > Ni(II) for **1, 5,** and **6**; Cu(II) > Zn(II) > Co(II) > Ni(II) for **2** and **3**; and Cu(II), Zn(II) > Co(II) > Ni(II) for **4**. Similar selectivity orders were observed in competitive transport of these transition metal cations across emulsion liquid membranes in which a high utilization degree of the carrier was demonstrated. The di-(p-alkylphenyl)phosphoric acids **3-6** are found to be efficient carriers for proton-coupled transport of transition metal cations from a weakly acidic aqueous source phase across an emulsion liquid membrane. Multistage emulsion liquid membrane systems allow Co(II), Cu(II), Ni(II), and Zn(II) to be separated with Cyanex 272™ as the carrier or by a combination of di-(p-alkylphenyl)phosphoric acids **4, 5,** and **6.**

Acknowledgments.

Financial support of this work by the CPBP 03.08 Program of the Polish Government is gratefully acknowledged. The authors wish to acknowledge the assistance of Dr. Richard A. Bartsch in the preparation of the revised version of this manuscript.

References.

1. Draxler, J.; Furst, W.; Marr, R. *J. Membr. Sci.* 1988, *38*, 281.
2. Visser, H. C.; Reinhoudt, D .N.; de Jong, F. *Chem. Soc. Rev.* **1994,** *23*, 75.
3. Li, N. N. *U. S. Patent* 3,410,794 (November 12, 1968).
4. Noble, R. D. *Sep. Sci. Technol.* **1987,** *22*, 731.
5. Noble, R. D.; Way, J. D. In *Liquid Membranes: Theory and Applications*; Noble, R. D.; Way, J. D., Eds.; ACS Symposium Series 347; American Chemical Society: Washington, DC: 1987; pp. 110-122.
6. Boyadzhiev, L. *Proc. ISEC '88*; Moscow, Russia, 1988, Vol. 2; p 56.
7. Tavlarides, L. L.; Bae, J. H.; Lee, C. K. *Sep. Sci. Technol*, **1987,** *22*, 581.
8. Strzelbicki, J; Charewicz, W. *Wiad. Chem.* **1977,** *31*, 639.
9. Noble, R. D.; Way, J. D. In *Liquid Membranes: Theory and Applications*;, Noble, R. D., Way, J. D., Eds.; ACS Symposium Series 347; American Chemical Society: Washington, DC: 1987; pp. 1-26.
10. Acharya, S.; Nayak, A. *Hydrometallurgy* **1988,** *19*, 309.
11. Marczenko, Z. *Separation and Spectrophotometric Determination of Elements*; Ellis Horwood, Ltd.: Chichester, Great Britain, 1986.
12. Szymanowski, J.; Prochaska, K. *J. Coll. Interfacial Sci.* **1988,** *125*, 649.
13. Strzelbicki, J.; Schlosser, S. *Hydrometallurgy* **1989,** *23*, 67.
14 Strzelbicki, J.; Charewicz, W. A. *Hydrometallurgy* **1984,** *13*, 89.

Chapter 14

Sugar Separation Using Liquid Membranes and Boronic Acid Carriers

Bradley D. Smith

Department of Chemistry and Biochemistry,
University of Notre Dame, Notre Dame, IN 46556

Boronic acids are shown to facilitate the transport of a range of hydrophilic sugars through lipophilic membranes. The results suggest that boronic acids have promise as transport carriers in membrane-based sugar separations. Most of the work involves Bulk Liquid Membranes (BLMs), for which the various chemical and physical factors that control transport are examined in detail. The use of boronic acids in large-scale separations will require changing to more practical liquid membrane systems, such as Supported Liquid Membranes (SLMs) with hollow fiber geometries. Boronic acids can also transport sugars through lipid bilayers. The possibility of developing a sugar separations method based on liposomes is discussed.

This chapter describes a method of transporting sugars through lipophilic membranes using boronic acid carriers. Potential applications of this technology range from industrial-scale separations to drug delivery. The focus of this article is on separations. While the utility of liquid membranes in metal cation separations is well-recognized, there are very few examples of liquid membranes being used in molecular separations (1). Boronic acids are currently the only synthetic compounds known to selectively facilitate the transport of hydrophilic sugars through lipophilic membranes (2). In principle, sugar-selective carriers should be useful in a variety of membrane-based separation problems ranging from sugar refining to the purification of fine chemicals and pharmaceuticals (3). This article summarizes our recent progress in this developing research area.

Sugar Complexation Using Boronic Acids.

Boric acid complexes of sugar compounds have been investigated for well over one hundred years (4). By the end of the nineteenth century it was known that addition of boric acid to aqueous sugar solutions changed the optical rotation of the sugars, increased the acidity of the boric acid, and increased the solution's electrical conductivity (5). Analogous complexation studies with boronic acids were apparently not conducted until the middle of the twentieth century. In 1954, Kuivila reported

0097–6156/96/0642–0194$15.00/0
© 1996 American Chemical Society

that phenylboronic acid formed trigonal boronate ester precipitates with sugars such as mannitol (*6*). In 1959, Lorand and Edwards proved that phenylboronic acid was not a Brønsted acid but rather a Lewis acid (*7*). While doing so, these workers determined boronate association constants for a number of diol-containing compounds in aqueous solution.

The covalent associations between boronic acids and diol-containing compounds can be summarized in the following way. In anhydrous aprotic solvents, boronic acids readily condense with diols to form trigonal boronate esters, **1**. In aqueous solution, the trigonal boronates are unstable and either hydrolyze back to starting compounds or ionize to form anionic, tetrahedral boronates, **2**. Since a boronate ester is more acidic than its parent boronic acid, the predominant complexation product between phenylboronic acid (pKa = 8.9), and a diol at pH 7 is boronate **2**. A salient point is that although covalent bonds are formed, the associated activation energy is low, so the process is rapid and reversible. This reversible interaction has been exploited extensively as a diol protecting group strategy in organic synthesis (*8*), and is the basis of a chromatographic method for separation of polyols (*9*).

Boronic Acids As Sugar Transport Carriers.

To date, most of the boronic acid transport experiments have used Bulk Liquid Membranes (BLMs) in a standard U tube apparatus. In this configuration an aqueous source phase is separated from an aqueous receiving phase by a dense organic layer (*e.g.*, dichloroethane) that is stirred using a magnetic stir bar (*2,10*). The main advantages of BLMs, compared to other membrane systems, are the ease of operation, low cost, and the less stringent requirement for the carriers to be highly lipophilic (*11*). Thus, BLMs are often used as the initial screen of transport ability for carrier candidates. However, BLMs are not practical on an industrial scale. For commercial applications, Supported Liquid Membranes (SLMs), where the organic layer is immobilized within the pores of a thin (~100 μm) polymer support, or Emulsion Liquid Membranes (ELMs), are more attractive (*1,11*). Since the goal of the separation is to reclaim the transported sugars, the use of SLMs in a hollow fiber geometry is probably the membrane system of practical choice (*12*). This means the final-generation carriers need to be compatible with hollow fiber SLMs.

Transport via Tetrahedral Boronates. Membrane transport using boronic acid carriers was first reported in 1986 by Shinbo and co-workers who discovered that a mixture of phenylboronic acid (PBA) and trimethyloctylammonium (TOMA) chloride transported reducing monosaccharides through BLMs (*13*). Transport was observed to be pH sensitive, in that uphill saccharide transport could be achieved from a basic source phase into an acidic receiving phase. The transport mechanism invoked to explain these observations is described in Figure 1. Subsequent studies by Czarnik (*14,15*) and our group (*16-20*) showed that other hydrophilic diol-containing compounds, namely nucleosides and aryl glycosides, could be transported by this ion-pair process. Under low extraction conditions the order of transport enhancements reflected the known order of boronic acid affinities for cyclic diols which is cis-α,β-diol > cis-α,γ-diol > trans-α,γ-diol >> trans-α,β-diol. Thus, the order of transport enhancements for nucleosides was ribonucleosides >> 2'-deoxyribonucleosides, and for glycopyranosides, galactoside > mannoside > glucoside > xyloside. In the case of monosaccharides, the transport enhancement order was fructose ~ mannose > galactose > glucose (*13*). In this series, however, the order of sugar selectivity is more difficult to rationalize because reducing sugars are known to isomerize in the presence of boronic acids making the identity of the sugar/boronate complex less certain (*18*). In particular, there is a strong bias towards complexation of hexoses as their furanose isomers (*21*).

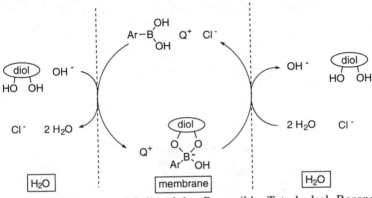

Figure 1. Transport Mediated by Reversible Tetrahedral Boronate Formation. pH < Boronic Acid pKa. Q^+ = Quaternary Ammonium Cation Such as Trioctylmethylammonium (TOMA).

Czarnik devised an elegant improvement on PBA-TOMA as a carrier admixture by synthesizing the lipophilic alkylated pyridylboronic acids **3** (*15*). Functioning as covalent versions of PBA-TOMA, carriers **3** were found to transport nucleosides up to eight-fold better (*e.g.*, a BLM containing 0.5 mM of carrier **3** enhanced uridine transport to over 100 times the rate of background diffusion). At pH 7, compounds **3** exist as zwitterions, thus the principal binding equilibrium is that shown in equation 1.

$$\begin{array}{c} ^-B(OH)_3 \\ \text{(pyridine ring)} \\ N^+{-}R \\ \textbf{3} \\ R = \text{n-}C_{18}H_{37},\ \text{steryl} \end{array} + \text{(diol, HO\ OH)} \rightleftharpoons \begin{array}{c} \text{(diol)} \\ O\quad O \\ ^-B\diagdown_{OH} \\ \text{(pyridine ring)} \\ N^+ \\ R \end{array} + 2\,H_2O \qquad (1)$$

Uphill Transport Carriers. Although zwitterionic carriers **3** are attractive because of their improved transport rates, they suffer from a potential drawback; their binding equilibria are pH independent. Unless a highly acidic receiving phase is employed (the pK_a for **3** is approximately 4) (*15*), these carriers cannot be used to transport sugars uphill using a pH gradient. The ability to transport uphill is an essential prerequisite if a practical sugar separation method is to be developed. As already stated, uphill transport driven by a pH gradient has been demonstrated for the ion-pair transport pathway shown in Figure 1. In certain separations it may be undesirable to deviate from neutral pH (*e.g.*, sample instability). Therefore, we have explored ways of driving uphill transport using electrochemical gradients other than pH. Our first attempt employed a fluoride ion gradient. Since the ability of F⁻ to form dative bonds with trigonal boron acids is similar to that of OH⁻, we wondered if F⁻ ions could substitute for OH⁻. In other words, could F⁻ ions promote formation of the anionic tetrahedral fluoroboronate **4** (equation 2), and induce transport via an ion-pair mechanism analogous to that described in Figure 1? This was indeed the case (*16*) Addition of KF (0.5 M) to a source phase buffered at pH 7 was found to increase the passive nucleoside transport ability of a PBA-TOMA carrier admixture by a factor of three. Moreover, uphill nucleoside transport was achieved in the direction of a F⁻ concentration gradient. Control experiments showed that KCl and KBr had no effect on transport.

$$\boxed{diol} + PhB(OH)_2 + F^- \rightleftharpoons \begin{array}{c} \boxed{diol} \\ O \quad O \\ B^- \\ Ph \quad F \end{array} + 2 H_2O \qquad (2)$$

4

We have also developed uphill transport systems that are driven by metal cation gradients. We reasoned that an alternative way of producing a lipophilic cation to ion-pair with a sugar-boronate anion would be to complex a metal cation inside a lipophilic ionophore (*17*). This idea lead to the design and synthesis of compound **5** as a functionally biomimetic sodium-saccharide cotransporter (*20*). A BLM containing 1 mM of **5** transported an aryl glucoside five times faster than the background rate, whereas an equimolar mixture of PBA and benzo-15-crown-5 produced negligible transport enhancement. Carrier **5**, however, was less than half as effective at glucoside transport as PBA-TOMA, reflecting among other things the inherent difficulty for a heterotopic receptor like **5** to simultaneously bind and cotransport two different solutes (equation 3). The putative binding equilibrium shown in equation 3 suggests that glycoside transport should be sensitive to Na⁺ ion concentrations in the aqueous phases. Moreover, uphill transport in the direction of a Na⁺ ion gradient was predicted to occur and subsequently found to be the case. Carrier **5** represents the first artificial sodium-saccharide cotransporter to mimic, functionally, the way nature uses the ubiquitous inward-directed Na⁺ gradient to actively transport sugars into cells (*17*).

Transport via Trigonal Boronates.   Under certain experimental conditions, significant BLM transport can be achieved using boronic acids alone.   For example, we found that boronic acids can mediate glycopyranoside transport by forming a reversible trigonal boronate ester with a glycopyranoside diol (Figure 2) (*18*).   The order of diol selectivity for this trigonal boronate transport pathway was observed to be cis-α,γ–diol > cis-α,β-diol ≈ trans-α,γ-diol >> trans-α,β-diol, which differs slightly from the selectivity of the tetrahedral boronate pathway.   As noted above, trigonal boronate esters are usually unstable in an aqueous environment, and thus are a minor presence.   However, at an aqueous/organic interface, a lipophilic boronate ester is able to partition into the organic phase, where it is protected from hydrolysis. Other factors that strongly affect the trigonal boronate transport pathway are the lipophilicity of the boronic acid carrier and the aqueous phase pH.   Shinkai and coworkers reported that highly lipophilic boronic acids are able to efficiently extract reducing monosaccharides into an organic layer (*22*).   Although liquid membrane transport was not the goal of their work, it seems likely that it would have occurred. The effect of pH on the trigonal boronate pathway is varied.   We have found that PBA alone is unable to transport ribonucleosides at neutral pH, while significant transport is observed at pH 4 (*23*).   On the other hand, galactopyranoside transport mediated by PBA was a maximum at pH 7 (*18*).   Most recently, Rotello has reported that a BLM containing 3 mM of PBA is able to enhance riboflavin transport by more than two hundred times over the background rate (*24*).   There is little doubt that the trigonal boronate transport mechanism is operating in this case, although the stoichiometry and regiochemistry of binding are still to be established.

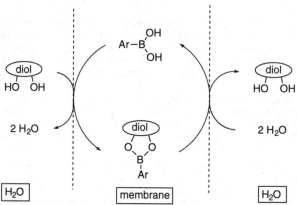

Figure 2.   Transport Mediated by Transient Trigonal Boronate Ester Formation.

Dopamine Transport.

Boronic acids have been shown to transport hydrophilic diol-containing compounds other than sugars.   For example, the crowned boronic acid **6** was designed as a selective carrier for dopamine transport (*25*).   A BLM containing 1 mM **6** transported dopamine 160 times faster than background diffusion (Table I).   An interesting design feature with carrier **6** is illustrated in equation 4.   Not only is it dopamine shape-selective, but the resulting complex **7** is charge balanced and does not need an accompanying counter-ion for transport.   This provided carrier **6** with a novel selectivity mechanism for dopamine transport which is reflected by the data shown in Table I.   Since the association of dopamine and carrier **6** was an acid-producing equilibrium, it was possible to use a pH gradient to drive the dopamine uphill into an acidic receiving phase.

$$ \text{6} \rightleftharpoons \text{7} + H_3O^+ \qquad (4) $$

6   **7**

Table I. Transport Rates for Catecholamines, Glycosides, and Uridine in the Presence and Absence of Carrier 6.

Entry	Transported compound[a]	Initial transport rate (10^{-8} M min^{-1})[b] no carrier	Initial transport rate (10^{-8} M min^{-1})[b] carrier **6**	Rate enhancement[c]
1	dopamine[d]	2.2	356	160
2	norepinephrine[d]	2	120	60
3	epinephrine[d]	3.4	6.8	2
4	tyramine[d]	70	70	1
5	aryl β-glucoside[e,f]	1.3	3.9	3
6	aryl β-mannoside[e,f]	1.6	5.3	2
7	uridine[e,g]	0.2	0.2	1

[a]Source phase: sodium phosphate buffer (100 mM, pH 7.4), sodium dithionite (10 mM); Organic phase: carrier **6** (1 mM) in chloroform; Receiving phase: sodium phosphate buffer (100 mM, pH 7.4), sodium dithionite (10 mM). [b]±15 %. [c]Transport rate in the presence of carrier **6** divided by the rate in the absence of carrier. [d]The source phase initially contained 41 mM catecholamine. [e]The starting concentration of transported species was adjusted to give a similar rate of background diffusion. [f]Source initially contained 1.36 mM glycoside. [g]Source initially contained 20 mM uridine.

Dopamine Transport Through SLMs. The goal of this ongoing project is to develop a dopamine carrier that will purify and concentrate body-fluid samples for clinical dopamine analysis (*25,26*). The most likely membrane system for such an application is a SLM, which means the carrier has to be highly lipophilic. In an attempt to satisfy this requirement, carrier **8** was synthesized and its SLM transport ability determined (*27*). Transport was measured through a liquid membrane of 2-nitrophenyl octyl ether supported by a flat sheet of Accurel polypropylene (area = 16 cm^2). In the absence of carrier, the flux from a source phase containing 50 mM dopamine at pH 7.4 was negligible. In the presence of carrier **8** (2 % wt) a dopamine flux of 5 x 10^{-7} mol/m^{2}s was observed. These preliminary results are highly encouraging and strongly suggest a high-flux dopamine transport device can be developed using SLM technology.

8

Chemical and Physical Factors That Control Transport Rate.

In an attempt to fully understand the BLM transport process, we undertook a detailed study of the factors that control glycopyranoside transport rates (19). We found that transport was dependent on the extraction ability of the boronic acid carrier. An extraction constant, K_{ex}, was calculated using the following expression:

$$G + B \underset{}{\overset{K_{ex}}{\rightleftharpoons}} GB \qquad (5)$$

(aq) (org) (org)

where: G = uncomplexed glycoside
B = uncomplexed boronic acid
GB = glycoside-boronate complex

A plot of transport rate versus log K_{ex} exhibited an approximate bell-shaped curve with maximal transport occuring when the carrier had an extraction constant, $K_{ex(max)}$ ~ 2.2 (Figure 3).

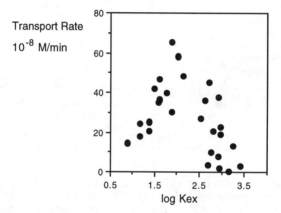

Figure 3. Plot of Glycoside Transport Rates vs. log K_{ex}.

We were able to show that of the various transport models which can explain this bell-shaped relationship, a diffusion-controlled process was the most likely. The most convincing evidence was that transport was dependent on the organic phase stirring rate (Figure 4) (19). The diffusion-controlled model, which is often seen in ionophore-mediated transport, assumes the kinetics of carrier complexation are rapid and that the rate-determining step is diffusion of the solutes through the unstirred layers (Nernst layers) of the three-phase system (10). Transport flux through the unstirred layers is in turn determined by the thickness of the layers (hence the dependence on organic phase stirring rate), the diffusion coefficient D from Fick's first law, and the carrier extraction equilibrium constant, K_{ex}. The observed bell-shaped correlation is rationalized in the following way (10). Transport is a multistep process involving extraction of the solute from the source phase, movement of the carrier/solute complex through the organic layer, and subsequent stripping of the complex into the receiving phase. Under conditions of weak extraction, transport is slow due to the low amounts of solute moving from the source phase into the organic layer. Under conditions of high extraction, it is the low solute concentrations moving from the organic layer into the receiving phase that is the rate-determining step. A corrollary of the diffusion-controlled process is that K_{ex} is the critical variable determining transport rate (10). Therefore, an analysis of the factors that control transport can be reduced to an analysis of the factors that change K_{ex} relative to $K_{ex(max)}$. In the case of glycopyranoside transport, the chemical and environmental factors that affect K_{ex} have been described in detail (*19*).

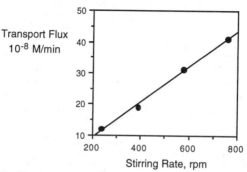

Figure 4. Glycoside Transport Rate vs. Membrane Stirring Rate.

Since transport is dependent on the rate of diffusion through the membrane, it is anticipated that transport rates should increase as the membrane becomes thinner. This correlation should continue, of course, until the rate of membrane diffusion becomes faster than the kinetics of sugar-boronic acid association. Since the effective membrane thickness associated with a stirred BLM is substantially greater than a SLM, it is expected that transport fluxes will be greatly increased when the above transport experiments are repeated with SLMs (*11,12*). Our most recent results with SLMs indicate this to be true (*27*).

Sugar Transport Through Lipid Bilayer Membranes.

Scattered within the transport literature are a handful of reports on the possibility of using liposomes (lipid bilayer vesicles) in separations (*28,29*). There are a number of attractive features associated with liposome membrane systems: (i) the bilayer constituents are highly biocompatible phospholipids, organic solvents are essentially eliminated; (ii) liposomes have a small size (100 nm diameter) and a very large surface area, while the walls are very thin (3 nm), thus transport rates can be very rapid; (iii) the technology and economics associated with manufacturing and storing liposomes has improved dramatically over the past ten years (*30*).

Transport Out of Liposomes. We have begun a study of the ability of boronic acids to transport sugars in and out of liposomes (*31*). Initially we focused on the experimentally easier problem of determining sugar efflux from liposomes. The experimental set-up for glucose efflux is described in Figure 5. Glucose (typically 300 mM) was encapsulated inside large unilamellar vesicles (LUVs) that were prepared by the rapid extrusion technique (*32*). It is worth noting that very similar results were obtained with the easier to prepare multilamellar vesicles (MLVs) composed of egg lecithin. A standard hexokinase/glucose-6-phosphate dehydrogenase enzyme assay was used for detection of the escaped glucose (*33*). The enzymes are unable to penetrate the liposomes, thus an absorbance reading at 340 nm, due to NADPH formation, results only when a glucose molecule is released from the liposome. As shown in Figure 6 and Table II, addition of various boronic acid compounds induced glucose leakage from the liposomes. Glucose efflux from the liposomes continued until the liposomes were completely empty. This active transport effect is attributed to the destructive assay which continually removes glucose from the system. Inspection of the data in Table II suggests efflux is dependent on both the lipophilicity of the boronic acid (as judged by hydrophobicity constant, π), as well as the acidity (as judged by pK_a). A large number of control experiments have been conducted to prove the glucose efflux is due to a selective transport process, and not a general increase in liposome permeability. In addition, the efflux experiments have been repeated with a range of sugar compounds. To date, the observed order of sugar efflux rates has been sorbitol > fructose > glucose > mannose > sucrose, which reflects the known order of boronate affinity for these sugars (7,34). An illustration of the transport selectivity that can be achieved with this approach is provided in Figure 7 (34). In this experiment, liposomes with equal amounts of sorbitol and sucrose encapsulated inside were teated with boronic acids. An enzymatic sorbitol assay showed that essentially all of the sorbitol was extracted from the liposomes within a couple of hours, whereas a sucrose assay of the same liposomes showed zero sucrose efflux. The selectivity is remarkable. In a single step, an equimolar mixture of these two sugars is completely separated

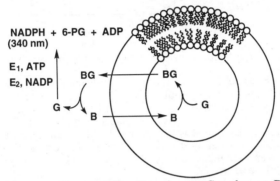

Figure 5. Liposome Glucose Efflux Experiment. G = glucose, B = boronic acid, BG = glucose-boronate complex, E_1 = hexokinase, E_2 = glucose-6-phosphate dehydrogenase, 6-PG = 6-phosphogluconate, NADP = nicotinamide adenine dinucleotide phosphate, NADPH = reduced form of nicotinamide adenine dinucleotide phosphate, ADP = adenosine diphosphate, ATP = adenosine triphosphate.

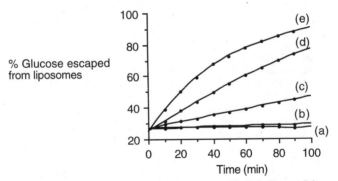

Figure 6. Percent of Glucose Escaped at pH 7.5 From Liposomes Containing 300 mM Glucose, After Treatment with: (a) no boronic acid, (b) phenylboronic acid, 1 mM, (c) phenylboronic acid, 5 mM, (d) 3,5-dichloro–phenylboronic acid, 1 mM, (e) 3,5-bis(trifluoromethyl)phenylboronic acid, 1 mM.

Table II. Relative Rates of Glucose Efflux at pH 7.5, pKa's, and Substituent Hydrophobicity Constants for Various Boronic Acid Derivatives.

Boronic acid	Relative rate of efflux (± 15 %)	pK_a	Sum of the π values[a]
Phenylboronic acid	1	8.9	0.0
3,5-Dichlorophenylboronic acid	15	7.4	1.42
3,5-Bis(trifluoromethyl)phenylboronic acid	30	7.2	1.76
4-Methylphenylboronic acid	1.9	9.3	0.56
4-Methoxyphenylboronic acid	0.8	9.3	-0.02
3-Methoxyphenylboronic acid	0.9	8.7	-0.02
4-*tert*-Butylphenylboronic acid	30	9.3	1.98
4-Carboxyphenylboronic acid	0.02	8.4	-4.36
1-Butaneboronic acid	0.3	10.4	-
Boric acid	10^{-4}	9.0	-

[a]Sum of hydrophobicity constants for aryl substituents excluding the boronic acid.

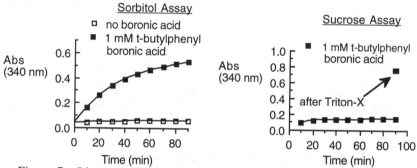

Figure 7. Liposome Efflux Rates for Liposomes Filled with 150 mM Sorbitol and 150 mM Sucrose. The absorbance at 340 nm corresponds to NADPH production produced by enzymatic assays.

Transport into Liposomes. Most recently, we have begun liposome influx studies using radiolabeled sugars. We have found that lipophilic arylboronic acids can facilitate sugar transport into empty liposomes. Figure 8 shows some typical preliminary results. ^{14}C-labeled liposomes were treated with ^{3}H-labeled glucose in the presence and absence of boronic acids. Every ten minutes an aliquot was removed and the amount of glucose associated with the liposomes (ratio of Glu:PL) was determined (*34*).

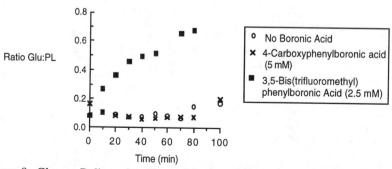

Figure 8. Glucose Delivery into Empty Liposomes Using Radiotracer Techniques.

Conclusions

Boronic acids can facilitate the transport of a range of hydrophilic sugar derivatives through various lipophilic membranes. Although the research program is still in its infancy, the results to date suggest that boronic acids show great promise as transport carriers in membrane-based sugar separations. Their use in large-scale separations will require moving to more practical membrane systems such as SLMs with hollow fiber geometries. This in turn means the carriers will have to be redesigned to meet the constraints of the membrane system (in the case of SLMs, an important requirement is very high carrier lipophilicity). Research efforts in this general direction are underway.

Acknowledgments

It is with sincere gratitude that I acknowledge the research efforts of Jeffrey T. Bien, Marie-France Paugam, Pamela R. Westmark, Gregory T. Morin, and Martin Patrick Hughes. This work was supported by a grant from the National Science Foundation and a Cottrell Scholar award of Research Corporation.

References

1.   *Liquid Membranes: Theory and Applications*.; Noble, R. D.; Way, J. D.; Eds.; ACS Symposium Series 347, American Chemical Society: Washington, DC; 1987.
2.   *Liquid Membranes: Chemical Applications*; Araki, T., Tsukube, H., Eds.; CRC Press: Boca Raton, LA; 1990.
3.   Binkley, B. W.; Wolfrom, M. L. *Adv. Carbohydr. Chem.* **1953**, *8*, 291-314.
4.   Vignon, L. *Compt. Rend.* **1874**, *78*, 148-149.
5.   Böeseken, *J. Adv. Carbohydr. Chem.* **1949**, *4*, 189-210.
6.   Kuivila, H. G.; Keough, A. H.; Soboczenski, E. J. *J. Org. Chem.* **1954**, *19*, 780-784.
7.   Lorand, J. P.; Edwards, J. O. *J. Org. Chem.* **1959**, *24*, 769-774.
8.   Ferrier, R. J. *Adv. Carbohydr. Chem.* **1978**, *35*, 31-80.
9.   Bergold, A.; Scouten, W. H., In *Solid Phase Biochemistry, Analytical and Synthetic Aspects*; W. H. Scouten, Ed.; Wiley: New York, NY; 1983, Ch. 4.
10.   Fyles, T. M., In *Inclusion Aspects of Membrane Chemistry*; Osa, T., Atwood, J. L., Eds.; Kluwer: Boston, MA; 1991, Chapter 2.
11.   Izatt, R. M.; Lamb, J. D.; Breuning, R. L. *Sep. Sci. Technol.* **1988**, *23*, 1645-1658.
12.   Visser, H. C.; Reinhoudt, D. N.; de Jong, F. *Chem. Rev.* **1994**, *23*, 75-82.
13.   Shinbo, T.; Nishimura, K.; Yamaguchi, T.; Sugiura, M., *J. Chem. Soc., Chem. Commun.* **1986**, 349-351.
14.   Grotjohn, B. F.; Czarnik, A. W., *Tetrahedron Lett.* **1989**, *30*, 2325-2328.
15.   Mohler, L. K.; Czarnik, A. W. *J. Am. Chem. Soc.* **1993**, *115*, 2998-2999.
16.   Paugam, M. -F.; Smith, B. D. *Tetrahedron Lett.* **1993**, *34*, 3723-3726.
17.   Paugam, M. -F.; Morin, G. T.; Smith, B. D. *Tetrahedron Lett.* **1993**, *34*, 7841-7844.
18.   Morin, G. T.; Paugam, M. -F.; Hughes, M. P.; Smith, B. D. *J. Org. Chem.* **1994**, *59*, 2724-2728.
19.   Morin, G. T.; Hughes, M. P.; M. -F. Paugam; Smith, B. D., *J. Am. Chem. Soc.* **1994**, *116*, 8895-8901.
20.   Bien, J. T.; Shang, M.; Smith, B. D. *J. Org. Chem.* **1995**, 60, 2147-2152.
21.   Norrild, J. C.; Eggert, H. *J. Am. Chem. Soc.* **1995**, *117*, 1479-1485.

22. Shinkai, S.; Tsukagoshi, K.; Ishikawa, Y.; Kunitake, T. *J. Chem. Soc. Chem. Commun.* **1991**, 1039-1041.
23. Bien, J. T.; Smith, B. D., unpublished results.
24. Lambert, E.; Breinlinger, E. C.; Rotello, V. M. *J. Org. Chem.* **1995**, *60*, 2646-2647.
25. Paugam, M. -F.; Valencia, L. S.; Smith, B. D., *J. Am. Chem. Soc.* **1994**, *116*, 11203-11204.
26. *Quantitative Analysis of Catecholamines and Related Compounds;* Krstulovic, A. M., Ed.; Ellis Horwood: Chichester, UK; 1986.
27. Paugam, M. -F.; Chrisstoffels, S.; De Jong, F.; Smith, B. D., unpublished results.
28. Walsh, A. J.; Monbouquette, H. G. *J. Membr. Sci.* **1993**, *84*, 107-121.
29. Powers, J. D.; Kilpatrick, P. K.; Carbonell, R., G. *Biotech. Bioeng.* **1989**, *33*, 173-182.
30. Lasic, D. D., *Liposomes; From Physics to Applications*, Elsevier: Amsterdam, 1993.
31. Westmark, P. R.; Smith, B. D. *J. Am. Chem. Soc.* **1994**, *116*, 9343-9344.
32. MacDonald, R. C.; MacDonald, R. I.; Menco, B. P. M.; Takeshita, K.; Subbarao, N. K.; Hu, L., *Biochim. Biophys. Acta* **1991**, *1061*, 297-303.
33. *Liposomes, a Practical Approach;* New, R. R. C., Ed.; IRL Press: Oxford, UK; 1990.
34. Westmark, P. R.; Smith, B. D., unpublished results.

APPLICATIONS IN CHEMICAL SEPARATIONS

Chapter 15

Recent Advances in Emulsion Liquid Membranes

W. S. Winston Ho[1] and Norman N. Li[2]

[1]Corporate Research, Exxon Research and Engineering Company,
Route 22 East, Clinton Township, Annandale, NJ 08801
[2]Research and Technology, Allied Signal Inc.,
50 East Algonquin Road, Des Plaines, IL 60017

This review covers recent advances in the theory for emulsion liquid membranes (ELMs) briefly and in ELM applications in more detail. In the theory, the state-of-the-art models for two types of facilitation for ELMs are discussed. In the applications, significant advances have been made recently. Commercial applications include the removal of zinc, phenol, and cyanide from wastewaters and in well control fluid. Potential applications include wastewater treatment, biochemical processing, rare earth metal extraction, radioactive material removal, and nickel recovery. The ELM systems for these applications are described.

Since their discovery by Li (1) over two decades ago, emulsion liquid membranes (ELMs) have demonstrated considerable potential as effective tools for a wide variety of separations (2-4). ELMs are essentially double emulsions, i.e., water/oil/water or oil/water/oil systems. ELMs are usually prepared by first forming an emulsion between two immiscible phases, and then dispersing the emulsion in a third (continuous) phase by agitation for extraction. The membrane phase is the liquid phase that separates the encapsulated, internal droplets in the emulsion from the external, continuous phase, as shown schematically in Figure 1. The effectiveness of ELMs is a result of two facilitated mechanisms called Type 1 and Type 2 facilitations. In Type 1 facilitation, the reaction in the internal phase of the ELM maintains a solute concentration of effectively zero. This is the minimization of the diffusing solute species in the internal phase. The reaction of the diffusing species with a chemical reagent in the internal phase forms a product incapable of diffusing back through the membrane. This type of facilitation can be illustrated by the extraction of phenol from wastewater (5) as shown schematically in Figure 2(a). In this figure, the phenol in the external aqueous phase dissolves in a membrane oil phase. Then, phenol diffuses across the membrane phase into the NaOH-containing internal phase, where it reacts with NaOH to form sodium

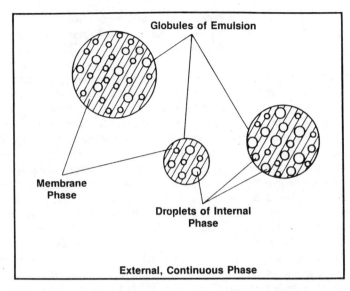

Figure 1. Schematic of an Emulsion Liquid Membrane System

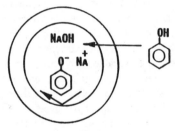

a. TYPE 1 FACILITATION

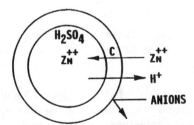

b. TYPE 2 FACILITATION (CARRIER FACILITATED TRANSPORT)

Figure 2. Schematic of Two Facilitated Mechanisms

phenolate. Since the ionic sodium phenolate is not soluble in the membrane oil, it is trapped in the internal phase. The reaction maintains the phenol concentration of effectively zero in the internal phase.

Type 2 facilitation is also called carrier facilitated transport. In this type of facilitation, the diffusing solute species is carried across the membrane phase by incorporating a "carrier" compound (complexing agent or extractant) in the membrane phase, and reactions involving the diffusing species and the carrier compound take place both at the external interface between the external and membrane phases and the internal interface between the membrane and internal phases. This facilitation can be illustrated by the removal of zinc from wastewater (4,6) as shown schematically in Figure 2b. In this figure, a zinc ion in the external aqueous phase reacts at the external interface with the carrier compound (extractant) C in the membrane phase to form a zinc complex. Then, the zinc complex diffuses across the membrane phase to the concentrated H_2SO_4-containing internal phase. The concentrated acid strips zinc from the membrane phase into the internal phase to become the zinc ion, and it donates protons to the carrier compound in the membrane phase. That is, protons are exchanged for zinc ions. The concentrated acid drives the stripping and maintains a low concentration of the zinc complex at the interface adjacent to the internal phase, giving a high driving force in terms of the zinc complex concentration difference between the external and internal interfaces and thus a high extraction rate. The concentrated acid allows the zinc ion to be effectively concentrated in the internal phase, resulting in high extraction capacity. As shown in this figure, anions in the external feed phase cannot complex with the carrier compound and are thus rejected.

For both Type 1 and Type 2 facilitations, simultaneous extraction and stripping take place in a single step rather than two separate steps as required for solvent extraction. The ELM feature of simultaneous extraction and stripping removes the equilibrium limitation inherent in solvent extraction. Therefore, complete removal of the solute from a feed can be achievable with single-step ELM extraction. In addition, the nonequilibrium feature results in a significant reduction of the extractant inventory required for the ELM extraction versus solvent extraction (2).

The theory and design for ELMs have been reviewed recently and extensively (2,3,7). This review covers recent advances in the theory briefly and in ELM applications in more detail.

Theory

The theory for ELMs may be classified into two categories: 1) diffusion-type mass transfer models for Type 1 facilitation; and 2) carrier facilitated transport models for Type 2 facilitation (2,7).

Diffusion-Type Mass Transfer Models for Type 1 Facilitation. The state-of-the-art model for Type 1 facilitation is the advancing front model (2,7,8). In this model, the solute is assumed to react instantaneously and irreversibly with the internal reagent at a reaction surface which advances into the globule as the reagent is consumed. A perturbation solution to the resulting nonlinear equations is obtained. In general, the zero-order or pseudo-steady-state solution alone often gives an adequate representation of the diffusion process.

Fales and Stroeve (9) extended the advancing front model to include the external phase mass transfer resistance outside the emulsion globules, which becomes significant (>10% error) if the Biot number is less than about 20. In addition to the external phase mass transfer resistance, Teramoto et al. (*10*) considered reaction reversibility in the internal phase.   Bunge and Noble (*11*) also considered the reversibility.   Chan and Lee (*12*) and Borwankar et al. (*13*) took into account the leakage of the internal phase into the external phase.   However, their resulting systems of equations including the reversibility and leakage were too complicated to develop even quasi-analytical solutions. Recently, Yan et al. (*14*) extended the advancing front model to include the external phase mass transfer resistance and an irreversible, first order reaction between the solute and the internal reagent, and they obtained a perturbation solution.

All of these models require the availability of the effective diffusivity of the solute within ELM globules.   Generally, the effective diffusivity is estimated from the Jefferson-Witzell-Sibbett equation (*2,7,8,15*).   In this equation, the ELM globule is considered to be an assembly of spheres of the internal phase embedded within cubes of the membrane phase.   Recently, Goswami et al. (*16*) pointed out that in order to fit cubic elements of the membrane phase, each with an embedded, internal phase droplet, within the spherical emulsion globule, the cubic elements must be distorted.   They introduced a shape factor to account for the distortion.

Carrier Facilitated Transport Models for Type 2 Facilitation.   The models take into account the diffusion of the carrier and carrier-metal complex in emulsion globules and reversible reactions at the external and internal interfaces.   Teramoto et al. (*17*) and Kataoka et al. (*18*) included the external phase mass transfer resistance and an additional mass transfer resistance in the peripheral thin membrane layer of the emulsion globule in their models.   The Teramoto et al. model also considered leakage. These models have complicated equations and many parameters.   Teramoto et al. evaluated their model parameters experimentally, which was quite tedious.

Lorbach and Marr (*19*) simplified the system of model equations by the use of the typical facilitated transport assumption of a constant sum of the free and complexed carrier concentrations.   They further simplified their system of equations with constant pH in their external phase and they eliminated the resistance for the peripheral thin membrane layer.   Their model has become the state-of-the-art model for Type 2 facilitation.   However, they used 4 reaction parameters, i.e., 2 reaction rate constants for forward and backward reactions and 2 apparent equilibrium constants for extraction and stripping.   The two apparent equilibrium constants were different.   In addition, they determined the external phase mass transfer coefficient experimentally and used the effective diffusivity as an adjustable parameter.   In principle, the reaction rate constants should be sufficient to define the reaction and equilibrium constants. Thus, accurate reaction rate constants are crucial to the success of the modeling.   The external phase mass transfer coefficient may be estimated reasonably accurately or it may be eliminated since the external mass transfer resistance is generally negligible for typical ELM systems (*2,7*).   Effective diffusivity may also be estimated (*2,7*).   These appear to be the areas of improvement that can be made to the modeling.

Ortner et al. (*20*) attempted to extend the Lorbach and Marr model to countercurrent column operation.   This resulted in a nonlinear partial differential equation system, which was complicated and was solved numerically.

Recently, Yan (*21*) further simplified the system of model equations for batch extraction by assuming an irreversible, first-order extraction reaction between the

212   CHEMICAL SEPARATIONS WITH LIQUID MEMBRANES

solute and the carrier, an irreversible, first-order stripping reaction between the complexed carrier and the internal reagent, and constant distribution coefficients. With those simplifications and assumptions, he obtained a perturbation solution.

Applications

Significant advances have been made in the applications of ELMs. In addition to commercial applications, there are several potential applications.

Commercial Applications. Commercial applications include zinc removal from wastewater in the viscose fiber industry, phenol removal from wastewater, cyanide removal from waste liquors in gold processing, and in well control fluid.

Zinc Removal. Zinc removal from wastewater in the viscose fiber industry was first commercialized in 1986 at Lenzing, AG, Austria (4,6,7,22). Table I shows the typical ELM system for the zinc removal. As shown in this table, zinc can be removed from about 200 mg/L down to 0.3 mg/L with an extraction efficiency of greater than 99.5%. This zinc removal involves very slow stripping kinetics, i.e., the zinc ion forms a strong complex with the extractant di(2-ethylhexyl)dithiophosphoric acid (DTPA). The slow stripping kinetics are not suitable for solvent extraction. However, this kinetics problem is overcome by the very large internal area of the ELM (10^6 m^2/m^3). The removal of zinc has been studied recently for a zinc plant effluent (23) and wastewaters containing zinc (24) and other metals (25). As mentioned earlier, the zinc removal is a typical example for Type 2 facilitation.

Table I.   ELM System for Zinc Removal from Wastewater (4)

External Feed Phase: 200 mg/L Zn^{2+} in 6g/L H$_2$SO$_4$
Membrane Phase
Extractant:	Di(2-ethylhexyl)dithiophosphoric Acid (DTPA),	5 wt.%
Surfactant:	ECA 11522 Polyamine,	3 wt.%
Diluent:	Shellsol T (Paraffin),	92 wt.%
Internal Phase:	250 g/L H$_2$SO$_4$	
Efficiency:	>99.5% with Zn^{2+} removed down to 0.3 mg/L	

Phenol Removal.   Removal of phenol from wastewater was commercialized around 1986 at the Nanchung Plastic Factory in Guangzhou, China (5,7,22). Table II gives the typical ELM system for phenol removal. As shown in this table, phenol can be reduced from about 1000 mg/L to 0.5 mg/L with an extraction efficiency of greater than 99.95%. The surfactant LMS-2 used in this ELM system is of a copolymeric anion type and contains a C$_4$ alkene group (26). Its average molecular weight is 5000 with a viscosity of 8000 to 10000 cp (8 to 10 Ns/m^2) (at 25°C) and a specific gravity of 0.83 to 0.86. The internal reagent NaOH converts phenol to sodium phenolate and the phenolate is trapped in the internal phase. As mentioned earlier, phenol removal is a typical example for Type 1 facilitation.

Table II. ELM System for Phenol Removal from Wastewater (5)

External Feed Phase:	1000 mg/L Phenol	
Membrane Phase		
Surfactant:	LMS-2 (Anion Type),	3.5 wt.%
Diluent:	Kerosene,	89.8 wt.%
	Paraffin,	6.7 wt.%
Internal Phase:	5wt.% NaOH	
Efficiency:	99.95% with phenol removed down to 0.5 mg/L	

Cyanide Removal. Cyanide removal from waste liquors in gold processing is being commercialized at the Huang-hua Mountain Gold Plant, near Tian-jin, China (*7,22,27,28*). Table III shows the ELM system for the cyanide removal (*27*). Cyanide can be reduced from about 130 mg/L to 0.5 mg/L with an extraction efficiency of 99.6%, leading to pollution control and productivity increase. In addition to Extractant M and Lan 113-b used by Jin and Zhang (*27*), an amine extractant, e.g., Alamine (Henkel), and a polyamine surfactant, e.g., ECA 11522 or Paranox 100, could constitute an effective ELM system for the cyanide removal.

Table III. ELM System for Cyanide Removal from Waste Liquors (27)

External Feed Phase:	130 mg/L CN^- as in $Au(CN)_2^-$, pH 9-9.5	
Membrane Phase		
Extractant:	Extractant M (Protonated Cation Type) (or Alamine),	5 wt.%
Surfactant:	Lan 113-b (or Polyamine),	2 wt.%
Diluent:	Kerosene,	93 wt.%
Internal Phase:	0.8wt.% NaOH	
Efffciency:	99.6% with CN^- removed down to 0.5 mg/L	

Well Control Fluid. A well control fluid which has been commercially available from Exxon since 1985 (*29*) is a pumpable water-in-oil emulsion containing clay particles in the oil phase. The emulsion is thickened to a viscous, high-strength paste by high shear at the drill bit nozzles. The high shear ruptures the oil film separating the water droplets and clay particles. This allows direct contact between the water and clay particles, resulting in tremendous swelling of the clay particles and formation of a high-strength paste. The well control fluid prevents well blowout and seals loss zones in oil and gas wells.

Potential Applications. Potential applications for ELMs include: 1) wastewater treatment; 2) biochemical processing; 3) extraction of rare earth metals from dilute solutions; 4) removal of radioactive materials from nuclear waste streams; and 5) recovery of nickel from electroplating solutions.

Wastewater Treatment. Potential applications in wastewater treatment include the removal of heavy metals (Zn, Cd, Cu, Pb, and Hg), arsenic, acids (acetic acid and nitrophenols), and bases (ammonia) from wastewater streams. For the removal of heavy metals from wastewaters in metallurgical and incineration plants, Marr and

Draxler (*4*) have given the typical ELM system shown in Table IV. As demonstrated in their pilot plant operation, this ELM system can remove heavy metals very well with a high efficiency (about 99%) to achieve 0.2 mg/L Zn^{2+}, 0.02 mg/L Cd^{2+}, 0.007 mg/L Cu^{2+}, and 0.01 mg/L Pb^{2+} in the treated stream.

Table IV. ELM System for Removal of Heavy Metals from Wastewater (*4*)

External Feed Phase:	230 mg/L Zn^{2+}, 2.7 mg/L Cd^{2+}, 1.1 mg/L Cu^{2+}, 0.5 mg/L Pb^{2+}, pH 3.4	
Membrane Phase		
Extractant:	Di(2-ethylhexyl)thiophosphoric Acid (MTPA),	5wt.%
Surfactant:	ECA 11522 Polyamine,	3wt.%
Diluent:	Shellsol T (Paraffin),	92wt.%
Internal Phase:	250 g/L H_2SO_4	
Efficiency:	~99%, 0.2 mg/L Zn^{2+}, 0.02 mg/L Cd^{2+}, 0.007 mg/L Cu^{2+}, 0.01 mg/L Pb^{2+}	

The swelling behavior of similar ELM systems due to osmotic water transfer from the external feed phase to the internal phase has been reported recently by Ramaseder et al. (*30*). Increasing the polyamine surfactant concentration enhances membrane stability, but it increases swelling. The optimal surfactant concentration is about 3 wt.% in the membrane phase, which was the concentration used by Marr and Draxler (*4*) as shown in Table IV.

Wright et al. (*31*) recently reported a field test of a continuous ELM process using Acorga M5640 oxime extractant to recover Cu from mine solutions in an around-the-clock, 196-hour operation. No major equipment or mechanical failures were experienced, including an electrical coalescer to break the emulsion. Recently, Raghuraman et al. (*32*) demonstrated successful extractions of Pb and Cd from wastewaters by the use of ELM systems with di(2-ethylhexyl)phosphoric acid (D2EHPA) as the extractant. Using this extractant, Samar et al. (*33*) successfully carried out the extraction of Cd, Pb, and Hg from wastewaters. However, when Cyanex 301 (di(2,2,4-trimethylphenyl)dithiophosphinic acid) and Aliquat 336 (methyltricaprylammonium chloride) were used as the extractants, the stripping of Hg was impossible. By the use of Aliquat 336 as the extractant, Salazar et al. (*34*) conducted the extraction of Cr from aqueous solutions.

Although the removal of all heavy metals is generally needed for wastewater treatment, selective extraction of Zn^{2+}, Cd^{2+}, and Ni^{2+} was performed by Shiau and Jung (*25*). Their results show that the order of the degree of extraction is Zn > Cd > Ni for a low pH value of the external feed phase. However, at the high pH value with the presence of ammonia in the feed phase, the order is reversed to Ni > Cd > Zn. Recently, selective separation of palladium from wastewater of high iron concentration was conducted by Kakoi et al. (*35*).

Recently, Huang et al. (*36*) presented an ELLM system for the removal of arsenic from wastewater, which is shown in Table V. The extraction efficiency for arsenic was more than 95%, i.e., arsenic was reduced from 5.5 mg/L to less than 0.25 mg/L in a one-step extraction.

Table V. ELM System for Removal of Arsenic from Wastewater (*36*)

External Feed Phase:	5.5 mg/L As (III) (as in As(OH)$_3$) in 0.4M H$_2$SO$_4$	
Membrane Phase		
Extractant:	2-Ethylhexanol,	10 vol.%
Surfactant:	ECA 4360 Polyamine,	2 vol.%
Diluent:	Heptane,	88 vol.%
Internal Phase:	2M NaOH	
Efficiency:	>95%	

An ELM system for the removal of acetic acid from wastewater is shown in Table VI (*37*). Acetic acid can be removed quite well with an extraction efficiency of greater than 95%. Recently, Gadekar et al. (*38*) gave an ELM system for the removal of nitrophenols from wastewater, which is shown in Table VII. The wastewater contained 1000 mg/L of 4-nitrophenol, 1000 mg/L of 2-nitrophenol, and 500 mg/L of 2,4-dinitrophenol. These nitrophenols were removed effectively with a high extraction efficiency of greater than 98%. In the membrane phase, cyclohexanone was added to produce preferential micellization with the surfactant SPAN 80, thus minimizing the micellization of this surfactant with water and swelling of the emulsion.

Table VI. ELM System for Acetic Acid Removal from Wastewater (*37*)

External Feed Phase:	0.0861M CH$_3$COOH	
Membrane Phase		
Extractant:	Tri-n-butyl Phosphate (TBP),	1 wt.%
Surfactant:	E 644 Polyamine,	8 wt.%
Diluent:	Kerosene,	72 wt.%
	Paraffin,	19 wt.%
Internal Phase:	3M NaOH	
Efficiency:	>95%	

Table VII. ELM System for Removal of Nitrophenols from Wastewater (*38*)

External Feed Phase:	1000 mg/L p-Nitrophenol, 1000 mg/L o-Nitrophenol, 500 mg/L 2,4-Dinitrophenol	
Membrane Phase		
Surfactant:	SPAN 80,	2 vol.% of Emulsion
Diluent:	Kerosene,	75.5 vol.% of Emulsion
	Paraffin,	10 vol.% of Emulsion
	Cyclohexanone,	2.5 vol.% of Emulsion
Internal Phase:	0.7M NaOH,	10 vol.% of Emulsion
Efficiency:	>98%	

ELM removal of ammonia from wastewater has been studied considerably (*2,3*). Table VIII shows a typical ELM system for the ammonia removal (*39*). Ammonia can be removed from 1200 mg/L down to 6 mg/L with an extraction efficiency of 99.5%.

Table VIII. ELM System for Ammonia Removal from Wastewater (*39*)

External Feed Phase:	1200 mg/L NH$_3$, pH $\geq$ 12	
Membrane Phase		
Surfactant:	SPAN 80,	4 wt.%
Diluent:	Paraffin,	96 wt.%
Internal Phase:	20 wt.% H$_2$SO$_4$	
Efficiency:	99.5% with NH$_3$ removed down to 6 mg/L	

Extractions with microemulsion liquid membranes were investigated for the removal of acetic acid from water (*40*), copper from aqueous feeds (*41*), and mercury from wastewater streams (*42*). Also investigated for these membranes were electrical and chemical demulsification techniques (*43*).

Biochemical Processing. Potential ELM applications in biochemical processing (*44*) include the separation of aminoacids (L-phenylalanine), biochemicals (acrylic and propionic acids), and antibiotics (penicillin G) from fermentation broths. A typical ELM system for the recovery of L-phenylalanine from fermentation broths is given in Table IX (*45,46*). The recovery from a broth containing 12-35 g/L of L-phenylalanine can be about 80% with a single batch extraction. Hong et al. (*45*) obtained a recovery of about 99% with four serial batch extractions in simulating their proposed continuous process with mixer-settler extractors.

Table IX. ELM System for Recovery of L-Phenylalanine (*45,46*)

External Feed Phase:	12 - 35 g/L L-Phenylalanine, pH 2.5	
Membrane Phase		
Extractant:	Di(2-ethylhexyl)phosphoric Acid (D2EHPA),	20 wt.%
Surfactant:	Paranox 100 Polyamine,	5 wt.%
Diluent:	S-60NR (50 vol.% Olefins, 49 vol.% Paraffins,	
	1 vol.% Aromatics),	75 wt.%
Internal Phase:	1.5M H$_2$SO$_4$	
Efficiency:	~80%	

For the recovery of acrylic and propionic acids from a fermentation broth, O'Brien and Senske (*47*) studied an ELM system containing Paranox 100 polyamine surfactant, Solvent 100 Neutral diluent, 5 M NaOH internal phase, and no extractant. They obtained a recovery of about 70%.

For the extraction and subsequent derivatization of penicillin G from a fermentation broth, Scheper et al. (*48*) utilized an ELM system containing the enzyme pencillin acylase in the internal phase to convert the extracted penicillin G to the products 6-aminopenicillinic acid and phenylacetic acid. Recently, Lee and Lee (*49*) and Mok et al. (*50*) successfully performed the ELM extraction of this penicillin with sodium carbonate in the internal phase. Mok et al. (*50*) gave an ELM system for this extraction, which is shown in Table X. They obtained extraction efficiencies between 80% and 95% for emulsion/feed ratios ranging from 0.167 to 0.2 and an internal phase concentration of greater than 9 times the initial concentration in the external feed phase. In addition, extraction of peptides was investigated recently (*51*).

Table X. ELM System for Recovery of Penicillin G (50)

External Feed Phase:	0.02M Penicillin G in 0.408M Citrate Buffer, pH 5	
Membrane Phase		
Extractant:	Amberlite LA2 Secondary Amine,	0.01M
Surfactant:	ECA 4360J Polyamine,	5 wt.%
Diluent:	Kerosene,	95 wt.%
Internal Phase:	0.1M Na_2CO_3	
Efficiency:	80-95%	

Extraction of Rare Earth Metals from Dilute Solutions. ELM extractions of rare earth metals, such as lanthanum, europium, lutetium praseodymium, and neodymium, from dilute solutions have been studied recently (*2,3,52,53*). Tables XI and XII show ELM systems for the extractions (*54,55*). These ELM systems can extract the rare earth metals from dilute solutions effectively. Both the extractants, 2-(sym-dibenzo-16-crown-5-oxy)hexanoic acid and mono(2-ethylhexyl) 2-ethylhexyl phosphonate (PC-88A) are effective complexing agents. It has been found that the extraction rate for rare earth metals is strongly affected by the structure of surfactant. The extraction rate with a cationic surfactant, e.g., $2C_{18}\Delta^9GEC_2QA$ ($C_8H_{17}CH{=}CH_8H_{16}OOCCH(C_2H_4COOC_8H_{16}CH{=}CHC_8H_{17})NHCOCH_2N^+(CH_3)_2CH_2CH_2OH$ Cl$^-$, Table XII), was about 40 times faster than that with a nonionic surfactant, e.g., Span 80. This was due to the interaction between the cationic surfactant and the anionic carrier PC-88A at the interface (*53,55*). Recently, the separation of Eu^{3+} from Am^{3+} was investigated (*56*).

Table XI. ELM System for Extraction of Rare Earth Metals (54)

External Feed Phase:	13.9 mg/L La^{3+}, 15.2 mg/L Eu^{3+}, 17.5 mg/L Lu^{3+}, pH 6-7	
Membrane Phase		
Extractant:	2-(sym-Dibenzo-16-crown-5-oxy)hexanoic Acid,	0.01M
Surfactant:	SPAN 80,	5 vol.%
Diluent:	Toluene,	50 vol.%
	Mineral Oil,	45 vol.%
Internal Phase:	HNO_3, pH 2	
Efficiency:	>99%	

Table XII.   ELM System for Rare Earth Metal Extraction (55)

External Feed Phase:	0.0001M Pr^{3+}, 0.00035M Nd^{3+}, 0.0008M La^{3+}, pH2	
Membrane Phase		
Extractant:	Mono(2-ethylhexyl) 2-ethylhexylphosphonate (PC-88A),	0.05M
Surfactant:	$2C_{18}\Delta^9GEC_2QA$,	0.02M
Diluent:	n-Heptane	
Internal Phase:	0.5M H_2SO_4	
Efficiency:	>99% for Pr^{3+} and Nd^{3+} (good selectivity vs. La^{3+})	

Removal of Radioactive Materials from Nuclear Waste Streams. ELMs have potential for the removal of radioactive materials such as strontium, plutonium, cesium, uranium, and americium from nuclear waste streams. Eroglu et al. (57) reported an ELM system for the extraction of strontium, which is shown in Table XIII. They extracted about 92% of Sr^{2+} from a feed containing 100 mg/L Sr^{2+} by the use of D2EHPA as the extractant.

Table XIII. ELM System for Removal of Strontium (56)

External Feed Phase:	100 mg/L Sr^{2+}, pH 5.3	
Membrane Phase		
Extractant:	Di(2-ethylhexyl)phosphoric Acid (D2EHPA),	0.2M
Surfactant:	Span 80,	3 wt.%
Diluent:	Kerosene,	97 wt.%
Internal Phase:	HCl, pH 1.6	
Efficiency:	92%	

Shukla et al. (58) used dicylcohexano-18-crown-6 (DC18C6) as the extractant in bulk and supported liquid membranes to remove more than 90% of plutonium from a feed containing 50 mg/L of Pu^{4+}, 3 M HNO_3, and 0.05 M $NaNO_3$. Based on their work, incorporation of a surfactant, e.g., ECA 11522 polyamine, Paranox 100 polyamine, or SPAN 80, into the membrane system shown in Table XIV could constitute an effective ELM system for plutonium removal. Recently, Dozol et al. (59) employed crown ethers, e.g., n-decylbenzo-21-crown-7 (nDecB21C7), as the extractant to investigate the transport of Cs^+ through supported liquid membranes. Based on their work, an ELM system might be developed for the removal of Cs^+.

Table XIV. ELM System for Removal of Plutonium (58)

External Feed Phase:	50 mg/L Pu^{4+} in 3M HNO_3 / 0.05M $NaNO_3$	
Membrane Phase		
Extractant:	Dicyclohexano-18-crown-6 (DC18C6),	0.2M
Surfactant:	Polyamine or SPAN 80,	3 wt.%
Diluent:	Toluene or Paraffin,	97 wt.%
Internal Phase:	0.5M Na_2CO_3	
Efficiency:	>90%	

For the extraction of uranium, Hirato et al. (60) used an ELM system containing tri-n-octylamine extractant, SPAN 80 surfactant, kerosene diluent, and 1 M Na_2CO_3 internal phase. They extracted more than 95% of uranium from a feed containing 1.15 g/L of U(VI) as in the form of UO_2^{2+}.

For the removal of americium from nuclear waste streams, Muscatello and Navratil (61) employed dihexyl-N,N-diethylcarbamoylmethylphosphonate (DHDECMP) as the extractant without a diluent in a supported liquid membrane. They removed more than 95% of americium from the waste streams. The DHDECMP extractant and a surfactant, e.g., SPAN 80 or LMS-2, with 0.25M oxalic acid as the internal phase could constitute an effective ELM system for americium removal.

Recovery of Nickel from Electroplating Solutions. Marr and Draxler (*4*) have given a typical ELM system for the recovery of nickel from wastewaters in the electroplating industry. Table XV shows the ELM system. Nickel can be reduced from 400-6000 mg/L down to 1 mg/L in the external feed phase with an extraction efficiency of at least 99.8%. Recently, Juang and Jiang (*62*) investigated the recovery of nickel from a simulated electroplating rinse solution.

Table XV. ELM System for Nickel Recovery from Electroplating Solutions (*4*)

External Feed Phase:	400 - 6000 mg/L Ni^{2+} in H_2SO_4	
Membrane Phase		
Extractant:	Di(2-ethylhexyl)dithiophosphoric Acid (DTPA),	5 wt.%
Surfactant:	ECA 11522 Polyamine,	3 wt.%
Diluent:	Shellsol T (Paraffin),	92 wt.%
Internal Phase:	250 g/L H_2SO_4	
Efficiency:	$\geq$99.8% with Ni^{2+} removed down to 1 mg/L	

For ELM applications, conventional solvent extraction contactors, e.g., mixer-settlers and mechanically agitated columns, have been generally employed. Recently, Raghuraman and Wiencek (*63*) investigated ELM extractions in a microporous hollow-fiber contactor. The hollow-fiber contactor may improve the efficiency of extraction by decreasing membrane swelling and leakage, particularly for poorly formulated ELMs with stability problems.

Concluding Remarks / Perspectives

Perspectives for ELMs include the potential applications described above. Future needs and improvements for ELMs include: 1) the modeling of carrier facilitated transport discussed earlier (*2,7*); 2) the design and synthesis of surfactants for improved membrane stability (swelling and leakage) and extractant compatibility, e.g., the copolymeric anion surfactant LMS-2 is much more stable than the SPAN 80 surfactant for phenol removal (*26*), the anionic, sulfonic type surfactant EM-301 (sulfonated polyisobutylene) has lower swelling than SPAN 80 and E 644 surfactants (*64*), the surfactant LYF-G2 (sulfonated polybutadiene) has shown better emulsion stability than ECA 4360 polyamine and EM-301 for an aqueous HCl feed (*65*), and the maleic anhydride/1-octadecene copolymer surfactant PA 18 is compatible with the hydroxyoxime extractant whereas SPAN 80 and polyamine are not compatible (*66*); 3) the design and synthesis of extractants for improved extraction performance, e.g., an extractant with faster stripping kinetics than di(2-ethylhexyl)dithiophosphoric acid (DTPA) could improve the extraction rate of nickel (*4*) and 4-*t*-octylphenyl phenylphosphonate (4TOPPPA) has shown stronger extractability for cobalt and nickel than D2EHPA and PC-88A (*67*); and 4) the improved formulation of membrane compositions, e.g., addition of cyclohexanone to the membrane phase for preferential micellization with SPAN 80 to reduce swelling (*16*).

Literature Cited

1. Li, N. N., *U.S. Patent* 3,410,794 to Exxon, **1968**.
2. Ho, W. S.; Li, N. N. In *Membrane Handbook;* Ho, W. S.; Sirkar, K. K., Eds.; Chapman & Hall: New York, NY, **1992**; pp. 597-655.
3. Gu, Z. M.; Ho, W. S.; Li, N. N. In *Membrane Handbook;* Ho, W. S.; Sirkar, K. K., Eds.; Chapman & Hall: New York, NY, **1992**; pp. 656-700.
4. Marr, R. J.; Draxler, J. In *Membrane Handbook;* Ho, W. S.; Sirkar, K. K., Eds.; Chapman & Hall: New York, NY, **1992**; pp. 701-724.
5. Zhang, X.-J.; Liu, J.-H.; Fan, Q.-J.; Lian, Q.-T.; Zhang, X-T; Lu, T.-S. In *Separation Technology;* Li, N. N.;. Strathmann, H., Eds.; United Engineering Trustees: New York, NY, **1988**; pp. 190-203.
6. Draxler, J.; Marr, R. J.; Prötsch, M. In *Separation Technology;* Li, N. N.; Strathmann, H., Eds.; United Engineering Trustees: New York, NY, **1988**; pp. 204-214.
7. Ho, W. S.; Li, N. N. In *Preprints of First AIChE Separations Division Topical Conference on Separations Technologies: New Developments and Opportunities*; American Institute of Chemical Engineers: New York, NY, **1992**; pp. 762-767.
8. Ho, W. S.; Hatton, T. A.; Lightfoot, E. N.; Li, N. N. *AIChE J.* **1982**, *28*, 662.
9. Fales, J. L.; Stroeve, P. *J. Membr. Sci.* **1984**, *21*, 35.
10. Teramoto, M.; Takihana, H.; Shibutani, M.; Yuasa, T.; Hara, N. *Sep. Sci. Technol.* **1983**, *18*, 397.
11. Bunge, A. L.; Noble, R. D. *J. Membr. Sci.* **1984**, *21*, 55.
12. Chan, C. C.; Lee, C. J. *Chem. Eng. Sci.* **1987**, *42*, 83.
13. Borwankar, R. P.; Chan, C. C.; Wasan, D. T.; Kurzeja, R. M.; Gu, Z. M.; Li, N. N. *AIChE J.* **1988**, *34*, 753.
14. Yan, N.; Shi, Y.; Su, Y. F. *Chem. Eng. Sci.* **1992**, *47*, 4365.
15. Jefferson, T. B.; Witzell, O. W.; Sibbett, W. L. *Ind. Eng. Chem.* **1958**, *50*, 1589.
16. Goswami, A. N.; Sharma, A.; Sharma, S. K. *J. Membr. Sci.* **1992**, *70*, 283.
17. Teramoto, M.; Sakai, T.; Yanagawa, K.; Ohsuga, M.; Miyake, Y. *Sep. Sci. Technol.* **1983**, *18*, 735.
18. Kataoka, T.; Nishiki, T.; Kimura, S.; Tomioka, Y. *J. Membr. Sci.* **1989**, *46*, 67.
19. Lorbach, D.; Marr, R. J. *Chem. Eng. Process.* **1987**, *21*, 83.
20. Ortner, A.; Auzinger, D.; Wacker, H. J.; Bart, H. J. *Process Technol. Proc. (Comput.-Oriented Process Eng.)* **1991**, *10*, 399.
21. Yan, N. *Chem. Eng. Sci.* **1993**, *48*, 3835.
22. Cahn, R. P.; Li, N. N. In *Separation and Purification Technology;* Li, N. N.; Calo, J. M., Eds.; Marcel Dekker: New York, NY, **1992**; pp. 195-212.
23. Reis, M. T. A.; Carvalho, J. M. R. *J. Membr. Sci.* **1993**, *84*, 201.
24. Lu, G.; Li, P.; Lu, Q. *Water Treat.* **1993**, *8*, 439.
25. Shiau, C.-Y.; Jung, S.-W. *J. Chem. Technol. Biotechnol.* **1993**, *56*, 27.
26. Zhang, X.-J.; Fan, Q.-J.; Zhang, X.-T.; Liu, Z.-F. In *Separation Technology;* Li, N. N.; Strathmann, H., Eds.; United Engineering Trustees: New York, NY, **1988**; pp. 215-226; *Water Treat.* **1988**, *3*, 233.
27. Jin, M.; Zhang, Y. In *Proc. International Congress on Membranes and Membrane Processes*, Chicago, IL, August 20-24, **1990**, Vol. I; pp. 676-678.
28. Jin, M.; Wen, T.; Lin, L.; Liu, F.; Liu, L.; Zhang, Y.; Zhang, C.; Deng, P.; Song, Z. *Mo Kexue Yu Jishu* **1994**, *14*, 16.
29. Exxon. In *Chem. Eng. News* **1985**, (December 9), 28; *The Lamp* (an Exxon publication) **1985**, *67*, 17.
30. Ramaseder, C.;. Bart, H. J.; Marr, R. J. *Sep. Sci. Technol.* **1993**, *28*, 929.
31. Wright, J. B.; Nilsen, D. N.; Hundley, G.; Galvan, G. J. *Miner. Eng.* **1995**, *8*, 549.
32. Raghuraman, B. J.; Tirmizi, N. P.; Kim, B.-S.; Wiencek, J. M. *Environ. Sci. Technol.* **1995**, *29*, 979.

33. Samar, M.; Pareau, D.; Durand, G.; Chesné, A. In *Hydrometall. '94, Pap. Int. Symp.*; Chapman & Hall: London, UK, **1994**; pp. 635-654.
34. Salazar, E.; Ortiz, M. I.; Urtiaga, A. M. *Ind. Eng. Chem. Res.* **1992**, *31*, 1523.
35. Kakoi, T.; Goto, M.; Nakashio, F. *Solvent Extr. Res. Dev. Jpn.* **1995**, *2*, 149.
36. Huang, C. R.; Zhou, D. W.; Ho, W.S.; Li, N. N. Paper No. 89b Presented at AIChE National Meeting, Houston, TX, March 19-23, **1995**.
37. Yan, N.; Huang, S.; Shi, Y. *Sep. Sci. Technol.* **1987**, *22*, 801.
38. Gadekar, P. T.; Mukkolath, A. V; Tiwari, K. K. *Sep. Sci. Technol.* **1992**, *27*, 427.
39. Lee, C. J.; Chan, C. C. *Ind. Eng. Chem. Res.* **1990**, *29*, 96.
40. Wiencek, J. M.; Qutubuddin, S. *Sep. Sci. Technol.* **1992**, *27*, 1211.
41. Wiencek, J. M.; Qutubuddin, S. *Sep. Sci. Technol.* **1992**, *27*, 1407.
42. Larson, K. A.; Wiencek, J. M. *Environ. Progress* **1994**, *13*, 253.
43. Larson, K. A.; Raghuraman, B. J.; Wiencek, J. M. *J. Membr. Sci.* **1994**, *91*, 231.
44. Thien, M. P.; Hatton, T. A. *Sep. Sci. Technol.* **1988**, *23*, 819.
45. Hong, S.-A.; Choi, H.-J.; Nam, S.-W. *J. Membr. Sci.* **1992**, *70*, 225.
46. Itoh, H.; Thien, M. P.; Hatton, T. A.; Wang, D. I. C. *J. Membr. Sci.* **1990**, *51*, 309.
47. O'Brien, D. J.; Senske, G. E. *Sep. Sci. Technol.* **1989**, *24*, 617.
48. Scheper, T.; Likids, Z.; Makryaleas, K.; Nowottny, Ch.; Schügerl, K. *Enzyme Microb. Technol.* **1987**, *9*, 625.
49. Lee, S. C.; Lee, W. K. *J. Chem. Technol. Biotechnol.* **1992**, *55*, 251.
50. Mok, Y. S.; Lee, S. C.; Lee, W. K. *Sep. Sci. Technol.* **1995**, *30*, 399.
51. Hano, T.; Matsumoto, M.; Kawazu, T.; Ohtake, T. *J. Chem. Tecnol. Biotechnol.* **1995**, *62*, 60.
52. Nakashio, F.; Goto, M.; Kakoi, T. *Solvent Extr. Res. Dev. Jpn.* **1994**, *1*, 53.
53. Lee, C. J.; Wang, S. S.; Wang, S. G. *Ind. Eng. Chem. Res.* **1994**, *33*, 1556.
54. Tang, J; Wai, C. M. *J. Membr. Sci.* **1989**, *46*, 349.
55. Goto, M.; Kakoi, T.; Yoshii, N.; Kondo, K.; Nakashio, F. *Ind. Eng. Chem. Res.* **1993**, *32*, 1681.
56. El-Reefy, S. A.; El-Sourougy, M. R.; El-Sherif, E. A.; Aly, H. F. *Anal. Sci.* **1995**, *11*, 329.
57. Eroglu, I.; Kalpakci, R.; Gündüz, G. *J. Membr. Sci.* **1993**, *80*, 319.
58. Shukla, J. P.; Kumar, A.; Singh, R. K. *Sep. Sci. Technol.* **1992**, *27*, 447.
59. Dozol, J. F.; Casas, J.; Sastre, A. M. *Sep. Sci. Technol.* **1995**, *30*, 435.
60. Hirato, T.; Kishigami, I.; Awakura, Y.; Majima, H. *Hydrometallurgy* **1991**, *26*, 19.
61. Muscatello, A. C.; Navratil, J. D. In *Chemical Separations: International Conference on Separation Science and Technology*, New Tork, NY April 15-17, **1986**; Litarvan: Denver, CO, Vol. II; pp. 439-448.
62. Juang, R.-S.; Jiang, J.-D. *J. Memr. Sci.* **1995**, *100*, 163.
63. Raghuraman, B. J.; Wiencek, J. M. *AIChE J.* **1993**, *39*, 1885.
64. Li, W.-X.; Shi, Y.-J. *Sep. Sci. Technol.* **1993**, *28*, 241.
65. Liu, P.-Y.; Chu, Y.; Wu, Z.-S.; Yan, Z.; Fang, T.-R. *Sep. Sci. Technol.* **1995**, *30*, 2565.
66. Draxler, J.; Fürst; W.; Marr, R. J. *J. Membr. Sci.* **1988**, *38*, 281.
67. Kakoi, T.; Goto, M.; Sugimoto, K.; Ohto, K.; Nakashio, F. *Sept. Sci. Technol.* **1995**, *30*, 637.

Chapter 16

Hollow Fiber-Contained Liquid Membranes for Separations: An Overview

K. K. Sirkar

Department of Chemical Engineering, Chemistry, and Environmental Science, New Jersey Institute of Technology, University Heights, Newark, NJ 07102

The basic technique of hollow fiber-contained liquid membrane (HFCLM) permeation-based separation of gas mixtures and liquid solutions is reviewed and existing applications in organic acid/antibiotic recovery from fermentation broths, facilitated, countertransport and co-transport of heavy metals, facilitated CO_2-N_2 (CH_4) separation, and facilitated SO_2-NO_x separation from flue gas are surveyed. New applications of the HFCLM technique include separations of organic mixtures (including isomeric) through polar liquid membranes (including cyclodextrin-containing aqueous solutions), highly selective pervaporation of organics from water, and, for gas separation, a modified technique that can treat feed at 175-200 psig. The utility of HFCLM as a reaction medium as well as the separator in separation-reaction-separation and separation-reaction processes will also be illustrated.

The utility of a thin layer of liquid as a selective membrane for separations has been explored extensively over the last thirty years. Three techniques have been used traditionally to exploit a thin liquid layer as a membrane: emulsion liquid membranes (ELM) for separation of liquid solutions (*1*); supported liquid membranes (SLM) in the pores of a porous/microporous support membrane for the separation of liquid feeds; and immobilized liquid membranes (ILM) in the pores of a porous/microporous support membrane for separating a gas mixture. SLM and ILM are different names for the same liquid membrane technique and have been briefly reviewed by Majumdar *et al.* (*2*) and Boyadzhiev and Lazarova (*3*).

A more recent liquid membrane technique is the hollow fiber-contained liquid membrane technique which is sometimes abbreviated as the CLM technique or the HFCLM technique. This technique has overcome most of the shortcomings of the SLM/ILM techniques. First introduced during 1986-1988 (*4,5*), the CLM-based permeation technique is undergoing evolution. This paper will provide a brief

0097–6156/96/0642–0222$15.00/0
© 1996 American Chemical Society

illustration of the basic HFCLM technique and its earlier and recent applications. Recent developments in the HFCLM technique will then be presented together with the corresponding applications and a perspective on possible future developments.

Basic HFCLM Technique

In the basic HFCLM technique, the feed gas or liquid mixture flows through the lumen of a first set of microporous hollow fibers in the permeator shell. The sweep, strip or permeate stream (gas or liquid) flows through the lumen of a second set of identical hollow fibers in the same permeator shell. Individual hollow fibers of the first set are intimately commingled with the individual fibers of the second set. The thin gap in the shell side between the outside surfaces of the contiguous hollow fibers of the two sets is filled with a liquid that keeps the feed fluid and the permeate fluid streams apart and acts as the selective liquid membrane (Figure 1). The shell-side liquid is connected to an external reservoir of the same liquid under pressure. Any loss of this membrane liquid to the flowing fluid streams is counteracted by an automatic and continuous supply from the external reservoir.

The first applications employed symmetric microporous hollow fibers in which the fiber materials were hydrophobic or hydrophilic. The membrane liquid may or may not wet the pores spontaneously. To separate a gas mixture in a permeator employing symmetric hydrophobic microporous fibers with gas-filled pores and nonwetting aqueous liquid membranes (Figure 2a), the liquid membrane pressure (P_M) is maintained higher than those of the feed gas (P_{FG}) and the sweep gas (P_{SG}) to prevent their dispersion in the membrane liquid (4). However, the excess pressure of the liquid membrane over that of either gas stream must be less than the breakthrough pressure for the membrane liquid (which is a function of the membrane material, pore size, and the interfacial tension between the pore fluid (here, gas) and the membrane liquid). For other fibers and configurations, Figures 2b and 2c provide appropriate details.

The first gas separation studied involved removing CO_2 from a 40% CO_2-60% N_2 mixture using helium as the sweep gas (4). The feed gas pressure was 411.5 KPa and the sweep outlet pressure was 115.1 KPa. The membrane liquid employed was pure water or a 30 wt% K_2CO_3 solution at a pressure of 446 KPa. Excellent CO_2-N_2 separation factors were achieved (e.g., 38 with water and 180 with the K_2CO_3 solution). The liquid membrane thickness in a particular permeator was determined to be 0.01115 cm for one of the 5 ft. long permeators built out of 300 Celgard X-10 hydrophobic microporous polypropylene fibers in each set. Separation of CO_2 from N_2 or of CO_2 from CH_4 was next studied in larger permeators over extended periods (up to 750 hours) using He or N_2, respectively, as the sweep gas (6). This study demonstrated long term performance stability using either water or an aqueous diethanolamine (DEA) solution as the membrane.

In gas permeation through polymeric membranes, a sweep gas is not commonly employed. Only gas permeated through the membrane flows through the permeate channel. This mode of operation for a HFCLM permeator was studied by Guha et al. (7) for CO_2-N_2 separation. They demonstrated how one can pull a vacuum in the permeate channel and achieve high selectivity and stable performance

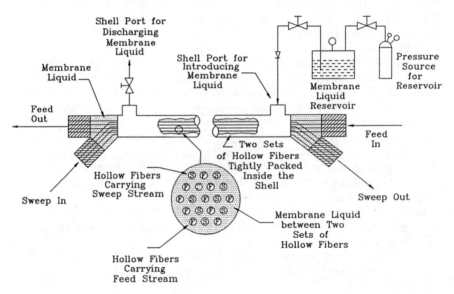

Figure 1. Basic hollow fiber-contained liquid membrane (HFCLM) permeator structure.

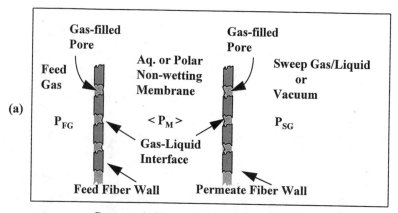

Symmetric Hydrophobic Substrate Fibers

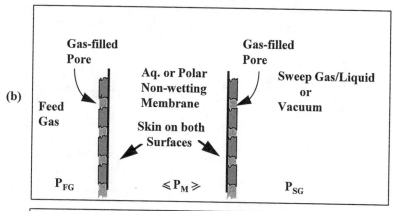

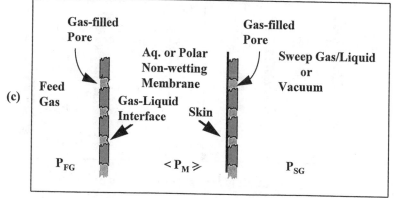

High Pressure Configurations with Composite Fibers

Figure 2. HFCLM configurations between two contiguous hollow fibers for gas separation. P_{FG}: feed gas pressure; P_M: liquid membrane pressure; P_S: strip solution pressure.

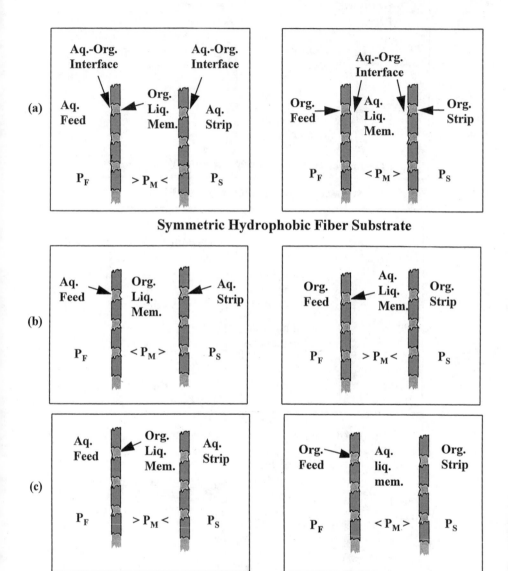

Figure 3. HFCLM configurations between the walls of two contiguous hollow fibers for liquid-liquid separation. P_F : feed pressure; P_M: liquid membrane pressure; P_S: strip solution pressure.

with a pure water membrane. An even higher selectivity ($\alpha_{CO_2-N_2} \cong 66$) was achieved by employing a sweep liquid of water in the permeate fibers. In such gas separations, the feed and the sweep gases were not humidified for the pure water liquid membrane (similarly for any pure liquid). For solutions (*e.g.*, 20 wt% aqueous DEA), it is preferable to humidify the feed gas unless the membrane is slowly circulated outside with water addition for long term stability. For a water sweep, no feed gas humidification is needed (*2*).

For aqueous solution separation by an organic liquid membrane, the microporous hollow fiber employed can be either symmetric hydrophobic or symmetric hydrophilic (*5*). If the fiber is symmetric hydrophobic, the organic membrane liquid wets the pores spontaneously (Figure 3a). To prevent dispersion, the membrane liquid pressure (P_M) is maintained lower than those of the aqueous feed (P_F) and the aqueous strip streams (P_S). The excess pressure of each aqueous stream over that of the organic membrane liquid must be less than the corresponding breakthrough pressure. If the feed and strip solutions are organic and the liquid membrane is either water or an aqueous solution or a polar organic immiscible with the feed and the strip solutions and the fibers are hydrophobic and wetted by the feed and the strip, then the membrane liquid pressure should be higher than those of the feed and the strip solutions (*8*) (Figure 3a). If a symmetric hydrophilic hollow fiber membrane with water in the pores or a hydrogel is employed, then the shell side organic liquid membrane pressure must be higher than those of the two flowing aqueous streams (*5*) for nondispersive operation (Figure 3b).

The first liquid separation studied (*5*) involved removal of phenol or acetic acid from an aqueous feed solution into an aqueous strip solution with or without 2% NaOH through a liquid membrane of decanol, xylene or methyl isobutyl ketone (MIBK). The fibers were either symmetric hydrophobic microporous Celgard X-10 of polypropylene (150 μm O.D., 100 μm I.D.) or hydrophilic regenerated cellulose fibers used for hemodialysis with water in the pores. The operation was stable even when there was substantial mutual solubility of the aqueous and organic phases.

The configuration in which the organic liquid membrane on the shell side will also occupy the pores of a symmetric hydrophilic fiber (Figure 3c) has not yet been employed in CLM. This configuration would be advantageous for those conditions where the solute prefers the liquid membrane over the feed solution. It is also necessary that the solute prefer the strip solution. In such a case, it is preferable to have the appropriate phase in the pores of the hydrophilic fiber. These ideas follow from a consideration of the various transport resistances present in CLM.

Considerations on Transport Rates

For a non-reactive system, the transport rate in the basic HFCLM technique is determined by five resistances in series: feed fluid phase boundary layer resistance, diffusional resistance through the fluid in the feed fiber pores, CLM resistance in the shell side, diffusional resistance in the fluid in the strip fiber pores, and strip fluid phase boundary layer resistance. Except for the CLM resistance, all others can be estimated generally without great uncertainty (5,8).

The diffusional resistance encountered by a species in the irregular shaped liquid film (i.e., the CLM) between a particular feed fiber and a contiguous strip fiber has been estimated by Majumdar et al. (9). This involved consideration of four equally probable square configurations of a total of four fibers which carry the feed gas and the sweep gas and numerically determining the net species transport rate from the feed fibers (fiber) to the sweep fiber (fibers) by solving the governing diffusion equations. An effective membrane thickness (EMT) was then determined. In such idealized fiber configurations, the EMT is a function of the fiber outside diameter (O.D.) and the fiber packing density (ε) and can be estimated from a graph which summarizes the numerical calculation results (9). This EMT is somewhat smaller than the fiber O.D. in a well-packed fiber bundle ($\varepsilon >0.33$) and decreases with increasing ε. Modules having a high fiber packing fraction and smaller O.D. will reduce the EMT substantially.

Since liquid membrane thickness influences the transport rate in many separations, one wonders about the difficulty of achieving the theoretically expected EMT in practice. Many smaller modules built in our laboratory have EMTs very close to those predicted theoretically. Larger modules have been developed in our laboratory as well as by Hoechst Celanese Corp. (Charlotte, NC). For example, a 2" diameter and 1 ft. long module containing 7500 Celgard X-10 fibers in each set has been built by Hoechst Celanese (LiquiCel®) and characterized by Basu and Sirkar (10). We have also built 1" diameter modules with 7008 Celgard X-10 fibers (6). The EMT values of these modules were higher than the theoretically predicted values (9). However, improved fiber bundle and module-making procedures developed by Hoechst Celanese Corp. (Charlotte, NC) (11) now provide values very close to the theoretically expected EMT. A semi-automatic fiber bundle making procedure is described in Majumdar et al. (12).

The CLM transport rates for aqueous solution separation have been compared with those for other liquid membranes (2) and polymeric membranes. Except in the cases where the membrane resistance dominates, the CLM transport rates are comparable to those in ELM for aqueous solution separation with a strip reaction (e.g., phenol removal from water with caustic in the internal phase (5)). Obviously the EMT value in a CLM is much larger than that in an ELM process. Use of CLM is suggested for aqueous solution separations where the solute partition coefficient between the organic liquid membrane and water is high. This renders the organic liquid membrane resistance negligible.

A new class of SLMs called supported polymeric liquid membranes (SPLMs) has been prepared by filling the pores of a microporous membrane with functional polymeric (oligomeric) liquids and crosslinking them (13; Ho, S.V.; Sheridan, P.W.; Krupetsky, E., J. Membr. Sci., in press). For solute phenol and polyglycols as a polymeric liquid membrane crosslinked with toluene diisocyanate, the mass transfer coefficient was twice that obtained by Sengupta et al. (5) for phenol transport through a CLM of MIBK (13). This author's conclusion that the CLM technique has much larger resistance seems unwarranted. For the phenol-MIBK-caustic strip system, Sengupta et al. (5) have shown that due to the high distribution coefficient, the CLM resistance with hydrophobic Celgard X-10 fibers was less than 1% of the total resistance. Most of the resistance was in the boundary layers of the strip and feed fibers. This conclusion is supported by the data in Table 6 of Ho *(13)* which shows that for the transport of *para*-nitrophenol with a 0.1 M caustic strip, the overall mass transfer coefficients in hollow fiber systems are two times smaller than those obtained with flat SPLMs. The comparisons which were carried out with CLMs for phenol transport were based on data from flat membranes where the boundary layer transport conditions are much more favorable.

Species transport rates in CLM devices for gas separation have also been compared with those obtained in polymeric devices. Majumdar *et al.* (*4*) have shown that the CO_2 permeance (= permeability coefficient/membrane thickness) for a CO_2-N_2 system with water as the liquid membrane was somewhat lower than that for commercialized polymeric systems although the selectivities were comparable (~ 38). When 30 wt% K_2CO_3 was used as the liquid, the permeance was about 3 times smaller. However, the selectivity was 180. For a CO_2-CH_4 system, Guha *et al.* (*6*) have obtained a CO_2 selectivity varying between 35 to 323 depending on the feed CO_2 partial pressure which compares well with values of 17-30 obtained in commercialized polymeric membrane systems. Since the EMT of their first ever large HFCLM devices were a few times larger than expected, the CO_2 permeance values were a few times smaller than in commercial polymeric systems. Low values of EMT are a necessity in gas permeation systems to achieve high permeance since, unlike liquid separations via CLM, the liquid membrane is essentially the only resistance. Improved module making (*11*) and smaller fiber outside diameter are required.

Additional Applications of the Basic HFCLM Technique

Liquid Separation.   Sorenson and Callahan (*14*) have studied the separation of chlorophenoxyacetic acid, a penicillin mimic from an aqueous feed at pH = 2.5 in a small HFCLM module (LiquiCel®) using an organic CLM of amyl acetate and an aqueous strip solution at pH = 7. This CLM process combined two sequential process steps, extraction of the antibiotic into an organic solvent and back extraction of the antibiotic into an aqueous strip solution. The extraction and back-extraction which were carried out traditionally in two devices were now achieved in one process with a HFCLM device. Basu and Sirkar (*15*) investigated the facilitated transport of citric acid through a HFCLM containing tri-n-octylamine in various organic solvents to simulate its recovery from a fermentation broth with either water or 0.5 M aqueous NaOH as the strip solution. A facilitated transport-based model accounting for interfacial reversible reaction kinetics and membrane diffusion was able to describe the almost complete solute recovery observed in a 48 cm long permeator. Yang *et al.* (*16*) have successfully recovered penicillin G (up to 90%) from aqueous solution with three HFCLM modules in series using a CLM of tributylphosphate containing Amberlite LA-2 as the penicillin complexing agent and a $NaHCO_3$ containing stripping solution.

Using a 20 cm long HFCLM permeator containing two sets of 180 Celgard X-10 fibers, Nguyen and Callahan (*17*) studied selective recovery of copper from zinc present in an aqueous ammonia solution by means of a liquid membrane of LIX 54 (phenyl alkyl beta-diketone, Henkel Corporation, Tucson, AZ). Nitrogen bubbling on the shell side of the CLM did not change the rate of copper extraction in this device suggesting that CLM resistance was not rate controlling.

Lamb et al. (18) investigated cation separations (e.g., facilitated K^+ transport over Na^+ and facilitated Ba^{2+} and Sr^{2+} transport over Ca^{2+}) using dicyclohexano-18-crown-6 as the carrier in a variety of membrane solvents, e.g., 2-octanone, 1-octanol, toluene, hexane, octanal, etc. These investigators employed the HFCLM configuration without a traditional shell so the liquid membrane could be stirred in a beaker and called it a dual module hollow fiber (DMHF) contactor. They found that the membrane system exhibited high fluxes, good control of individual phases, stability over long periods of time and allowed easy replenishment of the membrane solvent plus carrier without system downtime. The only problem encountered was the lack of control of the pressure of the liquid membrane in the beaker. This shortcoming is absent in the basic shell-and-tube configuration of HFCLM where the pressure of the liquid membrane can be independently controlled. Additional studies by Izatt et al. (19) in a DMHF contactor have focused on separating metallic cations like Ag^+ from other cations or K^+ from other alkali metal cations using proton-ionizable macrocycles in an appropriate organic solvent (2-octanone).

Sato et al. (20) have studied the preferential extraction of zinc from a feed solution of copper and zinc using 2-ethylhexyl phosphonic acid mono-2-ethyl hexyl ester in n-heptane and an aqueous HCl solution for stripping in a HFCLM permeator where both phases flowed cocurrently. Further, the liquid membrane was also made to move slowly and cocurrently. Guha et al. (21) have made an extensive study of separation and concentration of heavy metals like Cu^{2+}, Cr^{6+} and Hg^{2+} from their individual aqueous solutions by means of a number of HFCLM permeators. Both cotransport and countertransport schemes were experimentally studied. Countertransport of Cu^{2+} was successfully modeled. Heavy metal concentrations were reduced to sub-ppm levels for Cu^{2+} by countertransport with a counter pH gradient and to ppb level for Hg^{2+} by cotransport in a single short (24-48 cm long) module. The heavy metal was highly concentrated in the strip solution for recovery and recycle. The system performance was generally found to be stable and highly efficient. Results of studies on long term stability will be communicated soon.

Gas Separations. Majumdar et al. (22) have investigated separation of SO_2 from a simulated flue gas mixture of SO_2, CO_2, O_2 and N_2 through a HFCLM of pure water or 1 N $NaHSO_3$ solution in short (~ 44 cm) permeators containing 300 feed and 300 sweep hydrophobic microporous fibers (Celgard X-10, 100 μm I.D., 150 μm O.D.). Up to 95% of the SO_2 was removed using a sweep gas (He) or vacuum. Some runs involved simultaneous removal of NO in the feed gas using water as the liquid membrane and a NO-complexing agent Fe^{2+}EDTA in dilute concentrations (0.01M). Runs were carried out primarily at room temperature, but also at 70°C for as long as six days. A NO removal efficiency of 56% was achieved as SO_2 was being removed simultaneously. A simple permeation model for SO_2 employing an effective permeability of SO_2 from ILM studies (23) was able to describe the separation performance of the HFCLM permeator. The notion of an effective permeability was used since SO_2 transport through water is facilitated by the reaction $SO_2 + H_2O = HSO_3^- + H^+$. There is no effective model for facilitated transport through the complex membrane geometry of a HFCLM permeator.

In a flat membrane variation called flat-plate contained liquid membrane (FPCLM), Pakala *et al.* (*24*) studied and modeled SO_2 separation from flue gas using liquid membranes of aqueous sodium citrate or aqueous sodium sulfite solutions. The measured SO_2 fluxes were predicted well by a nonequilibrium boundary layer analysis for SO_2 transport. The SO_2 fluxes for sodium citrate films were a few times higher than that for sodium sulfite as the reagent concentration was increased form 0.0 to 0.667 M. They also studied the same system in a HFCLM module which, unfortunately, had a large EMT. Therefore, the percentage of SO_2 removal was much less than that in Majumdar *et al.* (*22*).

Recent Developments in the HFCLM Technique

Recent developments in the HFCLM technique fall under the following broad areas: Gas separation at higher pressures;
Liquid separation via a HFCLM-based pervaporation processes;
HFCLM-based perstraction processes;
HFCLM-based isomer separation processes;
HFCLM-based membrane reactors; and
HFCLM based on unequal number of fibers in each set.
These developments follow naturally from the basic configuration of a liquid membrane contained between porous polymeric membranes. Any pure liquid or liquid solution or liquid suspension can be used as the liquid membrane regardless of whether it wets the porous hollow fiber substrate (required in SLM; otherwise a complex exchange process is needed (*25-26*)) or not. Further, a variety of hollow fibers may be used to expand the capabilities of the technique. In addition, the CLM can be used as a reaction medium as well as a separating membrane. Further, the CLM may be used to separate a mixture to provide the right feed to a reaction system. All such concepts will be briefly illustrated here for a basic configuration of two microporous/porous (or nonporous) fibers next to each other with a layer of liquid in between.

Gas Separation at Higher Pressures. In the basic gas separation technique using aqueous liquid membranes and symmetric microporous hydrophobic fibers, the excess pressure of the liquid membrane over the gaseous strip/sweep/permeate streams can at the most be around 100 psi for currently available hydrophobic microprous polypropylene membranes (minimum pore size 0.03 μm) without membrane liquid breakthrough especially on the sweep side. Smaller pore sizes can enhance this value. However, a smaller pore size is also generally accompanied by a reduction in porosity and an increase in tortuosity. This substantially increases the resistance to transport.

Papadopoulos and Sirkar (*27*) employed symmetric microporous hydrophobic polypropylene hollow fibers with a thin nonporous plasma-polymerized skin of silicone on the outside surface. For species like N_2, O_2, CO_2, H_2S, SO_2 which have high permeability through a thin silicone skin, the extra skin resistance on top of the liquid membrane resistance is limited.   Yet, it eliminates liquid membrane breakthrough when P_M exceeds P_{FG} or P_{SG} by 100 psi. Unless the silicone coating is ruptured or the composite fibers break, the membrane liquid remains contained

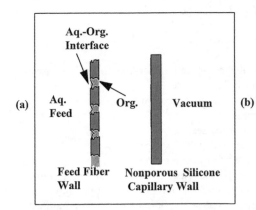

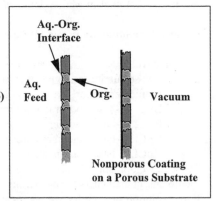

Hydrophobic Feed Fiber

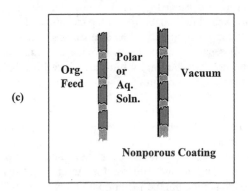

Hydrophobic or Hydrophilic Feed Fiber

Figure 4. HFCLM configurations for pervaporation separation.

between two sets of coated fibers (Figure 2b).   A variation of this employs feed fibers without any nonporous coating (Figure 2c).   It is recommended that the liquid membrane pressure remain equal to or larger than the feed gas pressures to prevent bubbling.   The premeate pressure is generally much lower than the feed side pressure.   Separation of CO_2-N_2 mixtures was achieved over a feed pressure range of 50-200 psig using a CLM of pure water or an aqueous 20 wt% DEA solution. Stronger fibers will allow higher pressure operation.

Liquid Separation via a HFCLM-based Pervaporation Processes. Pervaporation-based separation of a feed liquid mixture is usually carried out by means of a nonporous polymeric membrane subjected to vacuum on the permeate side.   The feed may be a dilute aqueous solution of volatile organic species, an organic solvent containing small amounts of water, or a liquid mixture of volatile organics (*28*). Yang *et al.* (*29*) have employed the basic HFCLM configuration of a symmetric hydrophobic microporous hollow feed fiber, an aqueous feed containing small amounts of a volatile organic compound (VOC), toluene or trichloroethylene (TCE), and an essentially nonvolatile organic solvent as the CLM.   However, they have used nonporous silicone capillaries as the second fiber set with vacuum in its bore (Figure 4a).   The silicone capillary functions as the pervaporation membrane.   Note that the organic CLM rejects water strongly so that the feed seen by the silicone capillary is substantially different from the actual feed.   The result is a much higher separation factor (*e.g.*, 11,000 for toluene over water compared to 1,600 for silicone capillary-based conventional pervaporation) and a marginally increased (by about 10%) transport resistance due to the feed fiber substrate and the HFCLM added to the feed boundary layer and the silicone membrane.   The increase in VOC selectivity is essentially due to a decrease in water flux.   Considerable improvement can be made over the observed flux values by employing a thin nonporous silicone coating on a porous substrate as the pervaporation membrane instead of the silicone capillary (Figure 4b).

This HFCLM pervaporation system (Figure 4a) was run continuously for 11 days. Performance was unchanged which suggests a stable membrane.   The liquid membrane to be used should have extremely low water solubility and be virtually nonvolatile to prevent contamination of the condensed permeate.   The solvent decanol used (*29*) is not ideal for water purification. There are many other suitable candidates.   However, the concept is demonstrated.   An area of great potential involves the use of complexing agents in such a liquid membrane and achieving facilitated transport.   An additional opportunity lies in pervaporation of an organic mixture using a polar or an aqueous solution as the CLM and an appropriate nonporous substrate for the strip fiber (Figure 4c).

HFCLM-based Perstraction Processes.   Perstraction, the combination of permeation and extraction, is selective permeation of a species from a liquid feed through a membrane into a strip or sweep liquid which selectively extracts the permeating species (Sirkar (*30*) has a brief review).   Cahn and Li (*31*) report separation of toluene and n-heptane through an aqueous ELM and kerosene as the strip liquid.   Papadopoulos and Sirkar (*32*) explored the separation of a 2 vol% isopropanol-n-heptane mixture through a highly polar liquid membrane in a HFCLM permeator in which dodecane, the high boiling strip liquid, flowed countercurrently (Figure 3a). The polar liquid membrane was pure water or 3 wt% water in sulfolane. In this HFCLM application, the feed and strip streams spontaneously wetted the pores of hydrophobic fibers. The polar liquid membrane was truly contained on the shell side between the fibers as suggested by Sengupta et al. (8). Papadopoulos and Sirkar (32) were able to selectively remove as much as 90% of the isopropanol into the sweep liquid in a 30.5 cm long permeator. The selectivity of isopropanol removal achieved with respect to n-heptane varied with the liquid membrane (447 for water). The highest $K_o a$ value obtained was 10.14 h^{-1} (compare 11.6 h^{-1} for toluene n-heptane system (31)).

HFCLM-based Isomer Separation Processes. Armstrong and Jin (33) studied liquid membrane permeation through an aqueous SLM in a cellulose filter placed in a batch cell. The feed consisted of a 50-50 mixture of organic isomeric solute systems (structural, stereoisomers, etc.) in an organic solvent with the same solvent present in the permeate side. They incorporated β-cyclodextrin (they also tested α- and γ-) in the aqueous solution to develop selectivity for one isomer over the other. Considerable selectivity was achieved initially. With time the selectivity was lost due to the inherent nature of the batch system.

Mandal et al. (in preparation) incorporated β-cyclodextrin in an aqueous solution (0.7 M β-cyclodextrin, 7.5% NaOH, 37.5% urea, pH = 12) and employed it as a CLM in a permeator containing 300 symmetric hydrophobic feed fibers and 300 similar strip fibers. The feed was an equimolar (0.005 M) mixture of structural isomers o-nitroaniline and p-nitroaniline in 80% octanol-20% heptane with the same solvent mixture as the strip liquid (Figure 3a). They obtained a selectivity of almost 5 for p-nitroaniline over the ortho isomer. The operation was stable. The longest run continued for a period of almost two days with the feed solution and the strip solvent in countercurrent flow. This successful separation indicated that employing a reflux arrangement at the end of the extraction cascade, it would be possible to continuously separate structural isomers into high purity (99%) .

In this example, the isomers had significant water solubility. Further, water has a small selectivity for p- over o-nitroaniline. Mandal et al. (in preparation) also studied HFCLM separation of stereoisomers cis- and trans-stilbene which have very low water solubility. The solubility and the observed selectivity in the β-cyclodextrin liquid was significant (selectivity of almost two for the cis- over the trans-isomer). Further, the separation was stable and continuous which opens a way for continuous separation of isomers as opposed to chromatographic separations.

HFCLM-based Membrane Reactors. A number of different types of reactor structures and reaction processes have been studied. Consider first HFCLM as a separator feeding a reactant to the permeate side. Basu and Sirkar (10) separated diltiazem (D) from an aqueous reaction mixture containing $NaHCO_3$ and NaCl through a HFCLM of decanol and fed it into an aqueous strip solution of L-malic acid to form diltiazem malate (D-M). The permeate side aqueous solution contained an overwhelming excess of L-malic acid. This solution was recirculated to build up the D-M concentration to many times over that possible in one continuous pass. A different example of HFCLM as a separator feeding a reactant is provided by

Tompkins *et al.* (*34*) who employed an organic CLM and separated aromatics from a waste stream having a high salt concentration into a salt-free bioreactor feed stream for biodegradation of the aromatic pollutant.

The second type of HFCLM-based membrane reactor is quite novel. It was developed by Guha *et al.* (*35*) for destruction of organic pollutants in wastewater with ozone employing an inert perfluorocarbon medium which has an O_3 solubility of 16 times that in water. One set of porous fibers of polytetrafluoroethylene had wastewater flowing in the fiber bore. The other set of nonporous fibers were silicone capillaries through which an ozonated oxygen stream flowed. The shell side was filled with the perfluorocarbon medium. This medium was supplied with ozone from one side and from the other side pollutants (the pollutants were partitioned from wastewater (and concentrated for hydrophobic species) into this reaction medium). The pollutants were degraded. The degradation products were partitioned into the wastewater as well as the gaseous stream. The perfluorocarbon reaction medium also serves to limit the amount of ozone (if any) leaking into the wastewater just as it prevented VOCs from leaking into the gaseous stream. Thus, the CLM performs the following functions: partitioning and concentration of individual reactants into it as a reaction medium; partitioning of reaction products into the two flowing streams; selective membrane to prevent leakage of both reactants to the other stream.

HFCLM Based on Unequal Number of Fibers in Each Set. In the basic HFCLM technique, the surface area of the feed fiber set is essentially equal to that of the strip fiber set. In some membrane permeation processes, the feed side boundary layer resistance and/or the feed-liquid membrane interfacial reaction resistance controls the transport rate; in others the corresponding quantities on the strip side controls. Since mass transfer rates are proportional to the product of the mass transfer coefficient and the interfacial area, one way to increase the transfer rate is to increase the interfacial area for the controlling resistance. Guha *et al.* (*21*) built permeators with one set having twice the number of fibers as the other set. For the countertransport removal of Cu^{2+} from a wastewater and its concentration in the strip solution where the feed side boundary layer and the interfacial reaction were found to control the transport rate, they demonstrated that the mass transport rate was considerably higher when the feed/strip fiber ratio was 2 instead of 1 or 1/2. Although such a configuration may impose constraints for fabrication, it opens up an additional dimension to HFCLM performance.

Perspectives on the HFCLM Technique

The versatile HFCLM technique can perform a variety of separations for gas mixtures and liquid solutions. It has been found to be stable for the gas and liquid separations studied. Larger modules (2" diameter and a total of 15000 fibers) have been built. Production methods appear to be in place to achieve EMT values close to the theoretical limits (*11*). The liquid membrane may be organic or aqueous and the feed and strip phases may be organic, aqueous or gaseous.

Several aspects of HFCLM-based separations need further development. The ability to make larger modules (4", 8" diameter) with many more fibers and low

EMT values has to be demonstrated. To reduce the change in permeation rates over very long terms (four months or longer) for multicomponent liquid membranes, we need to understand the relation between the liquid membrane composition profile along the permeator, the external membrane reservoir composition and the shell side liquid membrane entry locations. For mass transfer rate maximization, one has to optimize fiber types (asymmetric, hydrophilic, *etc.*) and the numbers for each set of fibers (*21*). The latter will complicate the module making process. Asymmetric fibers as well as hydrophilic fibers are especially relevant because of mass transfer advantages that can be achieved. For liquid separations with hydrophilic fibers, a particular configuration, namely organic liquid in pores (Figure 3c) has not been explored. Prasad and Sirkar (*36*) have pointed out that aqueous organic interfaces can be immobilized at the membrane pores with the organic phase in hydrophilic membrane pores if the external aqueous phase is at a higher pressure. For aqueous feed systems where hydrophilic fibers are favored for other reasons (*e.g.*, enzyme immobilization) and the organic phase favors the solute, the first set of fibers should then be hydrophilic with an organic phase in the pores. Correspondingly, the strip fiber should be hydrophobic with an organic phase in the pores to reduce strip side resistance for the same solute. For higher pressures (> 200 psig) in gas separations, smaller and stronger fibers are needed. Smaller fibers will also yield lower EMT. Although a variety of driving forces have been employed, temperature gradient as a driving force between the two sets of fibers has not yet been studied. The critical question will be the degree of thermal transport through the liquid membrane along the permeator length and the consequent driving force reduction due to change in the temperatures of the two flowing streams.

Acknowledgments

The contributions of R. Basu, J. S. Cha, A. K. Guha, S. Majumdar, D. Mandal, T. H. Papadopoulos, R. Prasad, A. Sengupta, D. Yang, Z-F. Yang and C. H. Yun are acknowledged. Support for research was obtained from DOE PETC, ERRC at NJIT, EPA NHSRC, GRI, Hoechst Celanese Corp., HSMRC at NJIT, Merck & Co., NYGAS and NYSERDA. A. S. Kovvali drew Figures 2 to 4.

Literature Cited

1.   Ho, W.S. Winston; Li, N.N. In *Membrane Handbook*; Ho, W.S.; Sirkar, K.K., Eds.; Chapman and Hall: New York, NY, 1992; pp. 597-655.

2.   Majumdar, S.; Sirkar, K.K.; Sengupta, A. In *Membrane Handbook*; Ho, W.S.; Sirkar, K.K., Eds.; Chapman and Hall: New York, NY, 1992; pp. 764-808.

3.   Boyadzhiev, L.; Lazarova, Z. In *Membrane Separations Technology. Principles and Applications*; Noble, R.D.; Stern, S.A., Eds.; Elsevier: Amsterdam, 1995; pp. 283-352.

4.   Majumdar, S.; Guha, A.K.; Sirkar, K.K. *AIChE J.* **1988**, *34*, 1135-1145.

5.   Sengupta, A.; Basu, R.; Sirkar, K.K. *AIChE J.* **1988**, *34*, 1698-1708.

6.   Guha, A.K.; Majumdar, S.; Sirkar, K.K. *J. Membr. Sci.* **1991**, *62*, 293-307.

7.   Guha, A.K.; Majumdar, S.; Sirkar, K.K. *I&EC Res.* **1992**, *31*, 593-604.

8.   Sengupta, A.; Basu, R.; Prasad, R.; Sirkar, K.K. *Sep. Sci. Technol.* **1988**, *23*, 1735-1751.
9.   Majumdar, S; Guha, A.K.; Lee, Y.T.; Sirkar, K.K. *J. Membr. Sci.* **1989**, *43*, 259-276.
10. Basu, R.; Sirkar, K.K. *J. Membr. Sci.* **1992**, *75*, 131-149.
11. Carroll, R.H.; Barber, T.A.; Reed, B.W. *US Patent No.* 5,169,529, 1992.
12. Majumdar, S.; Guha, A.K.; Lee, Y.T.; Papadopoulos, T.H.; Khare, S.; Sirkar, K.K. *Liquid Membrane Purification of Biogas*, Report 91-9. New York State Energy Research and Development Authority: Albany, NY, 1991.
13. Ho, S.V. In *Industrial Envirnomental Chemistry*; Sawyer, D.T.; Martell, A.E., Eds.; Plenum Press: New York, NY, 1992, pp. 229-245.
14. Soreson, B.V.; Callahan, R.W. *Proc. International Congress on Membranes and Membrane Processes*; NAMS; Chicago, IL, August 20-24, 1990, Vol. 1, pp. 695-697.
15. Basu, R; Sirkar, K.K. *AIChE J.* , **1991**, *37*, 383-393.
16. Yang, Z.F.; Rindfleisch, D.; Scheper, T.; Schügerl, K. Proc. 3rd International Conference on Effective Membrane Processes, New Perspectives; BHR Group: Bath, UK, May 12-14, 1993, pp. 49-57.
17. Nguyen, K.V.; Callahan, R.W. *Polym. Mat. Sci. Eng. ACS Proceedings*, Miami, FL, 1989, 61.
18. Lamb, J.D.; Bruening, R.L.; Linsley, D.A.; Smith, C.; Izatt, R.M. *Sep. Sci. Technol.*, **1990**, *25*, 1407-1419.
19. Izatt, R.M.; Lamb, J.D.; Bruening, R.L.; Wang, C.; Edge, N.; Bradshaw, J.S. *Sep. Sci. Technol.*, **1993**, *28*, 383-395.
20. Sato, Y.; Kondo, K.; Nakashio, F. *J. Chem. Eng. Japan*, **1990**, *23*, 23-29.
21. Guha, A.K.; Yun, C.H.; Basu, R.; Sirkar, K.K. *AIChE J.*, **1994**, *40*, 1223-1237.
22. Majumdar, S.; Sengupta, A.; Cha, J.S.; Sirkar, K.K. *I&EC Res.*, **1994**, *33*, 667-675.
23. Sengupta, A.; Raghuraman, B.; Sirkar, K.K. *J. Membr. Sci.*, **1990**, *51*, 105-126.
24. Pakala, N.R.; Varanasi, S.; LeBlanc, S.E. *I&EC Res.*, **1993**, *32*, 553-563.
25. Bhave, R.R.; Sirkar, K.K. *J. Membr. Sci.*, **1986**, *27*, 41-61.
26. Bhave, R.R.; Sirkar, K.K. In *Liquid Membranes Theory and Applications*; Noble, R.D.; Way, J.D., Eds.; ACS Symposium Series No. 347; American Chemical Society: Washington, DC, 1987, pp. 138-151.
27. Papadopoulos, T.H.; Sirkar, K.K. *J. Membr. Sci.*, **1994**, *94*, 163-181.
28. Fleming, H.L.; Slater, C.S. In *Membrane Handbook*; Ho, W.S.; Sirkar, K.K., Eds.; Chapman and Hall: New York, NY,1992, pp. 105-159.
29. Yang, D.; Majumdar, S.; Kovenklioglu, S.; Sirkar, K.K. *J. Membr. Sci.*, **1995**, *103*, 195-210.
30. Sirkar, K.K. In *Membrane Handbook*; Ho, W.S.; Sirkar, K.K., Eds.; Chapman and Hall: New York, NY, 1992, pp. 904-912.
31. Cahn, R.P.; Li, N.N. In *Membrane Separation Processes*; Meares, P., Ed.; Elsevier: Amsterdam, 1976, 327-349.
32. Papadopoulos, T.H.; Sirkar, K.K. *I&EC Res.*, **1993**, *32*, 663-673.
33. Armstrong, D.W.; Jin, H.L *Anal. Chem.*, **1987**, *59*, 2237-2241.

34.   Tompkins, C.J.; Michaels, A.S.; Peretti, S.W. AIChE Annual Meeting Abstracts, San Francisco, CA, November 14, 1994, Paper No. 213b.
35.   Guha, A.K.; Shanbhag, P.V.; Sirkar, K.K.; Vaccari, D.; Trivedi, D. *AIChE J.*, **1995**, *41*, 1998-2012.
36.   Prasad, R.; Sirkar, K.K. In *Membrane Handbook*; Ho, W.S.; Sirkar, K.K., Eds.; Chapman and Hall: New York, NY, 1992, pp. 726-763.

Chapter 17

Facilitated Transport of Carbon Dioxide Through Supported Liquid Membranes of Aqueous Amine Solutions

Masaaki Teramoto, Qingfa Huang, and Hideto Matsuyama

Department of Chemistry and Materials Technology,
Faculty of Engineering and Design, Kyoto Institute of Technology,
Matsugasaki, Sakyo-ku, Kyoto 606, Japan

A series of experiments on the facilitated transport of CO_2 through supported liquid membranes containing various amines such as monoethanolamine (MEA), diethanolamine (DEA) and ethylenediamine hydrochloride (EDAHCl) was performed. The feed gas was a mixture of CO_2 and CH_4. Permeation rate of CO_2 through these membranes is in the order, MEA<DEA<EDAHCl. This tendency is discussed quantitatively on the basis of the facilitated transport theory using the physicochemical properties of amines such as the reaction rate constant and chemical equilibrium constant, K_{eq}. It was confirmed that too high value of K_{eq} lowers the permeation rate of CO_2. If $\tau^{1/2}L$ where L is the membrane thickness and τ is the tortuosity factor of the microporous support membrane is used as the effective diffusional path length, experimentally observed permeation rates of CO_2 can be simulated by the facilitated transport theory. A method for estimating the solubilities of CO_2 in the membrane solutions from the permeation rates of CH_4 is also proposed.

Recently, the application of membrane separation techniques to the separation of CO_2 has been attracting attention due to the low energy consumption compared to traditional separation methods such as gas absorption and adsorption. Although many polymeric membranes have been developed for CO_2 separation, the separation factor of CO_2 over N_2 and CH_4 is lower than about 60 (*1*).

To overcome the problem of low selectivity, use of facilitated transport membranes has been proposed (*2,3*) and many works on the CO_2 facilitated transport were well summarized in review articles (*4,5*). As the amine type carriers of CO_2, monoethanolamine (MEA) (*6-8*), diethanolamine (DEA) (*7,9,10*) and triethanolamine (*7*) have been investigated. Most of the data were obtained by the "tracer transport experiments" where the tracer flux of $^{14}CO_2$ through the membrane of aqueous amine solution which had been equilibrated with a known partial pressure of CO_2 was measured (*6,7*). However, very few quantitative studies have been performed for the "net transport experiments" where the partial pressure of CO2 in the downstream side was kept lower than that in the upstream side (*9,10*).

0097–6156/96/0642–0239$15.00/0
© 1996 American Chemical Society

In this study, experiments on the simultaneous permeation of CO_2 and CH_4 through aqueous MEA, DEA and ethylenediamine hydrochloride (EDAHCl) membranes are performed and the permeation rates of CO_2 are quantitatively discussed on the basis of the approximate solution of facilitated transport (11). A method is also proposed for applying the facilitated transport theory developed for a liquid membrane consisting of a liquid membrane phase alone to the analysis of facilitated transport through a supported liquid membrane prepared by impregnating a microporous polymer support having tortuous pore structure with a carrier solution.

The Approximate Solution of Facilitation Factors

Reaction Kinetics and Basic Equations. The reaction of CO_2 with primary and secondary amines RR'NH (R, R'; functional group or hydrogen), such as MEA and DEA, is expressed as follows (12):

$$CO_2(A) + 2RR'NH(B) = RR'NCOO^-(E) + RR'NH_2^+(F) \qquad (a)$$

The reaction rate is generally expressed by Eq.(1) (12).

$$r = -(dC_A/dt) = k_1\{C_A C_B - C_E C_F/(K_{eq}C_B)\}/\{1+k_2/(k_3 C_B)\} \qquad (1)$$

For the CO_2-MEA system, the relation $k_2/(k_3 C_B)<<1$ holds while for the CO_2-DEA system the term $k_2/(k_3 C_B)$ cannot be neglected when C_B is low. The differential mass balance equations in the membrane and the boundary conditions are expressed as follows:

$$D_A d^2 C_A/dx^2 = r \quad (2), \qquad\qquad D_B d^2 C_B/dx^2 = 2r \qquad (3)$$

$$D_E d^2 C_E/dx^2 = -r \quad (4), \qquad\qquad D_F d^2 C_F/dx^2 = -r \qquad (5)$$

$$x = 0: \ C_A = C_{A0}, dC_B/dx = dC_E/dx = dC_F/dx = 0 \qquad (6)$$

$$x = L: \ C_A = C_{AL}, dC_B/dx = dC_E/dx = dC_F/dx = 0 \qquad (7)$$

The conservation of the carrier B in the membrane is expressed as

$$\int_0^L (C_B + C_E + C_F)dx = C_{BT}L \qquad (8)$$

The facilitation factor F is defined as the ratio of the permeation flux in the presence of carrier to that of physical permeation in the absence of carrier, and is expressed as follows:

$$F = \{-D_A(dC_A/dx)_{x=0,L}\}/(D_A C_{A0}/L) = -(da/dy)_{y=0,1} \qquad (9)$$

The Approximate Solution. Recently, Teramoto (13) developed an approximate solution for the facilitation factor in facilitated transport membranes where a reaction A(permeant) + B(carrier) = C(complex) occurs. Very recently, this approximation method was extended to the facilitated transport system where reaction (a) occurs in the membrane (11). It was confirmed that this approximation method provides sufficiently accurate facilitation factors over the entire range from the physical diffusion region to the chemical equilibrium region.

According to this approximation method, the facilitation factor F can be calculated by solving the following simultaneous algebraic equations.

$$F = \gamma_0 \frac{(1+\dfrac{f_0}{r_E q K b_0^2})(1-a_L)+(\cosh\gamma_0 -1)(1-\dfrac{e_0 f_0}{K b_0^2})}{\sinh\gamma_0 + \gamma_0 f_0/(r_E q K b_0^2)}$$

$$= \gamma_L \frac{(1+\dfrac{f_L}{r_E q K b_L^2})(1-a_L)+(\cosh\gamma_L-1)(\dfrac{e_L f_L}{K b_L^2} - a_L)}{\sinh\gamma_L + \gamma_L f_L/(r_E q K b_L^2)}$$

$$= 1-a_L-q(b_0-b_L)/2 = 1-a_L+r_E q(e_0-e_L) = 1-a_L+r_F q(f_0-f_L) \qquad (10)$$

$$(b_0+b_L+e_0+e_L+f_0+f_L)/2 = 1 \qquad (11)$$

$$e_0 = f_0, \qquad (12), \qquad e_L = f_L \qquad (13)$$

Equations 12 and 13 represent electrical neutrality in the membrane. Here, a_L, b_0, b_L, e_0, e_L, f_0, f_L are the dimensionless concentrations at the boundaries of the membrane, and γ_0 and γ_L are defined as follows:

$$\gamma_0 = \delta[\{b_0+(f_0/r_E q K b_0)\}/(1+m/b_0)]^{1/2} \qquad (14)$$

$$\gamma_L = \delta[\{b_L+(f_L/r_E q K b_L)\}/(1+m/b_L)]^{1/2} \qquad (15)$$

where δ is defined as

$$\delta = L(k_1 C_{BT}/D_A)^{1/2} \qquad (16)$$

The six unknown concentrations at the two boundaries of the membrane can be determined from Equations 10-15 for the given values of dimensionless parameters $a_L = C_{AL}/C_{A0}$, $K = K_{eq}C_{A0}$, $m = k_2/(k_3 C_{BT})$, $q = D_B C_{BT}/(D_A C_{A0})$, $r_E = D_E/D_B$, $r_F = D_F/D_B$ and $\delta = L(k_1 C_{BT}/D_A)^{1/2}$. Then the facilitation factor F can be calculated from these concentrations.

Application of the Approximate Solution to Supported Liquid Membranes. The above approximate solution was developed for a facilitated transport membrane comprised of a liquid phase alone. This solution can be easily applied to the analysis of the facilitated transport through a supported liquid membrane consisting of a microporous polymer support and a liquid membrane phase constrained in the pores of the support. In this case, the differential mass balance equation for A is expressed as follows:

$$D_{eA} d^2 C_A/dx^2 = \varepsilon r \qquad (17)$$

Here ε is the porosity of the support membrane and D_{eA} is the effective diffusivity defined by the following equation.

$$J_A = -D_{eA}(dC_A/dx) \tag{18}$$

The effective diffusivity is related to the molecular diffusivity D_A using the porosity ε and the tortuosity factor τ of the support membrane as follows:

$$D_{eA} = D_A \varepsilon/\tau \tag{19}$$

Equation 19 is considered as the definition of τ. Then, the following dimensionless equation is obtained from Equation 17.

$$d^2a/dy^2 = \tau L^2(k_1 C_{BT}/D_A) \{ab - ef/(Kb)\}/(1 + m/b)$$

$$= \tau\delta^2\{ab - ef/(Kb)\}/(1 + m/b) \tag{20}$$

The dimensionless form of Equation 2 is expressed by the following equation.

$$d^2a/dy^2 = L^2(k_1 C_{BT}/D_A) \{ab - ef/(Kb)\}/(1 + m/b)$$

$$= \delta^2\{ab - ef/(Kb)\}/(1 + m/b) \tag{2'}$$

Comparison of Equation 20 with Equation 2' suggests that instead of the parameter δ defined by Equation 16, the parameter δ' defined by

$$\delta' = \tau^{1/2} L(k_1 C_{BT}/D_A)^{1/2} \tag{21}$$

should be used in calculating the facilitation factor by use of Equations 10-15. This indicates that the effective diffusional path length L_e is expressed by the following equation using the membrane thickness L and τ.

$$L_e = \tau^{1/2}L \tag{22}$$

Then, the permeation flux of A is expressed by the following equation.

$$J_A = -D_{eA}(dC_A/dx)_{x=0,L} = -(\varepsilon/\tau)(D_A C_{A0}/L)(da/dy)_{y=0,1} = -(\varepsilon/\tau)(D_A C_{A0}/L)F \tag{23}$$

Here, F is the facilitation factor corresponding to the value of δ'.

The same equation can be derived on the basis of the parallel pore model which pictures transport as occurring through a number of parallel capillaries of the same size (*14*). The detailed derivation for this is described elsewhere (Teramoto, M. et al., *Ind. Eng. Chem. Res.* in press.).

It is noted that when the permeation occurs in the physical diffusion region ($\delta = 0$) or in the chemical equilibrium region ($\delta = \infty$), the facilitation factor F does not depend on δ or δ'. In other words, F is independent of the effective diffusional path length in these regions.

Experimental

The experimental apparatus employed in this study was the same as used in a previous study (*15*). The permeation cell consisted of two compartments for a feed and a sweep gas (thickness, 3.7mm; membrane area, 7.92cm^2). Two types of microporous membrane, i.e., Durapore® VVLP (Millipore Ltd.) and H010A membranes (Advantec Co., Ltd.) were used as the supports (Table I). The thickness of the membrane was measured with a micrometer.   The porosity of each membrane was obtained from the weight of water in the pores and the volume of the membrane (= thickness x area).   MEA, DEA and EDA were purchased from Wako Pure Chemical Industries, Ltd., Osaka, Japan.   EDAHCl was prepared by adding HCl to equimolar EDA.

The support membrane was soaked in an aqueous amine solution for more than 2 hours.   After the solution on the surface of the support was blotted by a cellulose filter paper, the membrane was used for permeation experiment.   The feed gas was a mixture of CO_2 and CH_4 and the sweep gas was helium.   Both gas streams were supplied to the permeation cell at atmospheric pressure after presaturation with water. The partial pressure of CO_2 in the feed gas was changed from 0.046 to 0.97 atm and the flow rate of the feed gas was in the range from 200 to 300 cm^3/min and that of the sweep gas was 50 cm^3/min.   The partial pressure of CO_2 in the sweep gas was less than 2 % of that in the feed gas.   The sweep gas from the cell was analyzed by a gas chromatograph equipped with a thermal conductivity detector (Shimadzu, GC-8APT, column: activated carbon).   The permeation rates of CO_2 and CH_4 were obtained from the partial pressures in the feed and the sweep gases and the sweep gas flow rate.   The temperature was in the range from 298 K to 318 K.

Table I   Properties of the Support Membranes

membrane	pore diameter/µm	thickness/µm	porosity (ε)	tortuosity (τ)
VVLP [a]	0.1[d]	100[c]	0.63[c]	2.91[c]
H010A [b]	0.1[d]	30[c]	0.71[d]	1.95[c]

[a]Millipore Ltd., Durapore® VVLP hydrophilic membrane, poly(vinylidene difluoride). [b]Advantec Co., Ltd., hydrophilic poly(tetrafluoroethylene). [c]Measured in this study. [d]Reported by the supplier.

Results

Determination of the Tortuosity Factor of the the Support Membrane.   The tortuosity factors of the support membranes were determined from the permeation rate of pure CO_2 through the supported liquid membranes of pure water at 298 K. The permeation flux J_{CO2} is expressed by the following equation.

$$J_{CO2} = \varepsilon D_{CO2,w} H_{CO2,w}(p_{CO2,F}-p_{CO2,S})/(\tau L) \tag{24}$$

Here, $D_{CO2,w}$ is the diffusivity of CO_2 in water and $H_{CO2,w}$ is the Henry constant of CO_2-water system.   They are correlated as follows (*16*):

$$D_{CO2,w} = 2.35 \times 10^{-6} \exp(-2119/T) \ (m^2/s) \tag{25}$$

$$H_{CO2,w} = 3.59 \times 10^{-5} \exp(2044/T) \ mol/(dm^3 atm) \tag{26}$$

The values of τ are listed in Table I.   Similar values of τ were obtained utilizing the experiment on the permeation of CH_4 through water impregnated membranes.

Permeation Rates of CO2 and CH$_4$ through Aqueous MEA, DEA and EDAHCl Membranes.

MEA Membranes.   Figure 1 shows the effects of the partial pressure of CO_2 in the feed gas, $p_{CO2,F}$, and the MEA concentration on the permeation rates of CO_2 and CH_4 and the separation factor α (CO_2 /CH_4), the ratio of R_{CO2} to R_{CH4}.   It is seen that for each MEA concentration the permeation rate of CH_4 which is transported by the simple solution-diffusion mechanism is almost constant irrespective of $p_{CH4,F}$ (=

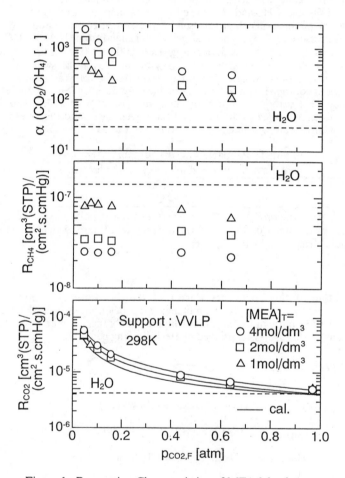

Figure 1. Permeation Characteristics of MEA Membranes.

1-water vapor pressure (0.03 atm) - $p_{CO2,F}$). It is also seen that R_{CH4} decreases considerably as the MEA concentration increases. One reason for this is that with increasing MEA concentration, the diffusivity of CH_4 in the membrane solution decreases due to the increase in the viscosity. Another reason is that as the MEA concentration increases, the solubility of CH_4 in the membrane solution decreases since the ionic strength of the membrane solution is enhanced with an increase in MEA concentration due to the formation of ionic species, e.g., carbamate ion and protonated MEA, as will be described later. On the other hand, the permeation rate of CO_2 decreases with an increase in $p_{CO2,F}$ since CO_2 is transported by the carrier-mediated transport mechanism. R_{CO2} does not increase appreciably with increasing the MEA concentration because the favorable facilitation effect of high carrier concentration is compensated by the decreases in both the solubility of CO_2 and the diffusivities of the chemical species in the solution of high MEA concentration. Therefore, a higher separation factor $\alpha(CO_2/CH_4)$ was obtained at lower CO_2 partial pressure and higher MEA concentration. The value of α at $[MEA]_T = 4$ mol/dm³ and $p_{CO2,F} = 0.046$ atm was about 2000.

The separation factor for pure water membrane was 29.9. This value agreed with 29.0 which was calculated from the solubilities and the diffusivities of these gases in water ($D_{CH4,W} = 1.70 \times 10^{-5}$ cm²/s (*17*), $H_{CH4,W} = 1.34 \times 10^{-3}$ mol/(dm³atm) (*18*)).

$$\alpha(CO_2/CH_4)_W = (H_{CO2,W}D_{CO2,W})/(H_{CH4,W}D_{CH4,W}) \qquad (27)$$

DEA and EDAHCl Membranes. The effect of $p_{CO2,F}$ on R_{CO2}, R_{CH4} and $\alpha(CO_2/CH_4)$ obtained with the DEA and EDAHCl membranes is shown in Figure 2. The data obtained with the MEA membranes are also shown for comparison. It is seen that R_{CO2} increases in the order of MEA<DEA<EDAHCl.

Effect of Temperature. The effect of temperature on the permeation rate of CO_2 through the MEA membrane is shown in Figure 3. It is seen that R_{CO2} is higher at higher temperature. R_{CO2} at 318 K is about three times that at 298 K. This is because the increases in both the reaction rate and diffusivity with temperature have a favorable effect on R_{CO2} although the solubility of CO_2 is lower at higher temperature. Another reason is that the chemical equilibrium constant K_{eq} decreases with temperature and this makes the release of CO_2 at the sweep side of the membrane faster as will be discussed later.

Effect of Membrane Thickness. Permeation rates were measured through a single membrane, a two-membrane laminate and a three-membrane laminate. H010A membranes were used as the support membrane. The effect of total membrane thickness on the permeation rate of CO_2 is shown in Figure 4. It is seen that as anticipated, the permeation rate decreases with increasing the membrane thickness.

Simulation of the Permeation Rate of CO_2

Physicochemical Properties of Facilitated Transport Systems. Physicochemical properties of the amines studied are listed in Table II. The diffusivities in the liquid membrane solutions of various amine concentrations were estimated from the values in water using the following equation (*19*).

$$D_{i,M}/D_{i,W} = (\mu_M/\mu_W)^{-2/3} \qquad (28)$$

The viscosity of the membrane solution in the pores of the support, μ_M, cannot be measured. Here, the viscosities of amine solutions saturated with pure CO_2 at 1atm were measured with an Ostwalds viscometer and these values were taken as the viscosities of the membrane solutions. The viscosities are listed in Table III. This

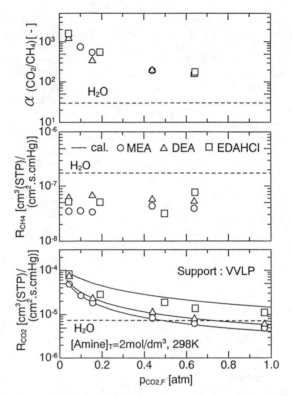

Figure 2. Comparison of Permeation Characteristics of MEA, DEA and EDAHCl Membranes.

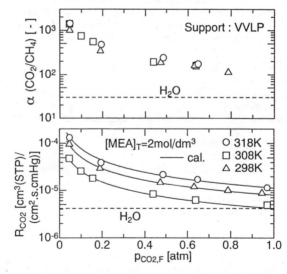

Figure 3. Effect of Temperature (MEA Membrane).

approximation may be reasonable for the MEA and DEA membranes since in the simulation of the permeation rates of CO_2 through these membranes, the average amine concentration in the membrane calculated by the present approximation method was found to be less than 16% of the initial amine concentration C_{BT}. As shown in Table III, these viscosities are higher than μ_{Am}'s, the corresponding values of the aqueous amine solutions. The diffusivities of carbamate ion and protonated amine were assumed to be equal to that of the amine.

Table II   Properties of MEA, DEA and EDAHCl

amine	T/K	$k_1/M^{-1}s^{-1}$	$k_2/k_3/M^{-1}$	K_{eq}/M^{-1}	D_W/cm^2s^{-1}
MEA	298	5.92×10^3 [a]	0	1.75×10^5 [d]	1.16×10^{-5} [f]
MEA	303	1.01×10^5 [a]	0	7.53×10^5 [d]	1.49×10^{-5} [g]
MEA	313	1.68×10^5 [a]	0	3.20×10^5 [d]	1.75×10^{-5} [g]
DEA	298	1.41×10^3 [b]	1.18 [b]	4.52×10^3 [d]	0.82×10^{-5} [f]
EDAHCl	298	1.35×10^3 [c]	0	1.38×10^3 [e]	1.18×10^{-5} [h]

[a]Reference *21*. [b]Reference *22*. [c]Measured by a chemical absorption method using semi-continuous stirred vessel with a flat gas-liquid interface. [d]Reference *23*. [e]Reference *24*. [f]Reference *19*. [g]Estimated from the value at 298 K (*19*) using the Wilke-Chang equation. [h]Estimated by the Wilke-Chang equation.

Table III   Estimation of the Solubility of CO_2 in the Membrane Solutions

amine	T/K	C_{BT}/M	$R_{CH4,M}/R_{CH4,W}$	μ_{Am}/μ_W	μ_M/μ_W	$H_{CO2,M}/H_{CO2,W}$
MEA	298	1.0	0.55	1.19	1.29	0.66
MEA	298	2.0	0.27	1.42	1.67	0.41
MEA	298	4.0	0.17	2.14	2.98	0.43
MEA	303	2.0	0.46	1.39	1.66	0.71
MEA	313	2.0	0.50	1.67	1.76	0.80
DEA	298	1.0	0.59	1.42	1.50	0.81
DEA	298	2.0	0.42	2.12	2.39	0.83
DEA	298	4.0	0.20	5.73	7.80	0.94
EDAHCl	298	2.0	0.31	1.40	1.50	0.44

Solubilities of CO_2 in Membrane Solutions.   The solubility of CO_2 in the liquid membrane solutions was estimated as follows. Since CH_4 permeates through the membrane by a simple solution-diffusion mechanism, its permeation rate is proportional to both the solubility and the diffusivity in the membrane solution. Therefore, the following equation holds for the ratio of the permeation rate through the aqueous membrane solution to that through a pure water membrane.

$$R_{CH4,M}/R_{CH4,W} = (H_{CH4,M}/H_{CH4,W})(D_{CH4,M}/D_{CH4,W})$$

$$= (H_{CH4,M}/H_{CH4,W})(\mu_M/\mu_W)^{-2/3} \qquad (29)$$

Then, the solubility ratio $H_{CH4,M}/H_{CH4,W}$ can be calculated from the experimantally measured permeation rates of CH_4 and the viscosities of the aqueous membrane solutions using Equation 29.   It should be noted that $H_{CH4,M}$ is not the Henry constant of CH_4 for the aqueous amine solution but that for the aqueous membrane solution which is in contact with CO_2 and contains unreacted amine and reaction products, i.e., carbamate ion and protonated amine.

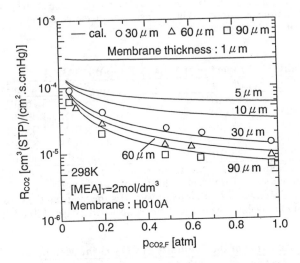

Figure 4. Effect of Membrane Thickness (MEA Membrane).

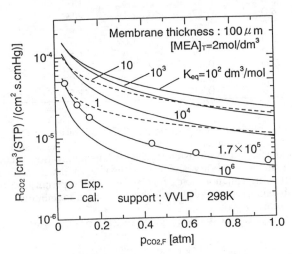

Figure 5. Effect of K_{eq}. The physicochemical parameters except K_{eq} are the same as those of MEA.

The solubility ratio $H_{CO2,M}/H_{CO2,W}$ is not necessarily equal to $H_{CH4,M}/H_{CH4,W}$. The solubilities of gases in electrolyte solutions can be estimated by the following equation (*20*).

$$\log(H/H_W) = -(i_g+i_++i_-)I \qquad (30)$$

Here, i_g , i_+ and i_- are the constants specific to the gaseous, cationic and anionic species, respectively, and I is the ionic strength. Since these values for the present system are not available, they were estimated from the solubility data of CO_2 and CH_4 in aqueous NaCl and KI solutions (*18*). By use of these data, the following relation was obtained:

$$i_{CO2} = i_{CH4} - 0.041I \qquad (31)$$

Then, the following relation holds.

$$H_{CO2,M}/H_{CO2,W} = (H_{CH4,M}/H_{CH4,W})10^{0.041I} \qquad (32)$$

From Equations 29 and 32, we obtain

$$H_{CO2,M}/H_{CO2,W} = (\mu_M/\mu_W)^{2/3}(R_{CH4,M}/R_{CH4,W}) \, 10^{0.41I} \qquad (33)$$

It was found in the calculation of the facilitation factor of CO_2 for the MEA membranes that more than 84% of the MEA initially incorporated in the membrane was converted to the ionic species, i.e., carbamate and protonated amine. The corresponding values for the DEA and EDAHCl membranes are 77 and 58%, respectively. Therefore, for the MEA and DEA membranes, the ionic strength of the membrane solution in contact with CO2 was approximately half the initial amine concentration and from the observed values of $R_{CH4,M}/R_{CH4,W}$ and the viscosity ratios, the values of $H_{CO2,M}/H_{CO2,W}$ were estimated by Equation 33 and are shown in Table III. Although Equation 33 does not hold for the EDAHCl membrane, this equation was used for the rough estimation of CO_2 solubility in the membrane solution.

Comparison of Experimental Results with Calculated Results. The calculated results for R_{CO2} obtained with the present approximation method are shown by the solid lines in Figures 1-4. It can be generally seen that the experimental data are satisfactorily simulated by the approximate solution.

As shown in Figure 2, R_{CO2} through the MEA membrane is lowest in spite of the fact that the reaction rate of CO_2 with MEA is faster than those with other amines (Table II). The reason for the low R_{CO2} through the MEA membrane is a too large value of K_{eq} which makes the stripping rate of CO_2 at the sweep side very slow.

To investigate the effect of K_{eq} on R_{CO2} , some hypothetical calculations were made for various values of K_{eq}. For the other physicochemical parameters, such as the reaction rate constant and diffusivities, the values for the CO_2-MEA system were used. As shown in Figure 5, the optimal value of K_{eq} which gives the highest permeation rate is about 100, suggesting that the value of K_{eq} for MEA is too large.

The reaction rate constant for the CO_2-EDAHCl system is the lowest among the three amine membranes (Table II). However, a rather small value of K_{eq} of this system (1380 dm³/mol) is favorable for obtaining a high permeation rate.

To examine the effect of membrane thickness, a hypothetical calculation was performed for the MEA membranes of various thickness and the calculated results are shown in Figure 4. It is seen that as the thickness decreases, R_{CO2} increases. It is also seen that when L = 1μm, R_{CO2} is almost independent of the CO_2 partial pressure since the CO_2 permeates through the thin membrane in the region of physical permeation (facilitation factor $F = 1$).   Therefore, when a very thin membrane is employed, an amine with much higher reactivity for CO_2 is necessary to obtain a high permeation rate.

Conclusion

Experiments on the separation of CO_2 from CH_4 by the supported liquid membranes containing aqueous amines such as monoethanolamine, diethanolamine and ethylenediamine hydrochloride were performed, and the data were discussed quantitatively on the basis of facilitated transport theory. The effects of the chemical properties of amines such as the reaction rate constant and chemical equilibrium constant and also the effect of the CO_2 partial pressure on the permeation rate of CO_2 could be interpreted by the proposed theory.   It was proposed to use $\tau^{1/2} L$ as the effective diffusional path length in the calculation of the facilitation factor, where L is the membrane thickness and τ is the tortuosity factor of the microporous support membrane. The permeation rates of CH_4 was found to provide useful information for evaluating the solubilities of CO_2 in the reactive membrane solutions.

Nomenclature

a C_A/C_{A0}
b C_B/C_{BT}
C   concentration (mol/m^3)
C_{BT} total carrier concentration (mol/m^3)
D   diffusivity (m^2/s)
D_e effective diffusivity (m^2/s)
e C_E/C_{BT}
F   facilitation factor
f C_F/C_{BT}
H   Henry constant $(mol/(m^3atm))$
I   ionic strength (mol/dm^3)
i_g, i_+, i_- salting-out parameter (dm^3/mol)
J   permeation flux $(mol/(m^2 s))$
K $K_{eq}C_{A0}$
K_{eq} equilibrium constant of reaction (a) (m^3/mol)
k_1 forward reaction rate constant $(m^3/(mol s))$
k_2/k_3 reaction rate parameter (mol/m^3)
L   membrane thickness (m)
L_e effective diffusional path length   (m)
m $k_2/(k_3C_{BT})$
p   partial pressure (atm)
q $(D_B/D_A)(C_{BT}/C_{A0})$
R   permeation rate $(cm^3(STP)/(cm^2 s\ cmHg))$
r   reaction rate $(mol/(m^3s))$
r_J D_J/D_B $(J = E, F)$
x   distance (m)
y   x/L
Subscripts
A CO_2
Am   aqueous amine solution
B   amine (MEA or DEA)
E   carbamate ion $(RR'NCOO^-)$
F   protonated amine $(RR'NH_2^+)$ or feed gas

L   value at x = L(receiving side interface)
M   membrane solution in contact with CO_2
S   sweep gas (receiving side)
T   total value
w   water
0   value at x = 0 (feed side interface)
Greek letters

γ_0 $L[\{(k_1 C_{B0}/D_A)+(k_1 C_{F0}/D_E K_{eq} C_{B0})\}/\{1+k_2/(k_3 C_{B0})\}]^{1/2}$
     $= \delta[\{b_0+(f_0/r_{Eq}Kb_0)\}/(1+m/b_0)]^{1/2}$

γ_L $L[\{(k_1 C_{BL}/D_A)+(k_1 C_{FL}/D_E K_{eq} C_{BL})\}/\{1+k_2/(k_3 C_{BL})\}]^{1/2}$
     $= \delta[\{b_L+(f_L/r_{Eq}Kb_L)\}/(1+m/b_L)]^{1/2}$

δ $L(k_1 C_{BT}/D_A)^{1/2}$

δ' $\tau^{1/2}L(k_1 C_{BT}/D_A)^{1/2}$ or $L_e(k_1 C_{BT}/D_A)^{1/2}$

ε porosity of support membrane

μ viscosity (Pa s)

τ tortuosity factor of support membrane

Literature Cited

1. Kesting, R.E.; Fritzsche, A.K. *Polymeric Gas Separation Membranes*; John Wiley: New York, 1993.
2. Ward, W. J.; Robb, W. L. *Science* **1967**, *156*, 1481.
3. LeBlanc, O. H.; Ward, W. J.; Matson, S. L. *J. Membrane Sci.* **1980**, *6*, 339.
4. Meldon, J. H.; Stroeve, P.; Gregoire, C. E. *Chem. Eng. Comm.* **1982**, *16*, 263.
5. Way, J. D.; Noble, R. D. *Facilitated Transport*; Ho, W. S. W.; Sirkar, K. K. Ed.; Membrane Handbook; Van Nostrand Reinhold, New York, 1992, pp.833-866.
6. Smith, D. R.; Quinn, J. A. *AIChE J.* **1979**, *21*, 197.
7. Donaldson, T. L.; Nguyen, Y. N. *Ind. Eng. Chem. Fundam.* **1980**, *19*, 260.
8. Meldon, J. H.; Paboojian, A.; Rajangam, G. *AIChE Symp. Ser.* **1986**, *82*, 114.
9. Guha, A.K.; Majumdar, S.; Sirkar, K. K. *Ind. Eng. Chem. Res.* **1990**, *29*, 2093.
10. Davis, R. A.; Sandall, O. C. *AIChE J.* **1993**, *39*, 1135.
11. Teramoto, M. *Ind. Eng. Chem. Res.* **1995**, *34*, 1267.
12. Danckwerts, P. V. *Chem. Eng. Sci.* **1979**, *34*, 443.
13. Teramoto, M. *Ind. Eng. Chem. Res.*, **1994**, *33*, 2161.
14 Youngquist, G.Y. *Ind. Eng. Chem.* **1970**, 52-63.
15. Matsuyama, H.; Teramoto, M.; Iwai, K. *J. Membrane Sci.* **1994**, *93*, 237.
16. Versteeg, G. F.; Van Swaaij, W. P. M. *J. Chem. Eng. Data* **1988**, *33*, 29.
17. Witherspoon, P. A.; Bonoli, L. *Ind. Eng. Chem. Fundam.* **1969**, *8*, 589.
18. Morrison, T. J.; Billett, F. *J. Chem. Soc.* **1952**, 3819.
19. Hikita, H.; Ishikawa, H.; Uki, K.; Murakami, T. *J. Chem. Eng. Data* **1980**, *25*, 324.
20. Van Krevelen, D.W.; P.J. Hoftijzer. *Chim. Industrr. XXIeme Congr. Int. Chim. Industr.* **1948**, 168.
21. Hikita, H.; Asai, S.; Ishikawa, H.; Honda, M. *Chem. Eng. J.* **1977**, *13*, 7.
22. Laddha, S. S.; Danckwerts, P. V. *Chem. Eng. Sci.* **1981**, *36*, 479.
23. Kent, R.L.; Eisenberg, B. *Hydrocarbon Processing* **1976**, Feb., 87.
24. Jensen, A.; Christensen, R. *Acta Chim. Scand.* **1955**, *9*, 486.

Chapter 18

Facilitated Transport of Carbon Dioxide Through Functional Membranes Prepared by Plasma Graft Polymerization Using Amines as Carrier

Hideto Matsuyama and Masaaki Teramoto

Department of Chemistry and Materials Technology,
Faculty of Engineering and Design, Kyoto Institute of Technology,
Matsugasaki, Sakyo-ku, Kyoto 606, Japan

New types of cation-exchange membranes were prepared by grafting acrylic acid and methacrylic acid to the substrates such as microporous polyethylene and polytetrafluoroethylene, and homogeneous poly[1-(trimethylsilyl)-1-propyne] by the use of the plasma graft polymerization technique. Various monoprotonated amines were immobilized by the electrostatic force in the ion exchange membranes as the carriers of CO_2 and the facilitated transport of CO_2 through the grafted membranes was studied. When the CO_2 partial pressure in the feed phase was 0.047 atm, the selectivity for CO_2 over N_2 reached more than 4700 with a CO_2 permeation rate of $1.0 \times 10^{-4} cm^3/(cm^2 \, s \, cmHg)$. These efficiencies are much higher than those of conventional polymer membranes and are also superior to efficiencies reported perviously for ion-exchange membranes containing carriers. These excellent results are probably attributable to the high ion-exchange capacity and the high water content of these ion-exchange membranes. The newly prepared membranes were stable. Membranes having tertiary amine groups as a fixed carrier were also prepared by plasma grafting of 2-(N,N-dimethyl)aminoethyl methacrylate. It was suggested that the dry membrane acts as a fixed carrier membrane for CO_2 facilitated transport in the dry condition and as a fixed reaction site membrane under the water-containing condition. The carrier transport mechanism is discussed for both the dry and wet membranes.

Facilitated transport membranes have been attracting attention since they have very high selectivity compared with conventional polymer membranes (1). This high selectivity is attributable to carriers which can react reversibly with permeant species. There are two types of facilitated (carrier) transport membranes. One is the mobile carrier membrane in which the carrier can diffuse in the membrane, and the other is the fixed carrier membrane in which the carrier cannot move.

Previous Work on Facilitated Transport of Gases.

Facilitated transport membranes for gas separation were prepared conventionally by impregnating the pores of microporous supports with the carrier solutions. This type of membrane is a typical mobile carrier membrane and is known as a supported liquid membrane (SLM) or immobilized liquid membrane (ILM).

Although remarkably high selectivities for gas separations have been reported for SLMs (2-6), the disadvantage of membrane degradation due to evaporation of the membrane solution and/or "washout" of the carrier is well recognized. One of the methods for preventing this degradation is the use of an ion-exchange membrane as the support for SLMs. LeBlanc et al. first reported the facilitated transport of CO_2 and C_2H_4 with ion-exchange membranes (7). The mechanism of facilitated transport of CO_2 using monoprotonated ethylenediamine ($EDAH^+$) as a carrier is shown in Figure 1. In this membrane, a cationic carrier, $EDAH^+$ was immobilized in a cation-exchange membrane by electrostatic force, which should reduce washout of the carrier. The work of LeBlanc et al. has stimulated research in facilitated transport using ion exchange membranes (8-15). Previous studies using ion exchange membranes are summarized in Table I.

Table I. Previous Work on CO_2 Facilitated Transport Using Ion-Exchange Membranes

	LeBlanc et al. (7)	Way et al. (8)	Langevin et al. (12)	Heaney et al. (11)
Ion-exchange membrane	membrane 1[a]	membrane 2[b]	membrane 3[c]	membrane 4[d]
Ion exchange capacity	5meq/g	0.91meq/g	2.2meq/g	0.91meq/g
Membrane thickness	110μm	200μm	100μm	294μm
Water content	11%	11%	47%	56%
CO_2 partial pressure in the feed gas	0.05atm	0.044atm	0.05atm	0.041atm
Permeation rate of CO_2 [$cm^3/(cm^2 \, s \, cmHg)$]	3.5×10^{-5}	6.5×10^{-6}	5.3×10^{-5}	7.1×10^{-5}
Permeability of CO_2 [Barrer]	3800	1290	5300	20900
Facilitation factor	13	9	21	10.5
Selectivity (CO_2/N_2)	380	850[e]	420	390[f]

[a] Sulfonated polystyrene grafted onto polytetrafluoroethylene.
[b] Perfluorosulfonic acid ionomer.
[c] Sulfonated strylene-divinylbenzene in fluororinated matrix.
[d] Heat-treated perfluorosulfonic acid ionomer (Nafion® N117).
[e] Estimated value from the selectivity of CO_2 over CH_4.
[f] Estimated value from the selectivity of CO_2 over CO.

There are several important points in designing efficient mobile carrier transport membranes with an ion-exchange membrane as supports. The first factor is the mobility of the permeant in the membrane. The transport mechanism for the ionic carrier through the ion-exchange membrane is different from that through the usual mobile carrier membrane because in the former the carrier moves under electrostatic interaction with the ion sites in the membrane matrix. This interaction usually leads to lower mobility of the complex (8). The mobility is related to the water content of the membrane and a swollen membrane with high water content favors high mobility (12). A hot glycerin treatment of the ion-exchange membrane is one way to improve mobilities (10,11,14). The morphology of the ion-exchange membrane, such as ion-exchange capacity and the distribution of the ion sites also affects the mobilities.

The second point is enhancement of the permeation rate by chemical reaction between the permeant and the carrier. Usually, high carrier concentration is favorable for obtaining both high permeability and selectivity. When an ionic carrier is used as in the case of ethylenediamine-mediated transport of CO_2, high ion-exchange capacity is desirable to retain the ionic carrier at high concentration in the membrane. The equilibrium constant of the complex between the permeant and the carrier is also very important. When a carrier with an extremely large equilibrium constant is used, the sorption of the permeant at the feed side interface occurs easily, while desorption at the sweep side interface becomes difficult, which results in a low permeation rate. A carrier with a moderate equilibrium constant is desirable (16).

The third factor is the stability of SLMs. When an aqueous carrier solution is used, the hydrophilic water-swollen membrane is suitable for retaining the membrane solution. Furthermore, high ion site concentration is desirable to prevent loss of the ionic carrier by washout.

In designing the membranes, these points must be considered. However, there have been very few studies which focused on the development of new ion-exchange membranes as the supports for facilitated transport membranes in gas separations. In most of pervious studies, commercial ion-exchange membranes have been used.

Purpose of This Work.

Recently, we prepared a microporous ion-exchange membrane by grafting acrylic acid (AA) onto a microporous polyethylene membrane by a plasma graft polymerization technique. We found that a facilitated transport membrane prepared by impregnating this membrane with $EDAH^+$ showed very high CO_2 permeability and selectivity for CO_2 over N_2. Also the membrane was stable (15). High efficiency of our membrane is considered to be due to both high water content and high ion-exchange capacity of the membrane.

This study is the extension of our previous study. Here, a variety of microporous ion-exchange membranes was prepared by plasma grafting various ionic monomers, such as acrylic acid (AA), methacrylic acid (MAA) and 2-(N,N-dimethyl)aminoethyl methacrylate (DAMA) on several kinds of substrate membranes such as microporous polyethylene (PE) membranes and polytetrafluoroethylene (PTFE) membranes. These membranes were evaluated as supports for facilitated transport of CO_2 using several kinds of monoprotonated amines as carriers. N,N-dimethylaminoethyl methacrylate (DAMA) grafted membranes prepared by a similar technique were also evaluated as a fixed carrier membrane for the facilitated transport of CO_2.

Experimental.

Membrane Preparation. The procedure and the apparatus used for plasma-graft polymerization are similar to those reported in the literatures(15,17,18). The substrate

(6x6 cm) contained in a glass tube was treated by a glow discharge plasma of the inductive coupling type, generated at a frequency of 13.56 MHz at 67 Pa under an argon atmosphere. The plasma treatment time and the power input were fixed at 60 sec and 10 W, respectively. Next, an aqueous monomer solution was introduced under vacuum into the glass tube containing the plasma treated substrate and graft polymerization was carried out in a shaking bath for a prescribed time. After graft polymerization, the membrane was washed and soaked in distilled water overnight to remove the residual monomer and the homopolymers. Then, the membrane was dried in an oven at about 50 °C. The degree of grafting was controlled by changing the grafting conditions, such as grafting time and monomer concentration.

Microporous polyethylene (PE) membranes with various pore diameters and porosities and microporous polytetrafluoroethylene (PTFE) membrane were used as substrates for the plasma graft polymerization (Table II). Besides these porous substrates, homogeneous poly[1-(trimethylsilyl)-1-propyne] (PTMSP), which has the highest gas permeability among polymeric materials, was used as the substrate. The poly[1-(trimethylsilyl)-1-propyne] was synthesized from 1-(trimethylsilyl)-1-propyne according to the literature procedure (19). Films were prepared by casting polymers from toluene solutions. Hereafter, the respective substrate membranes will be abbreviated as shown in Table II.

Table II. Membranes Used as Substrate.

materials	thickness[μm]	pore diameter[μm]	porosity[%]	abbreviation
polyethylene	13	0.03	60	PE1[a]
polyethylene	30	0.02	40	PE2[a]
polyethylene	30	0.02	37	PE3[a]
polyethylene	25	0.02	30	PE4[a]
polytetrafluoroethylene	60	0.10	57	PTFE[b]

[a] Tonen Chemical Co., Ltd., [b] Sumitomo Electric Co., Ltd.

A DAMA-grafted membrane was prepared by the similar method. The grafted membrane was tested without further incorporation of amine carrier.

Measurements of Membrane Properties. The amount of the graft polymer was calculated by the weight change of the membrane before and after the polymerization. The thickness of the membrane was measured by a micrometer. The ion-exchange capacity of the membrane was calculated by measuring the amount of K[+] in the membrane which had been soaked in an aqueous KOH solution (*15*). The cross section of the grafted membrane was observed by a microscope after staining it with methylene blue (*20*).

Gas Permeation. The permeation rates of gases were measured with a diffusion cell consisting of two compartments (depth of 3.7 mm) for a feed and a sweep gas. The plasma graft polymerization membranes were sandwiched between the two compartments. The feed gas, a mixture of CO_2 and N_2, and the sweep gas, helium, were supplied to the cell at atmospheric pressure. Both gas streams were saturated with water upstream of the cells. The permeation rates for CO_2 and N_2 were calculated from the outlet sweep gas composition, which was analyzed by a gas chromatograph equipped with a thermal conductivity detector (Shimadzu Co., Ltd., GC-8APT) and the sweep gas flow rate.

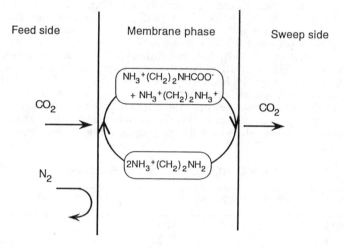

Figure 1. Schematic Diagram Showing the Facilitated Transport of CO_2.

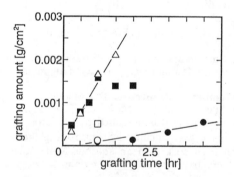

Figure 2. Relation between Grafting Amount and Grafting Time.
PE1 substrate: △ 3.85vol%(AA), ■ 3.85vol%(DAMA)
PTFE substrate: ● 3.85vol%(AA), ○ 5.66vol%(AA), □ 9.09vol%(AA)

Facilitated Transport by Mobile Carrier Membranes with Ion Exchange Membrane as Supports.

Graft Polymerization Characteristics. Figure 2 shows the relation between the acrylic acid (AA) grafting amounts and the grafting time. The grafting amount was defined as the mass of the grafted layer per unit area of the substrate. The grafting amounts increased with enhancement in both the grafting time and the monomer concentration. The grafting amounts for the PTFE substrate were lower than those for PE1 substrate. This is probably because the higher hydrophobicity of PTFE makes contact of aqueous monomer solution with the substrate more difficult.

Characteristics of Ion Exchange Membranes. The ion-exchange capacities of the AA grafted PE1 membrane (PE1-g-AA membrane) and PE1-g-MAA membrane were found to be 13.1 and 10.5 meq/g-dry grafted layer, respectively, and are nearly in accordance with the respective values of 13.9 meq/g and 11.6 meq/g, which are calculated from the molecular weight of acrylic acid (72.1) and methacrylic acid (86.1). The value of 13.1 meq/g-dry grafted layer means that the ion-exchange capacity per total weight of dry membrane is 8.0 meq/g when the grafting amount was 1.15×10^{-3} g/cm^2, the value for typical grafted membrane. This ion-exchange capacity is fairly high, compared with the values for the ion-exchange membranes listed in Table I.

Figure 3 shows cross sections of the AA grafted membranes. The black portions are the grafted layers stained with methylene blue. For the PE1 substrate membrane, the grafted layer formed mainly on the substrate surface at low grafting levels (Figure 3(a)). With an increase of the grafting amount, the grafted layer formed inside the pores of the substrate (Figure 3(b)), and further grafting brought about an increase in the total membrane thickness (Figure 3(c)). Yamaguchi et al. reported that a methyl acrylate grafted layer was formed inside the porous substrate even at low grafting levels (*18*). The difference between our results and theirs is probably attributable to the difference in the apparatus and procedures for the plasma graft polymerization and/or in the monomers used. Comparison of the PTFE-g-AA membrane with the PE1-g-AA membrane at a similar grafting level (Figure3 (b) and 3(d)) shows that the grafted layer formed more loosely inside the PTFE substrate. This is probably attributable to a larger thickness of the PTFE substrate and also to the hydrophobic property of PTFE.

The effects of the grafting levels on the membrane thicknesses and the water content are shown in Figure 4. The thicknesses of both PE1-g-AA and PTFE-g-AA membranes increased with increasing the grafting level. Water content in the grafted layer of the PE1-g-AA membrane was about 80% which is much higher than that of Nafion®117, which is only 11% (*21*). The high swelling and the high water content are probably attributable to the high hygroscopicity of poly(acrylic acid). The water contents of the PTFE-g-AA membrane was about 90% and higher than that of the PE1-g-AA membrane. This is explained as follows. In the observation of the cross section of the membrane with a microscope, we found that in the case of the PTFE-g-AA membrane, AA was grafted inside the pores of the PTFE membrane to lesser extent than in the case of PE-g-AA membrane. This suggests that a larger amount of water can be incorporated within the pores of the former membrane than the latter membrane.

Membrane Efficiencies in the Separation of CO$_2$ from N$_2$. The effect of the degree of grafting on the driving-force normalized flux of CO$_2$ R_{CO_2}, and the selectivity α of

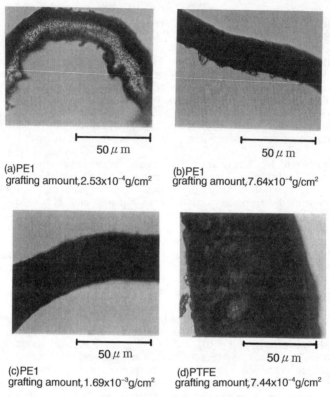

(a)PE1
grafting amount,2.53x10⁻⁴g/cm²

(b)PE1
grafting amount,7.64x10⁻⁴g/cm²

(c)PE1
grafting amount,1.69x10⁻³g/cm²

(d)PTFE
grafting amount,7.44x10⁻⁴g/cm²

Figure 3. Cross-sectional View of AA-grafted Membranes.

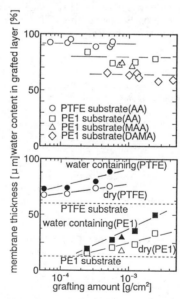

Figure 4. Effects of Grafting Amount on Membrane Thickness and Water Content.
○●□■ :acrylic acid, △▲ :metacrylic acid, ◇ :DAMA

CO_2 over N_2 are shown in Figure 5 for the case in which the membrane was impregnated with pure water. R_{CO2} and α are given by:

$$R_{CO2} = N_{CO2}/\Delta P \tag{1}$$

$$\alpha = R_{CO2}/R_{N2} \tag{2}$$

where N_{CO2} is the molar flux of CO_2 and ΔP is the pressure difference between the upstream and downstream side of the membrane. As the degree of grafting increased, R_{CO2} through the water-containing membrane decreased monotonously due to the increase in the membrane thicknesses, as shown in Figure 4, while the selectivity increased and reached a constant value. This suggests that the originally hydrophobic microporous polyethylene membrane became hydrophilic as the degree of grafting increased. When the degree of grafting was low, the grafted layer formed mainly on the substrate surface as shown in Figure 3, and all the pores were not covered with the grafted layer, which resulted in the low selectivity. The selectivity was about 63 and agrees with a result obtained with cellulose filter paper (Toyo Roshi Co., Ltd., No.4A) containing water, as shown by the dotted line in Figure 5. This value is close to 60, which was estimated from the solubility and diffusivity ratios of these gases (*15*).

Figure 6 shows the effect of the grafting level on R_{CO2} and α when ethylenediamine was immobilized in the grafted membrane as the carrier. Here, various AA-grafted membranes were used as the supports. In the case of the AA-grafted PE1 membrane, α increased with an increase of the grafting level and approached to a constant value, similar to the tendency shown in Figure 5. The constant value α was as high as 4700 and R_{CO2} was 1.0×10^{-4} cm^3/(cm^2 s cmHg).

The AA-grafted PTFE membrane showed a higher R_{CO2} compared with that of a PE1 membrane at the same grafting degree. This is because the pores of the PTFE substrate were more loosely filled with the grafted polymer, as shown in Figure 3, and the mass transfer resistance through the pores is smaller. The value of R_{CO2} for the PTFE membrane reached to about 2.0×10^{-4} cm^3/(cm^2 s cmHg) with $\alpha = 100$. When other PE substrates were used, the values of R_{CO2} were smaller than the R_{CO2} for the grafted PE1 membrane. The values of R_{CO2} for PE substrates decreased in order of PE1 > PE2 > PE3 > PE4. This order is in agreement with the porosity of the substrates as shown in Table II. Data obtained with a membrane grafted only on the one-side by sealing the other side with aluminum foil in the plasma treatment is also included in Figure 6. The data was hardly distinguishable from data for the usual grafted membrane.

The results using a poly[1-(trimethylsilyl)-1-propyne] (PTMSP) membrane are listed in Table III. Although the permeation rate of the substrate alone was as high as 1.51×10^{-3} cm^3/(cm^2 s cmHg) , the selectivity was only about 3. The permeabilities of CO_2 and N_2 were 27800 and 9600 Barrer, respectively, and were roughly in agreement with the respective values of 40500 and 9170 Barrer at 303K reported previously (*22*). When the AA-grafted PTMSP membrane was impregnated with water, the selectivity increased to 49. This value agrees approximately with the selectivity shown in Figure 5. When the grafted membrane was soaked in 25%(v/v) EDA solution, R_{CO2} became larger than that of the water-containing membrane, and reached to 2.2×10^{-4} cm^3/(cm^2 s cmHg) with $\alpha = 310$. This data is shown also in Figure 6.

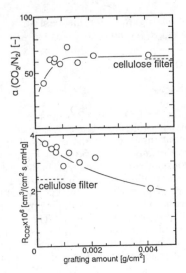

Figure 5. Effect of Grafting Amount on Permeation Rate and Selectivity.
The membrane was impregnated with pure water.
Reproduced with permission from reference 15. Copyright 1994 Elsevier
Science—NL.

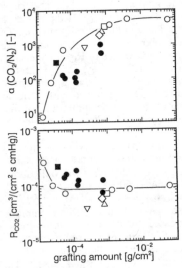

Figure 6. Effect of Grafting Amount on Permeation Rate and Selectivity.
monomer = acrylic acid, carrier = ethylenediamine, [EDA] = 25%(v/v),
feed CO_2 pressure = 0.047 atm, 298K
● PTFE, ○ PE1, ◇ PE2, △ PE3, ▽ PE4, □ PE1(one side grafted)
■ PTMSP

Table III. Results Using PTMSP Membranes.

	R_{CO2} [cm^3/(cm^2 s cmHg)]	$\alpha(CO_2/N_2)$ [-]
PTMSP alone[a]	1.51×10^{-3}	2.88
AA grafted PTMSP[b] (water)	1.00×10^{-4}	49
AA grafted PTMSP[b] (EDA 25%(v/v))	2.22×10^{-4}	310

The feed CO_2 pressure = 0.047 atm, 298K
[a] Membrane thickness of 20μm, [b] Grafting amount of 3.61×10^{-5} g/cm^2.

The effect of the EDA concentration in the aqueous solution in which the grafted membrane was soaked is shown in Figure 7. As [EDA] increases, both R_{CO2} and α increases sharply and the ionic sites are saturated with EDAH$^+$ when [EDA] reaches about 2 vol%. As [EDA] increases further, the EDA concentration in the membrane increases gradually since EDA is incorporated as free EDA which is not interacting with the ion sites of the membrane. The maximum facilitation factor obtained, which is defined as the ratio of the permeation rate in facilitated transport to that in physical transport, was 30. At low [EDA], the grafted membrane showed much a higher permeation rate of CO_2 and selectivity, compared to cellulose filter paper soaked in the same solution. This is due to the enrichment of EDA in the grafted membrane by ion exchange.

Figure 8 shows the effect of CO_2 partial pressure in the feed gas on R_{CO2} and α. Data reported by other researchers are also included in this figure. The original data by Way et al. were with respect to a CO_2/CH_4 system (8). These data were converted to values for a CO_2/N_2 system by considering the difference in the solubility and the diffusivity between N_2 and CH_4 (15,23,24). If the CO_2 transport obeys the facilitated transport mechanism, the CO_2 flux, not normalized to the driving force, increases proportionally with the CO_2 partial pressure in the low pressure region and reaches a constant value at a pressure where the carrier is saturated on the feed side (8). When the data are rearranged in the relation between the driving-force normalized flux and the CO_2 partial pressure, the flux decreases at the pressure where the carrier becomes saturated. This tendency was observed in Figure 8. This is a disadvantage in the facilitated transport membranes. However, in the removal of CO_2 from flue gases, the CO_2 partial pressure in the flue gases is usually 0.10-0.15 atm (25). Under such conditions, the value of α is more than 1000, as shown in Figure 8.

The PE1-g-MAA membrane showed a little lower R_{CO2} and α compared with the PE1-g-AA membrane. This may be due to lower ion-exchange site concentration and therefore lower carrier concentration in MAA-grafted membrane.

In addition to EDA, some other amines were tested as the carrier of CO_2. Typical results are shown in Table IV. As the chain length of the diamine increases, both the permeation rate and the selectivity decreases. With respect to amines having additional amine groups, such as diethylenetriamine and triethylenepentamine, a increase in molecular weight of the carrier also decreases both R_{CO2} and α. This may be due to lower diffusivity of larger molecules in the membrane. Thus, among the amines tested, ethylenediamine was found to be the best carrier of CO_2.

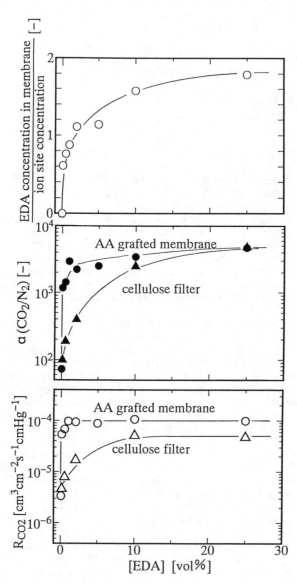

Figure 7. Effect of the EDA Concentration.
The grafting amount = 1.18×10^{-3} g/cm^2
Reproduced with permission from reference 15. Copyright 1994 Elsevier Science—NL.

Table IV. Effect of Carrier on R_{CO2} and α for the PE1-g-AA Membrane.

carrier	R_{CO2} [cm^3/(cm^2 s cmHg)]	$\alpha(CO_2/N_2)$ [-]	grafting amount [g/cm^2]
$H_2N(CH_2)_2NH_2$	8.26×10^{-5}	1900	1.33×10^{-3}
$H_2N(CH_2)_3NH_2$	1.88×10^{-5}	560	1.67×10^{-3}
$H_2N(CH_2)_4NH_2$	1.38×10^{-5}	320	1.91×10^{-3}
$H_2N(CH_2)_2NH(CH)_2NH_2$	2.86×10^{-5}	1160	1.67×10^{-3}
$H_2N(CH_2)_2NH(CH_2)_2NH(CH_2)_2NH_2$	0.93×10^{-5}	680	1.91×10^{-3}

Concentration of the carrier in the membranes were the same as those of the ion sites. The feed CO_2 pressure = 0.047 atm, 298K

Figure 9 shows a comparison of efficiencies for various membranes including both polymer membranes and liquid membranes in the separation of CO_2 over N_2. The ion-exchange membranes impregnated with EDA have very high permeability and selectivity compared with the polymer membranes. Our AA-grafted PE1 membrane showed the best membrane efficiency. Higher ion-exchange capacity, higher water content and lower membrane thickness are probably responsible for the good properties of our membrane.

Membrane Stability. A stability test was performed with the PE1-g-AA membrane impregnated with a 10 vol% EDA solution. When both the feed and the sweep gases were presaturated with water, no degradation was observed during the permeation experiment for two days. This stability is attributable to both the hygroscopicity of polyacrylic acid membrane and retainment of the carrier by electrostatic force.

Figure 10 shows the time-course of the PE1-g-AA membrane when the sweep gas was not saturated with water vapor. When the sweep gas flow rate was 113 cm^3/min, R_{CO2} decreased very quickly because the water content of the membrane diminished due to evaporation of water at the sweep side. The permeation rate at point A was about two orders of magnitude lower than the initial value. At point A, the sweep gas flow rate was decreased to 22 cm^3/min and at the same time, the temperature of the permeation cell was lowered from 298K to 288K. Then, R_{CO2} recovered to approximately the initial value quickly. When the cell temperature was raised to 298K at point B, the permeation rate decreased again. On the other hand, α is always very high at around 3000 under all conditions. These results indicate that in order to maintain a high permeation rate, control of temperature and the water vapor pressure in the gas streams is very important. In the removal of CO_2 from flue gas, the gas is saturated with water vapor. Therefore, the water can be supplied to the membrane from the feed side even if the sweep side is evacuated.

Facilitated Transport by a Fixed Carrier Membrane.

All of the results described above were obtained with mobile carrier membranes. Another type of membrane was prepared by grafting 2-(N,N-dimethyl)aminoethyl methacrylate (DAMA) onto a microporous polyethylene substrate (PE1). The relationship between the degree of DAMA-grafting and the grafting time is shown in Figure 2. Osada et al. reported that DAMA alone could not be graft-polymerized on a porous substrate by the low temperature plasma technique and that the addition of the easily graftable monomer, such as acrylic acid (AA) to DAMA, was necessary to

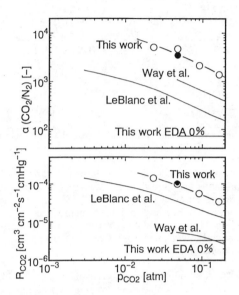

Figure 8. Effect of the Partial Pressure of CO_2 in the Feed Gas.
○ EDA = 25vol%, ● EDA = 10vol%, grafting amount = 1.18x10⁻³ g/cm²

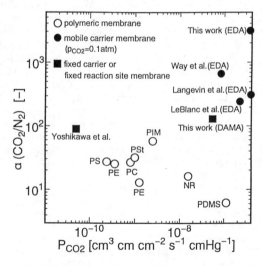

Figure 9. Relation between Permeability, P_{CO2} and $\alpha(CO_2/N_2)$.

achieve the graft polymerization (26). As can be seen in Figure 2, however, DAMA alone can be efficiently graft-polymerized on the porous PE substrate and the degree of grafting is equivalent to those of AA up to 60 min of graft time. The water content of PE1-g-DAMA membrane was about 60% as shown in Figure 4.

The effects of the degree of grafting on R_{CO2} and the selectivity α are shown in Figure 11. Data for both dry and water-containing membranes are included in the same figure. In the dry membrane, as the degree of grafting increased, R_{CO2} decreased monotonously due to the increase in the membrane thicknesses. The selectivities of the dry membrane, however, increased and reached a constant value. This constant selectivity suggests that all of the pores of the microporous substrate were efficiently filled with the grafted layers. The selectivity, however, was about 30.

When this membrane is impregnated with water, both R_{CO2} and α improved markedly, compared with those in the dry condition. The values of R_{CO2} were almost independent of the degree of grafting, probably because resistance in the grafted layer inside the substrate pores controlled the permeation rate and the diffusional resistance through the swollen grafted polymer layer on the surface was very small (18). The selectivity was about 130 when the CO_2 partial pressure of the feed gas was 0.047 atm. As far as we know, this selectivity is the highest among the values reported so far for polymeric membranes and fixed-carrier membranes, although our water containing membrane should be called a fixed reaction site membrane as described below. As described above, the selectivity of CO_2 over N_2 by the liquid membrane containing pure water was about 60. Therefore, the selectivity of 130 obtained in this work suggests that CO_2 transport through the PE1-g-DAMA membrane does not obey the simple solution-diffusion mechanism.

Figures 12 and 13 show the effects of CO_2 feed partial pressures, p_{CO2}, on R_{CO2} and α for dry and water-containing membranes, respectively. In both cases, as the CO_2 feed pressure increased, R_{CO2} decreased while R_{N2} was nearly constant. The decrease in R_{CO2} observed for both dry and water-containing membranes suggests that CO_2 permeates by the carrier transport mechanism in both conditions. However, the mechanism may be different for the two cases. In the dry membranes, the facilitated transport of CO_2 is expected to be attributable to the weak acid-base interaction between CO_2 and amine moiety, as suggested by Yoshikawa et al. (27). Therefore, the dry membrane is a fixed carrier membrane. On the other hand, tertiary amine groups in the wet membrane are considered to act as catalyst for the hydration of CO_2 as in the case of triethanolamine in a supported liquid membrane (28). The mechanism is schematically represented as follows:

$$CO_2 \longrightarrow R_3NH^+ + HCO_3^- \qquad (3)$$

In this case, CO_2 is transported as CO_2 and also as HCO_3^-. Therefore, the water containing membrane is not a fixed carrier membrane in the strict sense. The membrane should be called as a fixed reaction site membrane or a fixed catalytic membrane.

Efficiencies of the DAMA-grafted membrane are also shown in Figure 9. The efficiencies of the DAMA membrane are higher than those of the polymeric membranes, although inferior to those of the mobile carrier membranes. Compared

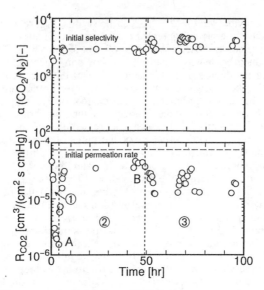

Figure 10. Time-course of Membrane Efficiencies.
Sweep gas was not saturated with water.
1 membrane and saturator at 298K, sweep gas flow rate = 113 cm^3/min
2 membrane at 288K, saturator at 298K, sweep gas flow rate = 22 cm^3/min
3 membrane and saturator at 298K, sweep gas flow rate = 22cm^3/min

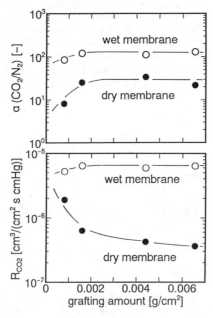

Figure 11. Effect of Grafting Amount of DAMA on the Permeation Rate and Selectivity.
The feed CO_2 partial pressure = 0.047 atm, 298K

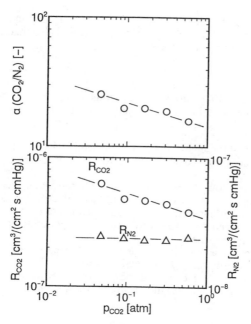

Figure 12. Effect of Partial Pressure of CO_2 in the Feed Gas for the Dry Membrane. The grafting amount = 1.59×10^{-3} g/cm^2, 298K

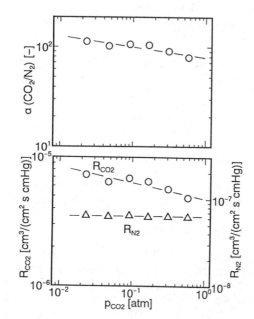

Figure 13. Effect of Partial Pressure of CO_2 in the Feed Gas for the Water-containing Membrane.
The grafting amount = 1.59×10^{-3} g/cm^2, 298K

with the mobile carrier membranes, the DAMA membrane probably has the advantage that there is no loss of carrier by washout and also that impregnation of the membrane with a carrier is unnecessary.

Conclusions.

Facilitated transport of CO_2 through ion exchange membranes prepared by the plasma graft polymerization was investigated using amines as the carrier. The effects of various experimental factors, such as grafting level, substrates, monomers and carriers were investigated.

It was found that these new ion-exchange membranes had the high ion-exchange capacities and high water contents, compared with those of the commercial ion-exchange membranes.

The membrane showed remarkably high CO_2 permeability and selectivity. The selectivity obtained was 4700 when the CO_2 partial pressure in the feed gas was 0.047 atm. These membrane efficiencies are much better than those of conventional polymer membranes and are also superior to those reported to date for ion-exchange membranes containing carriers.

We also prepared a new membrane with an amine moiety by plasma-grafting DAMA onto a microporous polyethylene substrate. The membrane obtained is suggested to act as a fixed carrier membrane for CO_2 facilitated transport in dry conditions and as a fixed reaction site membrane in wet conditions.

Literature Cited

1. Way, J. D.; Noble, R. D. In *Membrane Handbook;* Ho, W. S.; Sirkar, K. K., Eds.;Van Nostrand Reinhold; New York, 1992; pp.833-866.
2. Enns, T. *Science* **1967**, *155*, 44.
3. Ward, W. T.; Robb, W. L. *Science* **1967**, *156*, 1481.
4. Suchdeo, S. R.; Schultz, J. S. *Chem.Eng.Sci.* **1974**, *29*, 13.
5. Matson, S. L.; Herrick, C. S.; Ward, W. J. *Ind. Eng. Chem. Process Res. Dev.* **1977**, *16*, 370.
6. Guha, A. K.; Majumdar, S.; Sirkar, K. K. *Ind. Eng. Chem. Res.* **1990**, *29*, 2093.
7. LeBanc, O. H.; Ward, W.J.; Matson, S. L.; Kimura, S. G. *J. Membr. Sci.* **1980**, *6*, 339.
8. Way, J. D.; Noble, R. D.; Reed, D. L.; Ginley, G. M.; Jarr, L. A. *AIChE Journal* **1987**, *33*, 480.
9. Noble, R. D.; Pellegrino, J. J.; Grosgogeat, E.; Sperry, D.; Way, J. D. *Sep. Sci. Technol.* **1988**, *23*, 1595.
10. Pellegrino, J. J.; Nassimbene, R.; Noble, R. D. *Gas Separation and Purification*, **1988**, *2*, 126.
11. Heaney, M. D.; Pellegrino J. J. *J. Membr. Sci.* **1989**, *47*, 143.
12. Langevin, D.; Pinoche, M.; Selegny, E.; Metayer, M.; Roux, R. *J. Membr. Sci.* **1993**, *82*, 51.
13. Pellegrino, J.J.; Wang, D.; Rabago, R.; Noble, R. D.; Koval, C. *J. Membr. Sci.* **1993**, *84*, 161.
14. Eriksen, O.I.; Aksnes, E.; Dahl, I.M. *J. Membr. Sci.* **1993**, *85*, 99.
15. Matsuyama, H.; Teramoto, M.; Iwai, K. *J. Membr. Sci.*, **1994**, *93*, 237.
16. Teramoto, M.; Huang, Q.; Matsuyama, H. In *Facilitated Transport of Carbon Dioxide through Supported Liquid Membranes of Aqueous Amine Solutions;* Bartsch, R. A. and Way, J. D., Ed.; This volume.
17. Hirotsu, T. *Ind. Eng. Chem. Res.* **1987**, *26*, 1287.

18. Yamaguchi, T.; Nakao, S.; Kimura, S. *Macromolecules* **1991**, *24*, 5522.
19. Masuda, T.; Isobe, E.; Higashimura, T. *Macromolecules* **1985**, *18*, 841.
20. Lawler, J. P.; Charlesby, A. *Radiat. Phys. Chem.* **1981**, *15*, 595.
21. Koval, C. A.; Spontarelli, T.; Thoen, P.; Noble, R. D. *Ind. Eng. Chem. Res.* **1992**, *31*, 1116.
22. Nakagawa, T.,;Sekiguchi, M.; Nagai, K.; Higuchi, A. *Maku(Membrane)*, **1994**, *19*, 92.
23. Kagaku Binran (Handbook of Chemistry), Maruzen, Tokyo, **1984**, vol.2, pp. 158-159.
24. Witherspoon, P. A.; Bonoli, L. *Ind. Eng. Chem. Fundam.* **1969**, *8*, 589.
25. Haraya, K.; Nakajima, M.; Itoh, N.; Kamisawa, C. *Kakaku Kogaku Ronbunsyu* **1993**, *19*, 714.
26. Osada, Y.; Iriyama, Y.; Ohta, M. *Nihon Kagaku Kaishi* **1983**, pp. 831-837.
27. Yoshikawa, M.; Fujimoto, K.; Kinugawa, H.; Kitao, T.; Ogata, N. *Chemistry Lett.* **1994**, 243-246.
28. Donaldson, T. L.; Nguyen, Y. N. *Ind. Eng. Chem. Fundam.* **1980**, *19*, 260.

Chapter 19

Separation of Ethylene from Ethane Using Perfluorosulfonic Acid Ion-Exchange Membranes

Anawat Sungpet, Paul M. Thoen[1], J. Douglas Way, and John R. Dorgan

Chemical Engineering and Petroleum Refining Department, Colorado School of Mines, 1500 Illinois Street, Golden, CO 80401–1887

Facilitated transport of ethylene through Ag[+]-containing perfluorosulfonic acid ion-exchange membranes results in high separation factors for ethylene over ethane. Ethylene of higher than 99 percent purity was obtained from a 50:50 mixture of ethylene and ethane at 25 °C and feed and permeate pressures of 1 atmosphere. An ethylene permeability of over 2000 Barrer was obtained. Two different types of perfluorosulfonic acid membranes were studied to determine the importance of ionic site density and water content on membrane performance. The effect of temperature on the transport mechanism was also studied over the range of 5-35 °C. In addition, the transport data obtained at high pressure show carrier-saturation and membrane compaction phenomena.

Separation processes in petroleum and petrochemical industries are highly energy intensive. Among these processes, olefin separations consume the most energy, 0.12 Quad/year (1 Quad = a million billion BTU) (*1*). Ethylene purification alone required 9 trillion BTU in 1993 (*2*). Consequently, there is an economic incentive to develop alternative separation processes with lower energy consumption.

Membrane technology is a promising process for olefin/paraffin separation. It could be used in a hybrid process with existing cryogenic distillation processes or possibly replace distillation processes altogether to reduce energy consumption. Despite commercial applications in hydrogen separation, acid gas removal, and production of nitrogen (*3*), polymer membranes have not been applied to the separation of hydrocarbon mixtures, such as olefins from paraffins. The primary problems are lack of selectivity and low permeabilities.

Facilitated transport membranes are capable of providing very high separation factors while achieving reasonable fluxes (*4*). Unfortunately, this kind of

[1]Current Address: Golden Technologies Company, Inc. 4545 McIntyre Street, Golden, CO 80403

membrane has problems with deactivation of the carrier and loss of the solvent when used for gas separation. These two major problems must be solved before facilitated transport membrane processes can be applied to industrial separations.

The separation of ethylene from ethane by facilitated transport membranes containing Ag^+ has been studied by many investigators. Hughes et al. (*5,6*) were the first to demonstrate facilitated transport of olefins by immobilized Ag^+ solutions in porous supports. LeBlanc et al. (*7*) used sulfonated polyphenylene oxide membranes in the Ag^+ form and obtained a high separation factor for ethylene over ethane. Teramoto et al. (*8*) used an aqueous silver nitrate solution supported by a cellulose filter for the separation of ethylene from ethane. Kraus (*9*) described a water-free Nafion 415 (DuPont) treated with silver nitrate solution and glycerol. Kawakami et al. (*10*) studied the permeation of ethylene and propylene thorough supported liquid membranes of Rh^{3+}-poly(ethylene glycol) in glass microfiber filters. Teramoto et al. (*11*) proposed a flowing liquid membrane in which silver nitrate was used as a carrier for ethylene. Kanno et al. (*12*) used Nafion 417 (DuPont), equivalent weight 1100, incorporated with Ag^+. Davis et al. (*13*) developed a facilitated transport-distillation hybrid system for olefin separations. Yurkovetsky et al. (*14*) investigated the permeation of ethylene and ethane through Ag^+ exchanged sulfonated polysulfones and sulfonated poly(phenylene oxide) ion-exchange membranes. Ilinich et al. (*15*) discussed the transient increase of ethylene-ethane selectivity of poly(phenylene oxides) membranes. Eriksen et al. (*16*) studied facilitated transport of ethylene through Ag^+ form Nafion 117 as a function of the ethylene partial pressure up to 760 mm Hg. at 25 °C. The same authors (*17*) also studied glycerine-treated, water-swollen Ag^+-Nafion 117 for ethylene/ethane separation. Tsou et al. (*18*) developed a liquid membrane contactor system using aqueous $AgNO_3$ solution for ethylene/ethane separation. Finally, Richter et al. (*19*) described facilitated transport of ethylene across polyelectrolyte gels loaded with Ag^+.

In this present work, DuPont's Nafion 117 polymer and the experimental perfluorosulfonic acid (PFSA) membrane developed by the Dow Chemical Company (Product XUS-13204.10) were investigated for the separation of ethylene from ethane. The differences in chemical structures between the membranes strongly affects the membrane properties. The Dow membrane has a shorter repeat unit compared with Nafion 117 (Figure 1). Consequently, the Dow membrane has a higher ionic site density when dry. In addition, the Dow membrane is a more amorphous polymer because of the shorter repeat unit while Nafion 117 is more crystalline (*20*). This difference in the morphologies of these membranes leads to some very interesting variations in properties, such as water content and mechanical strength. The effect of water content on the facilitated transport of ethylene through the membranes was studied. In addition, the effect of run conditions, such as temperature and pressure, on the membrane performance have been evaluated.

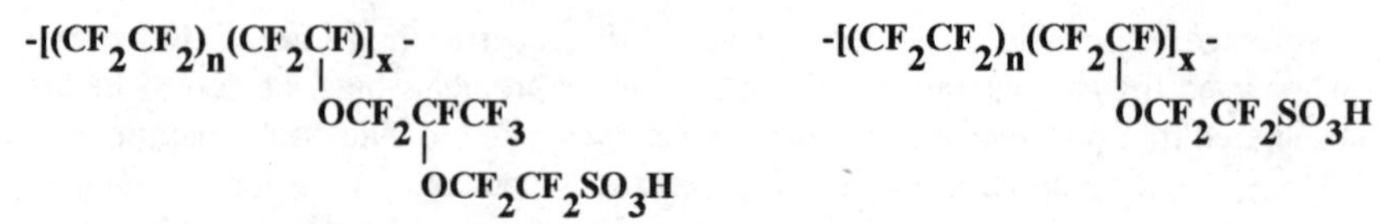

$$-[(CF_2CF_2)_n(CF_2\underset{|}{CF})]_x -$$
$$OCF_2\underset{|}{CFCF_3}$$
$$OCF_2CF_2SO_3H$$

$$-[(CF_2CF_2)_n(CF_2\underset{|}{CF})]_x -$$
$$OCF_2CF_2SO_3H$$

Nafion 117 Structure, n = 6-7   Dow Membrane Structure, n = 5-6

Figure 1. Nafion 117 and Dow Membrane Structures

Experimental.

Materials. Nafion 117 membrane with an equivalent weight of 1100 (Nafion 117) was purchased from C.G. Processing, Inc. (Rockland, DE). The membrane thickness is 170 μm when dry and 210 μm when fully hydrated. The experimental Dow membrane, with an equivalent weight of 803, was kindly provided by the Dow Chemical Company (Midland, MI). It is 95 microns thick when dry and 116 microns when hydrated. Silver nitrate, 99+%, A.C.S. reagent and sodium hydroxide, 97+ %, A.C.S. reagent were used. Purified and deionized water was used in all experiments. The ethylene and ethane purities were 99.5% and 99.0%, respectively. Helium of 99.999% purity was used as the flow system sweep gas and the carrier gas for gas chromatography. The compressed gases were used without further purification.

Membrane Preparation. The as-received membranes were cleaned by boiling at 113 °C in concentrated nitric acid for 2 hours. After slowly cooling to room temperature, then membrane was rinsed thoroughly with water and then exchanged into the Na^+ form by boiling at 95 °C in 1M NaOH for 1 hour. The membrane was allowed to cool to room temperature and then rinsed thoroughly with water. To obtain the Ag^+ form membrane, a Na^+-form membrane was immersed in a 1M AgNO3 solution for 12 hours and subsequently rinsed with water. The hydration temperature of the membranes prepared by this procedure is 113 °C. To study the effects of hydration temperature, some of the membranes were prepared by cleaning and ion exchange at room temperature. The hydration temperature of the membranes prepared by this procedure was 22 °C.

Transport Measurements. The permeability test apparatus and membrane cell are similar to those used by Way et al. (21). Pure ethylene and ethane were mixed in a gas flow system to obtain the desired feed composition. Helium was used as the sweep gas and both the feed and sweep gases were saturated with water before entering the membrane cell. The gas flow configuration in the membrane cell was cross flow. The humidifiers and membrane cell were placed in a water bath for temperature control. The temperature was kept at 25 °C, except for the temperature effect study. Permeate and retentate from the membrane cell were dried by passage over anhydrous calcium sulfate before being vented or injected into the gas chromatograph. The feed and sweep gas pressures were controlled by back pressure regulators.

Results and Discussion.

Water Content and Hydration Temperature Effects. As mentioned earlier, the Dow membrane is more amorphous than Nafion 117. This allows the Dow membrane polymer matrix to adsorb more water than Nafion 117. The temperature at which the membrane is hydrated also influences the swelling of the ionomer matrix. The highest temperature used during the membrane preparation procedure controls the water content of the membrane. The water content of the membrane was calculated by dividing the weight of water absorbed by the total weight of the hydrated membrane. The water content of a membrane is primarily controlled by the inherent structure of the ionic polymer (Table I).

Table I. Water Content (% by Weight) of Ag^+ and Na^+-Form Membranes

Hydration Temperature (°C)	Ag^+ Form Nafion 117	Ag^+ Form Dow	Na^+ Form Nafion 117	Na^+ Form Dow
22	12.9	28.3	15.1	36.3
113	19.6	34.9	23.5	44.7

Both the water content and swelling of the polymer matrix have a large effect on the local ionic concentration, i.e. moles of Ag^+ per volume of water (Table II). The local Ag^+ concentrations in the polar ionic cluster regions of the PFSA membranes were calculated by assuming complete Ag^+ exchange. The molar ion-exchange capacity and the volume of water in the hydrated membrane were determined and the local Ag^+ concentration was calculated.

Table II. Local Ag^+ Concentration (M)

Hydration Temperature (°C)	Nafion 117	Dow Membrane
22	7.89	4.14
113	4.45	3.83

Despite a lower local Ag^+ concentration, the Dow membrane showed a higher ethylene permeability (Figure 2). The ethane permeability of the Dow membrane, 13.1 ± 0.5 Barrer, was also higher than that of Nafion 117, 5.8 ± 0.3 Barrer.

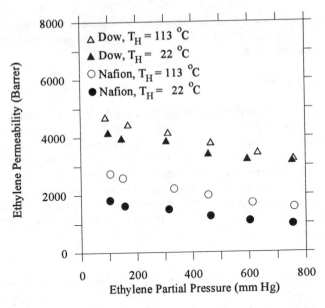

Figure 2. Ethylene/Ethane Separation Factors for Ag⁺-Form Nafion 117
and Dow Membrane at Various Hydration Temperatures

The same trend was also observed for the Na⁺ membrane. The average
ethylene and ethane permeabilities for Na⁺-form Nafion 117 (Table III) are lower
than those of Dow membrane (Table IV). This may be due to the higher water
content and the more open structure of the Dow membrane.

Table III. Ethylene and Ethane Permeabilities of Na⁺-Form Nafion 117

Hydration Temperature (°C)	Ethylene (Barrer)	Ethane (Barrer)
22	7.5	2.6
113	15.0	4.7

Table IV. Ethylene and Ethane Permeabilities of Na⁺-Form Dow Membrane

Hydration Temperature (°C)	Ethylene (Barrer)	Ethane (Barrer)
22	24.3	7.5
113	33.6	9.9

In contrast, Nafion 117 gave higher separation factors, defined as the ratio
of the ethylene permeability to the ethane permeability. Interestingly, the small
difference in water content due to a different hydration temperature does not have

a great effect to the separation factor (Figure 3). This is because the water content of a membrane influences both ethylene and ethane transport.

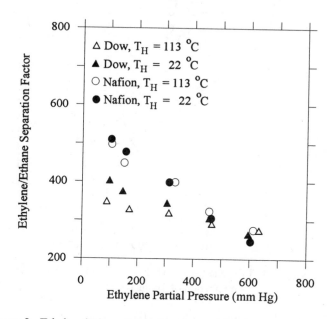

Figure 3. Ethylene/Ethane Separation Factor of Ag^+-Form Nafion 117 and Dow Membrane at Various Hydration Temperatures

   The facilitated permeability of a feed mixture of 50 mole% ethylene and 50 mole% ethane at atmospheric pressure measured by Eriksen et al. (*16*) was 1000 Barrer compared with a value 2100 Barrer observed in the present work. However, the separation factor observed by Eriksen et al. of 470 was higher than that reported in the present work.   Much higher ethylene permeabilities and separation factors were reported by Teramoto et al. (*8*).   In that study, an ethylene permeability of 3×10^5 Barrer with a C_2H_4/C_2H_6 separation factor of 1000 was obtained from a supported liquid membrane containing aqueous 4 M $AgNO_3$. The higher permeabilities observed by Teramoto are probably a result of higher ionic mobility in the supported liquid membrane.

Temperature Effects.   The temperature for each transport experiments was controlled with a water bath.   The membranes were tested at 5 , 25 and 35 °C. The results obtained from both Nafion 117 (Figure 4) and Dow membrane (Figure 5) show that the permeability increased as the temperature was raised.

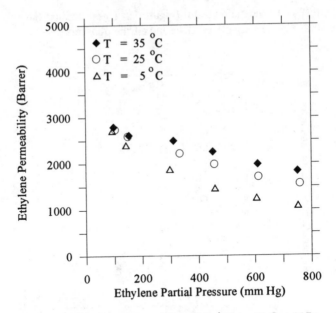

Figure 4. Ethylene Permeability of Ag⁺-Form Nafion 117
as a Function of Temperature

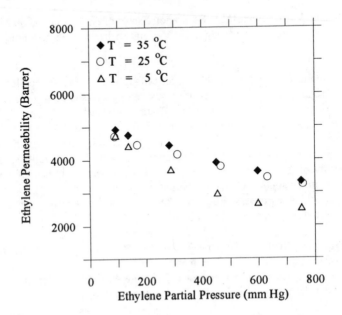

Figure 5. Ethylene Permeability of Ag⁺-Form Dow Membrane
as a Function of Temperature

For both membranes, the temperature effect was less pronounced when the partial pressure of ethylene was diminished. When the ethylene partial pressure was reduced to about 100 mm Hg, there was almost no change in the ethylene permeability with a 20 °C temperature increase. However, the same was not true for ethane. At constant temperature, the permeability of ethane was almost constant over the same range of partial pressure. The permeabilities of ethane at 5 , 25 and 35 °C were 3.7, 5.8 and 7.5 Barrer, respectively. This suggests that raising the temperature increases the diffusion coefficients of the gases in the membranes. The mathematical model developed by Way and Noble (*21*) was used to determine the diffusion coefficients of uncomplexed ethylene (D_A) and the Ag^+-ethylene complex (D_{AB}). The equilibrium constants (K_{eq}) for the reversible reaction between ethylene and Ag^+ were also calculated (Tables V and VI).

Table V. Calculated Diffusion Coefficients and Equilibrium Constants at Various Temperatures for Nafion 117

Temperature (°C)	D_A (cm²/sec)	D_{AB} (cm²/sec)	K_{eq} (M⁻¹)
5	9.7×10^{-7}	4.2×10^{-7}	253
25	3.4×10^{-6}	8.2×10^{-7}	214
35	4.5×10^{-6}	1.2×10^{-6}	168

Table VI. Calculated Diffusion Coefficients and Equilibrium Constants at Various Temperatures for Dow Membrane

Temperature (°C)	D_A (cm²/sec)	D_{AB} (cm²/sec)	K_{eq} (M⁻¹)
5	2.2×10^{-6}	9.4×10^{-7}	134
25	5.9×10^{-6}	1.9×10^{-6}	112
35	8.0×10^{-6}	2.0×10^{-6}	133

Surprisingly, the temperature does not have a great effect on the equilibrium constant for the reaction of Ag^+ and ethylene. All of the equilibrium constants shown in Tables V and VI are of the same magnitude. In contrast, a temperature increase from 5 to 35 °C enhances the ethylene diffusion coefficients by a factor of 3 to 4. Diffusion coefficients for the Ag^+-ethylene complex also increase by a factor of 2 to 3 over the same temperature range. This implies that the transport mechanism at low partial pressure of gas is controlled by gas solubility. More likely, a low concentration of gas at the gas-membrane interface limits the amount of gas adsorbed into a membrane.

These calculated diffusion coefficients and equilibrium constant for Nafion 117 are comparable to the values reported by Eriksen et al. (*16*) at 25 °C. Eriksen's effective diffusion coefficient for ethylene, as calculated from Fick's law, was 3.3×10^{-6} cm² sec⁻¹. The effective diffusion coefficient for the Ag^+-ethylene complex, D_{AB}, in their work is 2.0×10^{-7} cm² sec⁻¹. The disagreement in these values is attributed to the differences in the Ag^+ concentration in the Nafion 117 and the magnitude of the ethylene flux. The carrier concentration in Nafion 117 used for the calculation in the present work is half of the value employed by Eriksen et al. Moreover, the ethylene flux obtained from the Ag^+-form membrane is twice the value reported by Eriksen et al. Those authors reported an equilibrium constant, K_{eq}, of $390\ M^{-1}$.

Figures 6 and 7 present the influence of ethylene partial pressure and temperature on the ethylene/ethane separation factor. Both figures show the usual observation that permeability of the reactive gas increases with decreasing partial pressure. Consequently, the separation factor increases with decreasing ethylene partial pressure for both Ag^+-form Nafion 117 and Dow membranes.

For the most part, the separation factor decreases as the temperature increases. The transport model shows that the diffusion coefficient for ethylene is more sensitive to temperature variation than the diffusion coefficient of the Ag^+-ethylene complex. It seems reasonable that the diffusion coefficient of ethane is increased by a larger number than that of the complex. The decreasing separation factor may simply be explained by the fact that temperature favors the transport of ethane over the Ag^+-ethylene complex. When the ethylene partial pressure becomes small, the transport mechanism is controlled by gas solubility and the concentration of gas at gas-membrane interface. The amount of Ag^+-ethylene complex is decreased by the small amount of ethylene in the membrane and the effect of temperature becomes pronounced.

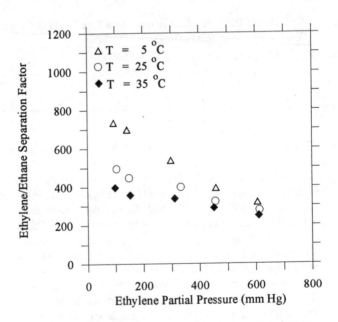

Figure 6. Ethylene/Ethane Separation Factor of Ag^+-Form
Nafion 117 as a Function of Temperature

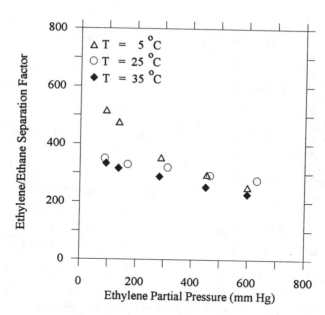

Figure 7. Ethylene/Ethane Separation Factor of Ag^+-Form
Dow Membrane as a Function of Temperature

Pressure Effects. Transport measurements with ethylene and ethane at partial pressures higher than 760 mm Hg were also performed. The partial pressures of ethylene and ethane were enhanced by increasing the feed side and sweep side total pressures. At a certain total pressure, the ethylene and ethane were premixed to a desired composition before entering the membrane cell. The pressure difference across the membranes was always maintained at zero. The temperature was 25 °C throughout the pressure effect measurements.

The ethylene flux obtained from Ag^+-form Nafion 117 increased consistently as the ethylene partial pressure increased (Figure 8). In contrast, the ethylene flux obtained from the Ag^+ form Dow membrane decreased when the feed and sweep total pressures were increased to 1.5 atmosphere (Figure 9).

Experimentally, it was found that the ethylene flux for the Ag^+-form Dow membrane started to decline when total pressure was higher than 1.3 atmosphere. However, this decline in flux was reversible. When total pressure was reduced to less than 1.3 atmosphere, the ethylene flux increased to the same value observed before the total pressure was increased to 1.3 atmosphere. A similar trend was also observed with ethane. The ethane flux obtained from Ag^+-form Nafion 117 increased consistently with an increasing partial pressure of ethylene regardless of the total pressure (Figure 10). However, a slight decrease of ethane flux with increasing total pressure was observed with the Ag^+-form Dow membrane (Figure 11).

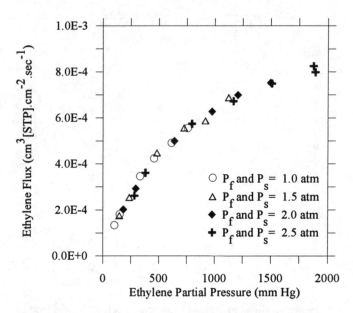

Figure 8. Ethylene Flux of Ag⁺-Form Nafion 117 as a Function
of Feed and Sweep Gas Total Pressures

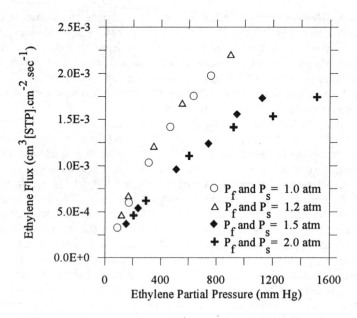

Figure 9. Ethylene Flux of Ag⁺-Form Dow Membrane as a
Function of Feed and Sweep Gas Total Pressures

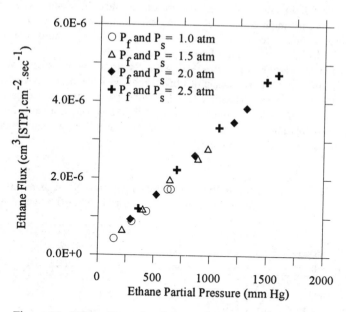

Figure 10. Ethane Flux of Ag⁺-Form Nafion 117 as a Function of Feed and Sweep Gas Total Pressures

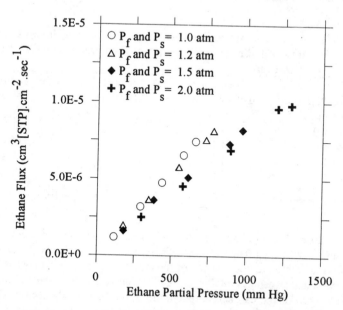

Figure 11. Ethane Flux of Ag⁺-Form Dow Membrane as a Function of Feed and Sweep Gas Total Pressures

The slight effect of total pressure on the flux of ethane, which is not a reactive species, implies that the total pressure influences the morphology of the Dow membrane. Since the Dow membrane is a totally amorphous ionic polymer, the diminished flux could result from membrane compaction.

Similar results were observed with Na^+-form membranes. Within the experimental error, the permeabilities of ethane and ethylene obtained from Na^+-form Nafion 117 were constant while those of Dow membrane decreased as the total pressures increased (Tables VII and VIII). However, a greater decrease in the ethane permeability was observed for the Ag^+-form Dow membrane compared to the Na^+-form Dow membrane.

Table VII. Average Ethane Permeabilities of Ag^+ and Na^+-Form Nafion 117 and Dow Membrane as a Function of Feed and Sweep Gas Total Pressures

Feed Side Total Pressure (atm)	Sweep Side Total Pressure (atm)	Ag^+ Form Nafion (Barrer)	Ag^+ Form Dow (Barrer)	Na^+ Form Nafion (Barrer)	Na^+ Form Dow (Barrer)
1.0	1.0	5.7	12.9	4.6	9.8
1.2	1.2	-	12.1	-	10.8
1.5	1.5	6.1	9.7	4.7	9.3
2.0	2.0	6.2	9.0	4.8	9.3
2.5	2.5	6.4	-	4.7	-

Table VIII. Average Ethylene Permeabilities of Na^+-Form Nafion 117 and Dow Membrane as a Function of Feed and Sweep Gas Total Pressures

Feed Side Total Pressure (atm)	Sweep Side Total Pressure (atm)	Nafion 117 (Barrer)	Dow Membrane (Barrer)
1.0	1.0	15.1	33.7
1.2	1.2	-	34.1
1.5	1.5	14.5	31.4
2.0	2.0	16.3	30.2
2.5	2.5	15.4	-

Figure 12 shows how the ethylene/ethane separation factor decreases with increasing ethylene partial pressure. These data were obtained by performing experiments where the composition of the ethylene/ethane binary feed was varied at four feed and sweep gas total pressures. The decrease of the separation factor is a consequence of the carrier saturation phenomena commonly seen in facilitated transport which causes the ethylene permeability to decrease with increasing ethylene partial pressure.

Since the ethane permeability is essentially constant, the separation factor, or ratios of permeabilities, decreases.

Note, however, that the ethylene/ethane separation factors for Ag^+-form Nafion 117 in Figure 12 are only a function of ethylene partial pressure, and are independent of the total feed pressure. In contrast, the separation factor data in Figure 13 appear to be more dependent on total feed gas pressure. Although the general trend of the data in Figure 13 is that separation factors decrease with increasing ethylene partial pressure, there is much more data scatter than the trend in Figure 12. This effect of total pressure may be due to structural changes in the hydrated Dow membrane produced by compaction at higher pressures. The water content of the Ag^+-form Dow membrane was almost twice that of the Ag^+-form Nafion membranes, suggesting that the Dow membranes are more "gel-like" and more deformable.

Conclusions. The separation of ethylene from ethane via facilitated transport of ethylene through perfluorosulfonic acid ion-exchange membranes has been studied. Nafion 117 and experimental PFSA membranes developed by Dow Chemical Company were used in this investigation. The different morphologies of these membranes lead to variation in their properties, such as water content and mechanical strength. A high water content in a membrane results in a higher permeability, but a lower ethylene/ethane selectivity. The hydration temperature of the membrane also has an effect on its water content. A membrane hydrated at 113 °C has a higher water content than one hydrated at 22 °C. However, the water content of a membrane is mainly controlled by the morphology of the membrane material. The effect of temperature on permeability and selectivity of the membranes was also studied. In general, permeabilities of ethylene and ethane increase and selectivity of ethylene over ethane decrease when temperature is raised. However, for a low partial pressure of ethylene on the feed side, the temperature has almost no effect on ethylene permeability. The total pressures also show an interesting result. The flux of ethylene thorough Ag^+ and Na^+-form Nafion 117 increases with increasing total pressure. In contrast, the flux of ethylene and ethane thorough the Ag^+ and Na^+-form Dow membrane drops when the total pressure surpasses 1.3 atm. The flux of ethane, which is not a reactive species, decreases as the feed and sweep gas total pressure increases at constant ethane composition. This suggests that membrane compaction at higher feed pressures causes structural changes in the membrane.

Acknowledgments. The authors gratefully acknowledge support from The Government of Thailand for A. Sungpet's graduate fellowship. Acknowledgment is made to the Donors of the Petroleum Research Fund, administered by the American Chemical Society, for the partial support of this research through Grant # 23187-G7. We would also like to thank Dr. Alan Cisar of the Dow Chemical Company for providing us with samples of PFSA membranes.

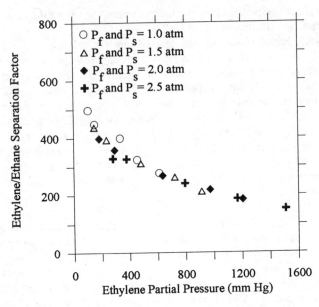

Figure 12. Ethylene/Ethane Separation Factor of Ag$^+$-Form Nafion 117
as a Function of Feed and Sweep Gas Total Pressures

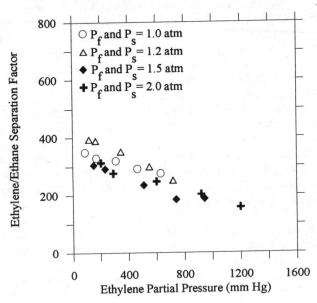

Figure 13. Ethylene/Ethane Separation Factor of Ag$^+$-Form Dow
Membrane as a Function of Feed and Sweep Gas Total
Pressures

Notation and Subscripts

P = total pressure
T = temperature
H = hydration
f = feed side
s = sweep side

Literature Cited

1. Humphrey, J. L.; Seibert, A. F.; Koort, R. A. *U. S. Department of Energy Report, 12920-1*,1991.
2. *Chem. Eng. News*, July 4, 1994.
3. Spillman, R.W. *Chem. Eng. Prog.* **1989**, *85*, 41.
4. Way, J.D.; Noble, R.D. *Membrane Handbook;* Van Nostrand Reinhold: New York, 1992; Chapter 44-Facilitated Transport, pp 833-866.
5. Hughes, R. D.; Steigelmann, E. F. *U.S. Patent 3,758,605*, 1973.
6. Hughes, R. D., Mahoney, J. A., Steigelmann, E. F., Paper read at the AIChE National Meeting, April 5-9, 1981, Houston, TX.
7. LeBlanc, O. H.; Ward, W. J.; Matson, S. L. *J. Membr. Sci.* **1980**, *6*, 339.
8. Teramoto, M.; Matsuyama H.; Yamashiro T.; Katayama Y. *J. Chem. Eng. Japan* **1986**, *19*, 419.
9. Kraus, M. A. *U.S. Patent 4,614,524*, 1986.
10. Kawakami, M.; Tateishi, M.; Iwamoto M.; Kagawa, S. *J. Membr. Sci.* **1987**, *30*, 105.
11. Teramoto, M.; Matsuyama, H.; Yamashiro, T.; Okamoto, S. *J. Membr. Sci.* **1989**, *45*, 115.
12. Kanno, T.; Fujita, S.; Kobayashi, M.; *J. Chem. Soc., Chem. Commun.* **1989**, 1854
13. Davis, J. C.; Valus, R. J.; Eshraghi, R.; Velikoff, A. E. *Sep. Sci. Tech.* **1993**, *28*, 463.
14. Yurkoestsky, A. V.; Bikson, B.; Kharas, G. B.; Watterson A. Paper read at the ACS Spring National Meeting, March 28-April 2, 1993, Denver, CO.
15. Ilinich, O. M.; Zamaraev, K. I. *J. Membr. Sci.* **1993**, *82*, 149.
16. Eriksen, O. I.; Aksnes E.; Dahl, I. M. *J. Membr. Sci.* **1993**, *85*, 89.
17. Eriksen, O. I.; Aksnes E.; Dahl, I. M. *J. Membr. Sci.* **1993**, *85*, 99
18. Tsou, D. T.; Blachman, M. W.; Davis J. C. *Ind. Eng. Chem. Res.* **1994**, *33*, 3209.
19. Richter, C.; Woermann, D. *J. Membr. Sci.* **1994**, *91*, 217.
20. Moore, R.B.; Martin, C. R. *Macromolecules* **1989**, *22*, 3594.
21. Way, J. D.; Noble, R. D.; Reed, D. L.; Ginley, G.M. *AIChE J.* **1987**, *33*, 480.

Chapter 20

Facilitated Transport of Unsaturated Hydrocarbons in Perfluorosulfonic Acid (Nafion) Membranes

Carl A. Koval, Debra Lee Bryant, Heidi Engelhardt, David Manley, Roberto Rabago, Paul M. Thoen[1], and Richard D. Noble

Departments of Chemistry and Biochemistry and of Chemical Engineering, University of Colorado, Boulder, CO 80309

In this paper, results from our groups and from others relating to separations of unsaturated hydrocarbons using Ag+-exchanged membranes are briefly reviewed. Next, some useful and interesting properties of Ag+-Nafion are summarized and equations defining common transport performance parameters are presented. This is followed by a brief description of membrane/distillation column hybrid processes as a motivation for developing better membrane materials. Since stability is a critical issue for all membrane separations, experimental data regarding the stability of Ag+-Nafion for both perstraction and pervaporation processes are presented. Finally, previously unreported performance data for close boiling alkene/alkane (olefin/saturate) mixtures and a variety of separations involving arenes (aromatic compounds) are described.

Due to their extensive use in the polymer industry and as solvents, there is a continuing need for better separation processes for alkenes and other unsaturated organic compounds from alkanes. Perfluorosulfonic acid (PFSA) membranes, such as Nafion (1), that have been ion-exchanged with silver(I) ion exhibit large transport selectivities for many unsaturated hydrocarbons with respect to saturates with similar physical properties. These selectivities are the result of reversible complexation reactions between the unsaturated molecules and Ag+ (2-4), which results in facilitated transport through the membranes (5).

---

[1]Current Address: Golden Technologies Company, Inc. 4545 McIntyre Street, Golden, CO 80403

0097–6156/96/0642–0286$15.00/0

Background and Motivation

The concept of using Ag+ in liquid membranes to promote facilitated transport of simple gaseous alkenes, specifically ethylene/ethane separations, began with papers by LaBlanc et al. (6) and Teramoto et al. (7,8). Interest in this process waned somewhat when it was discovered that Ag+ formed explosive side products with acetylene which was present in the feed stocks. Despite this potential problem, researchers at BP America developed a Ag+-based separation process for propene/propane separation (9).

For the past 8 years, our research groups have explored the use of Ag+-exchanged Nafion (1) membranes for the separation of various liquid phase unsaturates (10-21). Nafion (1) is a perfluorosulfonate ionomer membrane with outstanding chemical and thermal stability, as well as exceptional mechanical strength. Many studies have been performed on the chemical, morphological and structural properties of perfluorosulfonate ionomers (22-28). The chemical structure of Nafion consists of a Teflon-like backbone containing side chains that are ether linked and terminate in a sulfonate group (Figure 1, inset). Due to the extremely hydrophilic sulfonate groups and the very hydrophobic fluorocarbon backbone, the microstructure of Nafion consists of a series of ionic clusters interconnected by a network of channels (Figure 1). Nafion can absorb relatively large amounts (about 10-30% by mass) of water and other polar solvents due to the hydrophilicity of the ionic clusters. Data from X-ray and neutron scattering experiments indicate that the ionic clusters are approximately 50 Å in diameter while the channels that connect them are 10 Å wide.

Commercially available Nafion is 180 μm thick and has an equivalent weight (EW) of 1100 g/mol, indicating that most of the mass of the membrane is due to the fluorocarbon backbone. Nafion of 1100 EW is also commercially available as a solution. The casting of membranes from this solution has been studied and procedures have been developed make membranes with thicknesses as small as 2.5 μm (13,17,21,24,29-31).

Ion-exchange materials such as Nafion possess excellent properties as supports for immobilized liquid membrane studies. Although Ag+-exchanged Nafion membranes do not exhibit commercially viable permeabilities, these membranes possess many advantages over the use of microporous supports. Unlike microporous supports, an ionic carrier that is exchanged into an ion-exchange membrane cannot be lost from the membrane unless exchangeable ions are present in the phases contacting the membranes. Also, since the carrier loading is a function of the number of exchange sites within the membrane and not dependent on solubility, higher carrier loadings are possible. Finally, if the solvent is lost from the membrane, the ion-

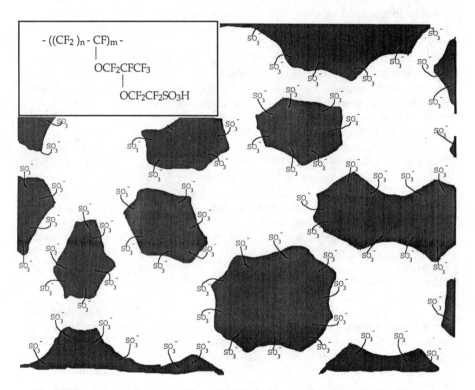

Figure 1.   Insert - Chemical Structure of DuPont's Nafion. Remainder - Cartoon Depicting Cluster-channel Network in Nafion membranes.   The dark sections represent regions containing mostly the fluorocarbon polymer backbone. The unshaded sections represent the ionic clusters which contain most of the absorbed water. The cations associated with the sulfonate groups have been omitted for clarity.

exchange membranes become a non-porous film preventing leaks from occurring.

The performance of Nafion membranes for separating various hydrocarbon mixtures can be characterized in several ways. Experimentally, fluxes (J, mol cm^{-2} s^{-1}) for individual solutes are determined using transport cells like those described below. [The results described in this paper utilize the same membrane preparation and other experimental procedures described in our earlier work (10-21)]. Nafion membranes can be ion-exchanged to produce either the Ag$^+$- or Na$^+$-containing forms by immersing the material in an aqueous solution containing the appropriate salt. The two forms absorb nearly the same amount of water. However, since Na$^+$ ion does not interact with organic molecules to any extent, flux through the Na$^+$-form provides information about transport due to physical solubility. The enhanced transport due to Ag$^+$ ion is usually expressed as a facilitation factor (F):

$$F = J^{total} / J^{diffusive} = J^{Ag+-Nafion} / J^{Na+-Nafion} \qquad (1)$$

where each flux is measured at the same driving force across the membrane. For nearly all membrane separations, the flux depends on the concentration driving force of a solute across the membrane. This dependence is often linear over certain concentration ranges and fluxes are sometimes normalized by dividing the observed flux by the concentration (or pressure) gradient. Also, over certain ranges of membrane thickness (L), many membrane materials exhibit fluxes for solutes that are inversely proportional to thickness. Therefore, thickness normalized fluxes (normalized flux = J x L) are often reported.

For separation of a mixture of solutes, a second important parameter is the separation factor (S). For a bicomponent mixture of solutes A and B:

$$S_{AB} = \frac{J_A}{J_B} \ X \ \frac{[B]_{feed} - [B]_{sweep}}{[A]_{feed} - [A]_{sweep}} \qquad (2)$$

where

S_{AB}^{obsd} = value observed using mixed feed solutions

S_{AB}^{ideal} = value calculated from two experiments involving single component feed solutions.

Since solutes with similar physical solubilities in a given membrane material tend to enhance each others solubility, it is often found that $S_{AB}^{obsd} < S_{AB}^{ideal}$. For the separation of styrene/ethylbenzene and for the separation of many diene/monoene mixtures, we have unexpectedly observed that $S_{AB}{}^{obsd} > S_{AB}{}^{ideal}$ due in part to competitive solubility effects (21). Unexpectedly large separation factors are also observed for diyne/monoyne pairs. For example, $S_{AB}{}^{ideal}$ for the separation of 1,7-octadiyne and 1-octyne (0.5 M feed solutions in isooctane, 20 μm cast Ag⁺-Nafion membrane) is 4.8 while the value of $S_{AB}{}^{obsd}$ is 31.

Various mathematical models have been developed to describe facilitated transport in liquid membranes (32-39). Under certain conditions (rapid complexation kinetics, large facilitation factors) the models allow calculation of the separation factor

$$S_{AB}{}^{calc} = \frac{[A]_0 \, K_A}{[B]_0 \, K_B} \cong \frac{K_A}{K_B} \tag{3}$$

where $[A]_0$ and $[B]_0$ are the physical solubilities of the solutes in the membrane and K_A and K_B are the equilibrium constants for complexation with the carrier, e.g. Ag⁺ ion. According to equation 3, separation factors can be estimated from solubility considerations and complexation data. With Ag⁺-Nafion or other ion-exchange materials in which the carrier is hydrated by water, it is tempting to make such predictions using data from the literature for aqueous solutions. However, experience with Ag⁺-Nafion membranes has often shown that $S_{AB}{}^{obsd} \neq S_{AB}{}^{calc}$. The unusual morphology and physical properties of Nafion are also partially responsible for the unanticipated separations that are observed and a number of studies have been directed at understanding the transport mechanisms in this material (17,18,21,40,41,42).

Potential Applications

Currently, virtually all industrial scale separations of hydrocarbons are performed by distillation (43,44). Distillation alone is inherently inefficient when the vapor/liquid equilibrium line is close to the operating lines in McCabe-Thiele diagrams. This occurs when components have similar volatilities, form azeotrope(s), or when high product purity is required. While it is unlikely that membrane separation processes will displace distillation, the use of membranes to enhance the performance of distillation columns is a possibility, if membrane materials with the necessary performance characteristics can be developed. Membrane/distillation column hybrid processes offer significant advantages in a number of situations:

- purification by distillation alone is impossible or requires a large number of theoretical trays;

- it is desirable to build new processes with reduced capital costs (fewer trays and/or lower operating-energy costs;
- it is desirable to modify an existing process to have lower operating costs, higher throughput, and/or higher purity product.

Recent theoretical work in our groups has been directed at developing design methodologies for membrane/distillation column hybrid processes (45-47). These simple design methods are based on minimizing the area under the operating lines in McCabe-Thiele diagrams and can be applied to any hybrid/distillation process. Using this method allows the design of the membrane unit to be decoupled from the design of the distillation column. It is possible to predict the best location of membrane with respect to the column and the potential reductions in theoretical trays required, the lowering of energy inputs and the increases in column throughput based on the membrane performance properties.

Stability Issues for Perstraction and Pervaporation Separations

Most of the results that we have reported for hydrocarbon separations involving Ag+-Nafion membranes are related to a process referred to as perstraction indicating that both the feed and sweep solutions are liquids. Transport cells for perstraction experiments are quite simple, as shown in Figure 2. An important issue regarding Ag+-Nafion facilitated transport flux measurements is the stability of fluxes and separation factors over extended periods of time. Typically, the facilitated alkene flux through a 20 μm Ag+-form membrane stabilizes within 30 minutes and flux experiments are generally run for several hours. To verify that the fluxes remain at the initial steady-state values for several days, a long-term alkene flux experiment was performed using a transport cell with very large feed and sweep volumes. A plot of sweep alkene concentration as a function of time is shown in Figure 3. The fluxes of both components are stable over one week and are approximately the same (±10%) as those observed over the first few hours of transport. Clearly, the facilitation effect is stable and short-term experiments generate accurate long-term steady-state flux values.

A pervaporation membrane separation system is one which contains a condensed phase feed solution and a vapor phase sweep. The two main pervaporation configurations involve either applying vacuum to the downstream side or flowing a sweep gas across the sweep interface. In both cases, rapid removal of permeate from the sweep interface and collection by condensation are the objectives. On a practical level, pervaporation is considerably more appealing than perstraction separations for several reasons. Condensed phase separations always result in the permeates being diluted into a receiving

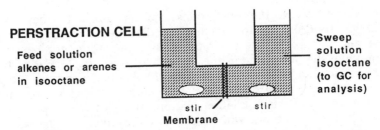

Figure 2.   Diagram of Transport Cell Used in Perstraction
Experiments.

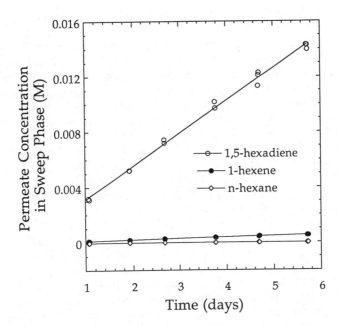

Figure 3.   Results for Long Term Perstraction Experiment.
Fluxes of 1,5-hexadiene, 1-hexene, and n-hexane (0.5 M each
in isooctane) through a 20 μm Nafion membrane. Feed and
sweep solutions were saturated with water.

phase. Therefore, if pure permeate is desired, an additional process of separating the permeate from the sweep solvent is required. The sweep solvent can usually be chosen such that the latter separation is easy, but materials, effort and energy must still be expended to achieve it. In contrast, a pervaporative separation generally collects only the permeates, undiluted in any receiving phase. Additionally, in pervaporation the permeates are continuously removed from the membrane sweep interface, hence the driving force is maintained at all times. With condensed phase systems, the permeate concentration in the sweep phase must be held below a certain level such that sufficient driving force is maintained. As is the case for perstraction systems, pervaporation separations potentially offer considerable energy savings compared to traditional separations such as distillation. They also have the ability to effectively separate mixtures that are difficult by distillation, such as close-boiling mixtures and azeotropes. Pervaporation systems have been studied since early in this century (48,49). However, work on identifying applications of the process for the oil and chemical industries was not undertaken until the 1960's (50). Since the 1980's a number of commercial pervaporation plants have come on line, mostly for the separation of azeotropic mixtures such as ethanol and water (51).

As mentioned previously, membrane transport can be described by three processes including partitioning of the permeate into the membrane phase, diffusion through the polymer matrix, and desorption from the membrane. Several differences distinguish pervaporation from perstraction processes. The first is that in pervaporation, the permeate undergoes a phase change on exiting the film. Therefore the heat of vaporization must be supplied from the surrounding medium, and a resultant temperature gradient may exist in the film. The various permeates may also have significantly different rates of vaporization. Finally, the lack of a swelling solvent in contact with the sweep interface may cause the membrane to contract near the sweep interface. The difference between condensed phase and pervaporation fluxes for a particular separation will depend critically on these parameters. In the case of Nafion films, the additional factor of membrane water content under pervaporation conditions is a dominant factor controlling the observed flux and separation factor.

Alkene pervaporation measurements were performed using the apparatus shown in Figure 4. The sweep reservoir could be varied to accommodate a dry flowing gas, a humidified flowing gas, or a vacuum. The membrane was secured in the cell and supported by a fiberglass mesh to prevent deformation from pressure gradients. To control the temperature of the membrane, the feed reservoir and membrane region were immersed in a water bath. Two collection U-tubes connected in parallel allowed continuous operation while collecting samples at

PERVAPORATION CELL

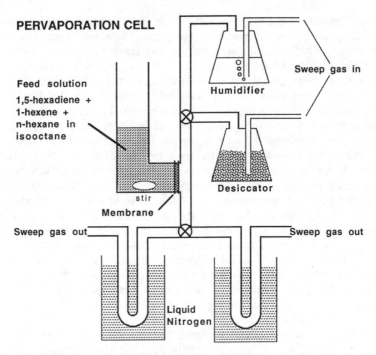

Figure 4.   Diagram   of   Transport   Cell   Used   in
Pervaporation Experiments.

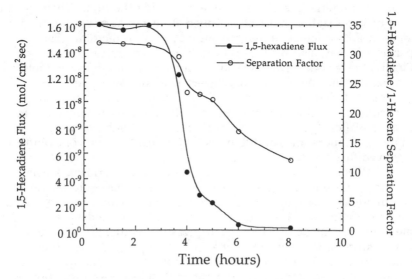

Figure 5.   Pervaporation of 1,5-Hexadiene and 1-Hexene
Through a 20 μm Nafion Membrane. Sweep gas is initially
water-saturated. At 2.5 hours, humidification is discontinued
and the sweep gas is directed through a desiccator.

various time intervals. After freezing out permeate samples in liquid nitrogen, each was diluted in isooctane and analyzed by gas chromatography.

The transport experiments involved evaluating the flux of three permeates (1,5-hexadiene, 1-hexene, and n-hexane) under the following conditions: a) humidified helium sweep, b) dry helium sweep, and c) vacuum sweep. The graph in Figure 5 shows the alkene fluxes and separation factor for a membrane which was initially swept by humidified helium. The initial diene flux was approximately 16×10^{-9} mol cm^{-2} sec^{-1} and the alkene separation factor was around 30. Comparison of these values with a membrane used in the perstraction mode reveals that the pervaporative flux is 50% less than the condensed phase flux and the separation factor is 35% smaller. The flux of n-hexane was virtually undetectable, indicating a minimum separation factor between 1,5-hexadiene and n-hexane of about 800. After 2.5 hours into the experiment, the sweep was directed through the desiccator instead of the humidifier; soon after both the diene flux and alkene separation factor began to fall. The diene flux fell two orders of magnitude before stabilizing and the alkene separation factor approached 10. After 8 hours the sweep gas was turned off and a vacuum of 1 torr was applied for several hours. The resultant diene flux was approximately 0.05×10^{-9} mol cm^{-2} sec^{-1} with an alkene separation factor of about 6.

The importance of water content on facilitated alkene transport is obvious from Figure 5. As water is withdrawn from the membrane, the facilitated flux falls by a factor of several hundred. In addition, because the separation factor decreases as the membrane loses water, it is clear that water plays a major role in the preferential transport of the diene compared to the monoene. The sensitivity of this separation to membrane water content is even more evident when comparing the initial pervaporative flux with the condensed phase flux. The sweep stream for the pervaporation experiment shown in Figure 5 had been passed through a humidifier, so it was expected that the sweep atmosphere was nearly 100% saturated in water vapor. However, as mentioned above, the diene flux through such a membrane was significantly lower than through a membrane in contact with a water-saturated isooctane sweep phase. It seems likely that in the pervaporation experiment, a thin layer of the membrane in contact with the humidified sweep gas was less than completely saturated and this slightly decreased level of hydration was responsible for both the reduced fluxes and reduced separation factors. The notion that the amount of water in Nafion membranes can differ significantly depending upon whether the membrane is in contact with a condensed solution or a saturated vapor has support in the literature. Pineri et al. found that a Nafion membrane in contact with water-saturated vapor contained overall less water and 35% less water just inside the membrane surface than a membrane which was immersed in liquid water at the same temperature(52). Therefore, even though the chemical potential of water in contact with a Nafion membrane surface is virtually the same in either the vapor or condensed phase, the amount of uptake and the distribution of water inside the membrane may differ dramatically. Simple absorption experiments can also be used to demonstrate that the presence of water in Ag^+-Nafion has a dramatic effect on the uptake of alkenes. If water is excluded from a feed solution containing 1-hexene in isooctane, the amount of 1-hexene absorbed into the membrane is diminished by more than a factor of 15 in comparison to the case where the feed solution is saturated with water. This difference persists even when the dry membrane is equilibrated with the feed solution for more than a week.

The results presented here demonstrate several points. Although the facilitated transport of alkenes in a pervaporative mode is not quite as productive as in a condensed phase mode, the separation is still highly effective for both alkene-alkene separations and alkene-paraffin separations. The requirement, however, is that the membrane must be swept with a water-saturated gas. This requirement at first may seem cumbersome due to consumption of a sweep gas and the presence of water with the collected permeate. However, the separation of water from hexenes is trivial and one can easily envision a system which recirculates a non-condensing sweep gas such as helium or nitrogen. The alkene-saturated water condensed from the sweep stream could also be recycled after thawing and separating from the organics.

Separations of Close Boiling Mixtures

Many experiments have demonstrated the ability of Ag^+-exchanged Nafion membranes to separate unsaturated hydrocarbons (alkenes and arenes) from saturated organics. This ability is potentially useful in the context of hybrid distillation-membrane processes and can be further demonstrated by performing separations of bi-component mixtures containing similar boiling points, i.e with relative volatilities close to 1. Table I is a compilation of several representative close boiling mixtures which would be difficult and expensive to separate by distillation alone. Each mixture consists of one alkane and one alkene. Therefore, the separation of the mixtures should be achievable based on their difference in reactivity with Ag^+.

Table II illustrates the effective separation of each close boiling mixture. Both Ag^+-exchanged and Na^+-exchanged membranes were used to see what effect the facilitation effect has on the separation factor. Since complexation does not occur in Na^+-exchanged membranes, separation is achieved from differences in permeate solubility and diffusivity. As can be seen, little separation is achieved for the close boiling mixtures based upon these physical properties.

Table I
Boiling Points and Relative Volatilities of
Several Alkene/Alkane Mixtures.

Mixture	Boiling point (°C)	Relative volatility
3-Methylpentane	63.3	
1-Hexene	63.3	1.009
3-Methylpentane	63.3	
2-Ethyl-1-pentene	64.7	1.0037
Isooctane	99.2	
2,4,4-Trimethyl-1-pentene	101.4	1.040
n-Heptane	98.4	
2,4,4-Trimethyl-1-pentene	101.4	1.045

Table II
The Fluxes, Separation Factors, and Facilitation Factors of Close Boiling
Mixtures Through 20 μm Thick, Cast 1100 EW Nafion Membranes.

Mixture[a]	Ag+-exchanged		Na+-exchanged		Facilitation factor
	Flux[b]	Sep. factor	Flux[b]	Sep. factor	
3-Methylpentane	0.017		0.012		
1-Hexene	5.7	330	0.047	3.9	121
3-Methylpentane	0.017		0.013		
2-Ethyl-1-butene	4.6	270	0.039	3.0	118
Isooctane	0.0061		0.0011		
2,4,4-Trimethyl-1-pentene	0.27	44	0.0034	3.1	79
n-Heptane	0.0047		0.0041	1.1	
2,4,4-Trimethyl-1-pentene	0.29	62	0.0036		80

[a] All mixtures are 0.50 M in each solute. The solvent is water-saturated isooctane, except for the isooctane/2,4,4-trimethyl-1-pentene mixture for which the solvent was water-saturated n-hexane.
[b] Units of mol cm^{-2} sec^{-1} x 10^{-9}

Facilitation of the alkene due to complexation by Ag+ leads to high facilitation factors and high separation factors for each mixture.

However, the extent of facilitation and the magnitude of the flux value are not constant for each alkene permeate. The facilitation factors are similar for both 1-hexene and 2-ethyl-1-butene signifying that the complexation reaction for these two molecules are very similar. The transport of 2,4,4-trimethyl-1-pentene exhibits both a lower flux and lower facilitation factor than for 1-hexene or 2-ethyl-1-butene. The reduced flux can be explained by the increase in molecular size causing lower physical solubility and slower diffusion in the hydrated membrane. These effects are also reflected in the fluxes through Na^+-form membranes. Alkane flux is dependent only upon solubility and diffusion in each membrane form. Similar alkane fluxes are observed for the Ag^+- and Na^+-exchanged membranes.

Separations of Arenes

Based on the data for complexation of arenes with Ag^+ ion in water, the separation of arenes from alkanes and of arene mixtures would seem an unlikely use of Ag^+-Nafion membranes (see equations 1-3). As an example, the equilibrium constants for 1-hexene, cyclohexene, and benzene with Ag^+ in water are 860, 79, and 2.4, respectively(2). Even though the trend in aqueous solubilities for these three components is opposite to the trend in K_{eq}'s, it is not obvious that large fluxes for solutes like benzene would be obtained. Nevertheless, for a bicomponent feed solution of benzene and methylcyclohexane (0.5 M each in isooctane), a separation factor S(benzene/methylcyclohexane) greater than 750 is obtained. An even more illustrative example is a tri-component feed solution containing benzene, cyclohexene and cyclohexane (0.5 M each) where the observed fluxes are 3.9×10^{-9}, 7.3×10^{-9} and $<0.01 \times 10^{-9}$ mol cm^{-2} s^{-1}, respectively. Even though the K_{eq}'s for benzene and cyclohexene differ by a factor of over 30, their fluxes through Ag^+-Nafion membranes differ by less than a factor of 2.

Since very large separation factors are obtained for arene/alkane mixtures, we also explored separations involving a variety of arene mixtures. The results shown in Table III illustrate that differences in flux values do occur for some aromatic mixtures. Potentially useful separation factors were found for the separation of benzene from alkyl arenes such as toluene, ethylbenzene, and cumene. For isomeric mixtures such as ethylbenzene/p-xylene, m-xylene/o-xylene, and p-xylene/o-xylene, little differences in flux values are found for either Na^+- or Ag^+-exchanged membranes.

From the Na^+-exchanged data it can be seen that there are not significant differences in the product of physical solubility and diffusivity for the various solute pairs. From the Ag^+-exchanged membrane experiments it can be seen that while the influence of complexation by Ag^+ does increase the flux values, it does not alter the separation factor. Therefore it can be concluded that the complexation constants for the solutes are similar. This assertion is supported by measurements that were made of solute solubility in the membranes (*14*).

Table III
Fluxes, Separation Factors, and Facilitation Factors of Arene Mixtures Through 20 μm Thick, Cast 1100 EW Nafion Membranes.

Mixture[a]	Ag+-exchanged		Na+-exchanged		Facilitation factor
	Flux[b]	Sep. factor	Flux[b]	Sep. factor	
Ethylbenzene	0.57		0.11		5.7
p-Xylene	0.57	1.0	0.11	1.0	5.7
m-Xylene	0.76		0.13		5.8
o-Xylene	0.93	1.2	0.14	1.1	6.6
p-Xylene	0.53		0.11	1.1	4.8
o-Xylene	0.62	1.2	0.10		6.2
Benzene	6.5	2.7	2.3	3.5	2.8
Toluene	2.4		0.66		3.6
Benzene	3.9	7.8	1.3	11	3.0
Ethylbenzene	0.50		0.12		4.2
Benzene	4.8	23			
Cumene	0.21				
Styrene	3.5	16	0.091	2.4	38
Ethylbenzene	0.22		0.038		

[a]All feed mixtures are 0.50 M of each solute. The solvent for each mixture is water-saturated isooctane.
[b]Units of mol cm-2 sec-1 x 10-9

The separations listed in Table III result from a combination of differences in solubility and diffusivity. A difference in physical solubility of benzene derivatives would be expected and is observed. In general, as the substituted side chain on benzene becomes longer its solubility in water is reduced due to its more organic nature. Therefore, if the separation was based solely on the differences in solubility, the order of decreasing flux values would be benzene > toluene > ethylbenzene > cumene. A difference in diffusion coefficients of the benzene derivatives would also be expected. In general, diffusion coefficients decrease as molecular size increases. Therefore, the order of decreasing diffusivity would be benzene > toluene > ethylbenzene > cumene. Both physical properties predict the trends seen for Na+-Nafion membranes. While incorporation of Ag+ ions into the membrane increases the magnitude of the flux values, it actually decreases the separation factor of the membrane. As can be seen from the facilitation factors, the longer the substituted chain on the benzene ring, the larger the facilitation factor becomes. Again, this trend is as expected. The electron donation to the ring is increased as the substituted chains lengthen, resulting in enhanced complexation to Ag^+.

Although Ag^+-Nafion membranes were unable to separate alkylarene isomers, the difference between ethyl and vinyl substituents on arenes does result in significant separation factors, e.g. the styrene/ethylbenzene separation. An even more subtle separation of this type is the separation of 3- and 4-divinylbenzene isomers (DVB) from 3- and 4-ethylvinylbenzene isomers (EVB). Commercially available DVB, which is important in polymer production, can contain over 40% EVB because these two substances are extremely difficult to separate by distillation. Perstraction of such a 58%/42% DVB/EVB mixture through a 25 μm Ag^+-Nafion membrane into isooctane produces a mixture containing less than 15% EVB. Better selectivity is obtained if the feed mixture is diluted into isooctane.

Conclusions

Experiments involving Ag^+-Nafion membranes clearly indicate that facilitated transport provides a mechanism for separating a wide variety of hydrocarbons. Although calculations indicate that membrane/distillation column hybrid processes could provide a potential market, less expensive membrane materials with better productivity and stability must be developed before commercialization of such processes will occur. Hopefully, the knowledge gained by studying facilitated transport of hydrocarbons in Ag^+-Nafion membranes will provide the necessary direction to promote the discovery of such membrane materials.

Acknowledgment

Various aspects of the research presented in this paper were supported by BP America, Chevron Research and Technology Co., the Ford Foundation, the National Science Foundation (CTS-9315090) and the NSF I/U CRC Center for Separations Using Thin Films.

References

1. Nafion® is a registered trademark of E. I. du Pont de Nemours.
2. Beverwijk, C. D. M.; Van Der Kerk, G.J.M. *Organometal. Chem. Rev. A.* **1970**, *5*, 215.
3. Hartley, F. R. *Chem. Rev.* **1973**, *73*, 163.

4. Winstein, S.; Lucas, H.J. *J. Am. Chem. Soc.* **1938**, *60*, 836.
5. Noble, R. D.; Koval, C. A.; Pellegrino, J. J. *Chem. Eng. Prog.* **1989**, *85*, 58.
6. LeBlanc, O. H.; Ward, W. J.; Matson, S. L.; Kimura, S. G. *J. Membr. Sci.* **1980**, *6*, 339.
7. Teramoto, M.; Matsuyama, H.; Yamashiro, T.; Katayama, Y. *J. Chem. Eng. Japan.* **1986**, *19*, 419.
8. Teramoto, M.; Matsuyama, H.; Yonehara, T. *J. Membr. Sci.* **1990**, *50*, 269.
9. Davis, J. *Sep. Sci. Technol.* **1993**, *28*, 463.
10. Engelhardt, H., Honors thesis, Univ. of Colorado, 1994.
11. Koval, C. A.; Spontarelli, T. *J. Am. Chem. Soc.* **1988**, *110*, 293.
12. Koval, C. A.; Spontarelli, T.; Noble, R. D. *Ind. Eng. Chem. Res.* **1989**, *28*, 1020.
13. Koval, C. A.; Spontarelli, T.; Thoen, P.; Noble, R. D. *Ind. Eng. Chem. Res.* **1992**, *31*, 1116.
14. Manley, D., M.S. thesis, Univ. of Colorado, 1994.
15. Manley, D. S.; Williamson, D.; Noble, R. D.; Koval, C. A. *Chem. Mat.* **1996**, *submitted.*
16. Rabago, R.; Noble, R. D.; Koval, C. A. *Chem. Mat.* **1994**, *6*, 947.
17. Rabago, R., Ph.D. thesis, Univ. of Colorado, 1994.
18. Rabago, R.; Noble, R. D.; Koval, C. A. *Ind. Eng. Chem. Res.* **1995**, *in press.*
19. Spontarelli, T., Ph.D. thesis, University of Colorado, 1989.
20. Thoen, P. M., Ph.D. thesis, University of Colorado, 1993.
21. Thoen, P. M.; Noble, R. D.; Koval, C. A. *J. Phys. Chem.* **1994**, *98*, 1262.
22. Gierke, T. D.; Munn, G. E.; Wilson, F. C. *J. Poly. Sci: Poly. Phys. Ed.* **1981**, *19*, 1987.
23. Kumar, S.; Pineri, M. *J. Polym. Sci. Part B: Polym. Phys.* **1986**, *24*, 1767.
24. Martin, C. R.; Moore, R. B. I. *Macromolecules* **1988**, *21*, 1334.
25. Moore, R. B.; Martin, C. R. *Macromolecules* **1988**, *21*, 1334.
26. Moore, R. B.; Martin, C. R. *Macromolecules* **1989**, *22*, 3594.
27. Mizutani, Y. *J. Membr. Sci.* **1990**, *49*, 121.
28. Halim, J.; Büchi, F. N.; Haas, O. *Electrochim. Acta.* **1994**, *39*, 1303.
29. Martin, C. R.; Moore, R. B. I. *Anal. Chem.* **1986**, *58*, 2569.
30. Martin, C. R.; Rhoades, T. A.; Ferguson, J. A. *Anal. Chem.* **1982**, *54*, 1639.
31. Gebel, G.; Alderbert, P.; Pineri, M. *Macromolecules* **1987**, *20*, 1425.
32. Kemena, L. L.; Noble, R. D.; Kemp, N. J. *J. Membr. Sci.* **1983**, *15*, 259.
33. Noble, R. D.; Way, J. D.; Powers, L. A. *Ind. Eng. Chem. Fundam.* **1986**, *25*, 450.
34. Noble, R. D. *Gas Sep. Pur.* **1988**, *2*, 16.
35. Noble, R. D. *J. Membr. Sci.* **1992**, *75*, 121.
36. Kirkkopru, A.; Noble, R. D. *J. Membr. Sci.* **1992**, *42*, 13.
37. Dindi, A.; Noble, R. D.; Yu, J.; Koval, C. A. *J. Membr. Sci.* **1992**, *66*, 55.

38. Dindi, A.; Noble, R. D.; Koval, C. A. *J. Membr. Sci.* **1992**, *65*, 39.
39. Jemaa, N.; Noble, R. D. *J. Membr. Sci.* **1992**, *70*, 289.
40. Leddy, J.; Vanderborgh, N. E. *J. Electroanal. Chem.* **1987**, *235*, 299.
41. Way, J. D.; Noble, R. D. *J. Membr. Sci.* **1989**, *46*, 309.
42. Whiteley, L. D.; Martin, C. R. *J. Phys. Chem.* **1989**, *93*, 4650.
43. Humphrey, J. L., ed. *Separation Technologies - Advances and Priorities. Final Report*; U.S. DOE: Washington, D.C., **1991**.
44. Humphrey, J. L., ed. *Separation Technologies - Marketing Factors. Final Report*; U.S. DOE: Washington, D.C., **1991**.
45. Moganti, S.; Noble, R. D.; Koval, C. A. *J. Membr. Sci.* **1994**, *93*, 31.
46. Stephan, W.; Noble, R. D.; Koval, C. A. *J. Membr. Sci.* **1995**, *99*, 259.
47. Pettersen, T.; Argo, A.; Noble, R. D.; Koval, C. A. *Gas Sep. Pur.* **1995**, *submitted*,
48. Kahlenberg, L. *J. Phys. Chem.* **1906**, *10*, 141.
49. Kober, P. A. *J. Am. Chem. Soc.* **1917**, *39*, 944.
50. Binning, R. C.; Lee, R. J.; Jenning, J. F.; Martin, E. C. *Industr. Eng. Chem.* **1961**, *53*, 45.
51. Aptel, P.; Challard, N.; Cuny, J.; Néel, J. *J. Membr. Sci.* **1976**, *1*, 271.
52. Pineri, M.; Thomas, M.; Escoubes, M. In *Third International Conference on Pervaporation Processes in the Chemical Industry*, Nancy, France, **1988**.

Chapter 21

Heavy Metal Ion Separation by Functional Polymeric Membranes

Takashi Hayashita

Department of Chemistry, Saga University,
Honjo Saga 840, Japan

Certain heavy metal cations such as Pb(II) and Cd(II) are extracted
into organic solvents as anionic metal halide complexes by liquid-
anion exchangers. Similar ion-association sorption is obtained by
anion-exchange polymers. The potential of these phenomena for
application in heavy metal ion separation by functional charged
membranes has been explored. Selective separation of Pb(II) and
Cd(II) chloride complexes was performed by both liquid membrane
containing liquid anion-exchangers as mobile carriers and solid
polymeric membrane containing anion-exchange sites as fixed carriers.
Donnan equilibrium of the system allows Pb(II) and Cd(II) chloride
complexes to be concentrated in the aqueous receiving phase. To
enhance permeation selectivity and efficiency, a novel polymeric
plasticizer membrane, which is composed of cellulose triacetate
polymer as a membrane support, o-nitrophenyl octyl ether as a
membrane plasticizer and trioctylmethylammonium chloride as an
anion-exchange carrier, has been developed. This type of membrane
exhibited a superior separation selectivity and efficiency to the
conventional anion-exchange membranes.

Selective separation of toxic heavy metal ions from waste solutions is frequently
required in hydrometallurgical processing (1). Solvent extraction is known to be an
useful method to separate such metal ions from solutions. To make a continuous
process of such a system, liquid membrane separations have been developed (2, 3).
However, the unstability of these systems often pose a problem to practical industrial
applications. With respect to the system stability, polymeric membrane separations
may be feasible. The problem for the polymeric membrane separations is their poor
transport selectivity toward heavy metal ions. Thus a development of the polymeric
membrane separation which can show a comparable metal separation ability with the
liquid membrane separation is an important issue to perform the practical heavy metal
ion separation (4).

Certain heavy metal ions complex with chloride, bromide, and iodide ions to
form metal halide complexes in solution, according to the following equation (5):

$$M^{2+} + X^- \underset{}{\overset{K_1}{\rightleftharpoons}} MX^+ + X^- \underset{}{\overset{K_2}{\rightleftharpoons}} \cdots \underset{}{\overset{K_{m-1}}{\rightleftharpoons}} MX_{m-1}^{(m-3)-} + X^- \underset{}{\overset{K_m}{\rightleftharpoons}} MX_m^{(m-2)-} \qquad (1)$$

Here, K_m represents the m-th successive formation constant between M^{2+} and the halide ion. These complexes are selectively extracted into an organic solution containing an liquid anion-exchanger such as lipophilic quaternary ammonium salts (6, 7). According to the differences in the formation constants of metal ions with halide ions and the interaction between metal halide complexes and extractant cations, the selective extraction of heavy metal ions has been observed. A similar sorption of metal halide complexes has been reported for anion-exchange polymers (8). Thus a membrane having anion-exchange sites is expected to show an effective recognition ability for heavy metal halide complexes.

Based upon the above concept, three different types of membrane separation, i.e., (1) liquid membrane separation which utilizes lipophilic anion-exchangers as mobile carriers, (2) polymeric membrane separation in which the anion-exchange sites function as fixed carriers, and (3) polymeric plasticizer membrane separation in which the membrane is composed of polymeric support, membrane plasticizer, and lipophilic anion-exchangers as a novel membrane material are discussed in relation to their transport efficiency and selectivity for separation of heavy metal chloride complexes.

Liquid Membrane Separation

Extraction of Heavy Metal Chloride Complexes.   The successive formation constants of some heavy metal ions with chloride ion are summarized in Table I (5). Since Pb(II), Cd(II) and Hg(II) are strongly complexed by chloride ion, selective extraction systems for the separation of these heavy metal ions can be designed (9-14).

Results for single metal extraction of Pb(II), Cd(II) and Zn(II) from aqueous solutions containing chloride or nitrate ions into chloroform solutions of tetra-n-heptylammonium chloride (THAC) or tetra-n-heptylammonium nitrate (THAN) are presented in Table II (14). The negligible extraction observed in the nitrate systems is attributed to a low complexing ability of nitrate for these heavy metal ions. On the other hand, high (89%) and selective extraction of Cd(II) was noted when the aqueous-phase chloride ion concentration was 0.020 M. When the chloride ion concentration was increased to 0.52 M, substantial amounts of Zn(II) and Pb(II) were extracted into the chloroform phase, and the extraction efficiency order was Cd(II) (100%) > Zn(II) (69%) > Pb(II) (44%).

For back-extraction of the heavy metal chloride complexes from the chloroform phase into 1.0 mM HCl, the efficiency was reversed: Pb(II) (99%) > Zn(II) (92%) > Cd(II) (75%). Thus a very high efficiency for back-extraction of Pb(II) was observed.

Table I. Successive Stability Constants of Metal Chloride Complexes

Metal Ion	Successive Stability Constants					(Media)
	$\log K_1$	$\log K_2$	$\log K_3$	$\log K_4$	$\log K_5$	
Ag(I)	3.04	2.00	0.00	0.26		
Cd(II)	1.58	0.65	0.12			(3 M NaClO$_4$)
Cu(II)	0.08					
Fe(II)	0.66					(1 M HClO$_4$)
Hg(II)	6.74	6.48	0.95	1.05		(0.5 M NaClO$_4$)
Ni(II)	-0.25	0.20				(3 M NaClO$_4$)
Pb(II)	1.23	0.53	0.39	-0.57	-0.28	(4 M NaClO$_4$)
Zn(II)	-0.49	0.59	0.09			(2 M NaClO$_4$)

Table II. Extraction of 1.0 mM Heavy Metal Cations from Aqueous Chloride or Nitrate Solutions into 0.10 M Chloroform Solutions of Tetraheptyl-ammonium Ions (THA$^+$)

X^-	$[X^-]$, M	Percent extraction, %		
		Pb(II)	Cd(II)	Zn(II)
Cl^-	0.020	5.2	88.8	2.8
Cl^-	0.52	43.9	100	69.2
NO_3^-	0.020	0.0	0.0	0.0
NO_3^-	0.52	1.7	2.5	0.0

(Reproduced with permission from Ref. 14. Copyright 1991, American Chemical Society)

When an ion pair is formed between THA$^+$ and an anionic metal chloride complex in the organic phase, the following equilibrium can be derived:

$$(M^{2+})_a + (Cl^-)_a + n(THAC)_o \overset{K_{ex}}{\rightleftharpoons} \{(THA)_n \cdot MCl_{n+2}\}_o \tag{2}$$

where subscripts a and o denote the aqueous and organic phase, respectively. The overall extraction constant, K_{ex}, is

$$K_{ex} = \frac{[(THA)_n \cdot MCl_{n+2}]_o}{[M^{2+}]_a[Cl^-]_a^2[THAC]_o^n} \tag{3}$$

which is represented logarithmically as

$$\log(\beta D) = 2\log[Cl^-]_a + n\log[THAC]_o + \log K_{ex} \tag{4}$$

where β is the coefficient for the chloride ion concentration when the overall stability constant is $\beta_i (= K_1 K_2 \cdots K_i)$. The distribution ratio (D) and β are defined by

$$D = \frac{[(THA)_n \cdot MCl_{n+2}]_o}{\Sigma [MCl_i^{2-i}]_a} \tag{5}$$

$$\beta = 1 + \Sigma \beta_i [Cl^-]_a^i \tag{6}$$

The β values at various chloride ion concentrations can be calculated from the successive formation constants. From equation 4, it is seen that for extraction according to equation 2, a plot of $\log(\beta D)$ versus $\log a_{Cl^-}$ would be linear with a slope of 2. The a_{Cl^-} represents activity of chloride ion calculated from Debuy-Huckel equation (*15*). The coordination number n can be determined from a plot of $\log (\beta D)$ against $\log [THAC]$ in the organic phase. For extraction of Cd(II) and Pb(II) into chloroform by THAC at several different aqueous-phase chloride ion

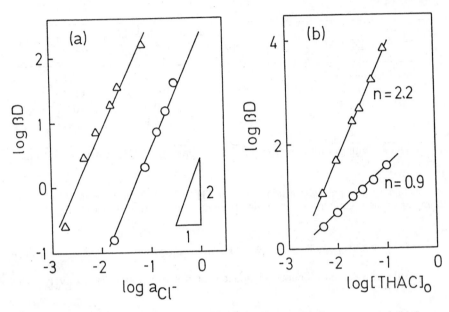

Figure 1. Relationship between Log (βD) Versus (a) Log a_{Cl^-} and (b) Log [THAC]$_0$. (o) Pb(II), (Δ) Cd(II). (Reproduced from Ref. 14. Copyright 1991 American Chemical Society.)

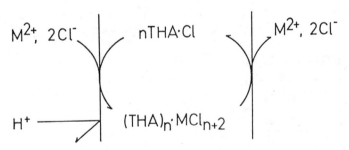

Figure 2. Transport Mechanism for Metal Chloride Complexes through the Liquid Anion-exchange Membrane. (Reproduced from Ref. 14. Copyright 1991 American Chemical Society.)

concentrations and organic-phase THAC concentrations, the plots are shown in Figure 1.

Both plots of log (βD) versus log a_{Cl^-} are linear with slopes of 2 (Figure 1a), which is consistent with the extraction equilibrium given in equation 2. Figure 1b shows that the slopes of log (βD) versus log [THAC]$_0$ are different for the extraction of Cd(II) and Pb(II). For Cd(II), the slope of 2 reveals a stoichiometry of (THA)$_2$CdCl$_4$ for the extraction complex, whereas for Pb(II) the slope of 1 shows a stoichiometry of THA·PbCl$_3$. By use of this specific extraction, a selective liquid membrane separation which combines extraction and stripping in a single system can be designed.

Bulk Liquid Membrane Separation. The permeation principle for the liquid membrane separation system is shown in Figure 2 (*14*). In this system, the liquid anion exchanger THAC functions as a mobile carrier. Heavy metal ions are selectively extracted at the membrane-source phase interface, transported through the organic membrane to the membrane-receiving phase. Transport is driven by concentration gradients of both metal and chloride ions. As indicated in Figure 2, the membrane is permeable only to heavy metal and chloride ions. At equilibrium, the Donnan equilibrium equation is

$$[M^{2+}]_s[Cl^-]_s^2 = [M^{2+}]_r[Cl^-]_r^2 \tag{7}$$

where the subscripts s and r denote the source and receiving phases, respectively. Thus, the metal ion species can be enriched in the receiving phase by a large concentration gradient of chloride ion across the membrane.

The dialysis cell shown in Figure 3 was utilized for the bulk liquid membrane separation. The source, membrane, and receiving-phase solutions were circulated from reservoirs through the three compartments. The compartments were separated from each other by cellulose dialysis membranes of area 6.6 cm^2. The aqueous source phase was a 250 mL of 0.50 M HCl with 1.0 mM Pb(II) and Cd(II). A 0.10 M chloroform solution (10 mL) of THAC was the membrane phase. The aqueous receiving phase (25 mL) was pure water.

Plots of the enrichment factor (EF, the metal ion concentration in the receiving phase divided by the metal ion concentration in the source phase) versus time and the chloride ion concentration in the source phase are shown in parts a and b of Figure 4, respectively. Selective permeation of Pb(II) over Cd(II) was observed (Figure 4a), which is the opposite of the extraction efficiencies for these two heavy metal ion species (Table II). The EF for both heavy metals increased with time and reached values of 4.5 and 0.9 for Pb(II) and Cd(II), respectively, after 26 h.

Figure 4b shows the influence of the source phase chloride ion concentration on EF after 24 h for both Pb(II) and Cd(II). The transport selectivity for Cd(II), which is noted at low chloride ion concentrations (4-24 mM), changes to Pb(II) transport selectivity when the chloride ion concentration is 0.10 M or above.

To prove the reason for the change in transport selectivity, the concentrations of Cd(II) and Pb(II) in the membrane phase were determined as a function of the chloride ion concentration in the source phase (Table III). Retention of Cd(II) in the membrane phase increased significantly as the chloride ion concentration was enhanced. On the other hand, no significant retention in the membrane phase was noted for Pb(II). This demonstrates that release of the Cd(II) chloride complex at the membrane-receiving phase interface is the rate-determining step for Cd(II) transport.

Emulsion Liquid Membrane Separation. The principle of the emulsion liquid membrane technique, which was discovered by Li (*17*), is based on the separation of internal (receiving phase) and external (source phase) aqueous solutions by an organic liquid membrane that contains an emulsifying agent (surfactant) and carrier

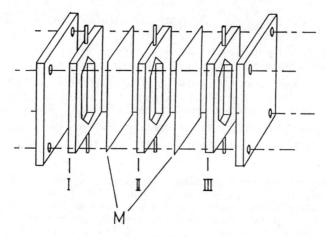

Figure 3. Bulk Liquid Membrane Cell: (I) Source Phase, (II) Membrane Phase, (III) Receiving Phase, (M) Cellulose Dialysis Membrane. (Reproduced from Ref. 14. Copyright 1991 American Chemical Society.)

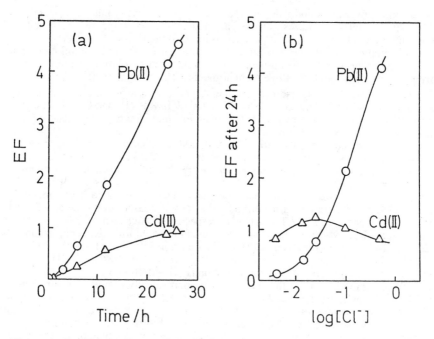

Figure 4. Enrichment Factor (EF) of Metal Ions as a Function of (a) Permeation Time and (b) Chloride Ion Concentration in the Source Phase Solution. (o) Pb(II), (Δ) Cd(II). (Reproduced from Ref. 14. Copyright 1991 American Chemical Society.)

Table III. Metal Ion Concentration in the Organic Membrane Phase as a Function of the Chloride Ion Concentration in the Aqueous Source Phase

[Cl⁻], M	Concentration,[a] mM	
	Cd(II)	Pb(II)
0.004	1.6	0.5
0.014	2.8	0.1
0.024	3.2	ND[b]
0.104	3.7	ND[b]
0.504	5.6	ND[b]

[a]After permeation for 24 h. [b]Less than 0.1 mM.
(Reproduced with permission from Ref. 14. Copyright 1991, American Chemical Society)

species in a W/O/W type emulsion. Rapid separation can be achieved due to the very thin liquid membrane (the skin of the emulsions droplet), which has a very large area (18-21).

Nakashio et al. reported the use of lipophilic dialkyldimethlylammonium bromides as surfactants in the formation of emulsion liquid membrane systems (22). The two long alkyl chains provided a good hydrophilicity-lipophilicity balance (HLB) to form a stable emulsion membrane. This finding led us to attempt the use of such quaternary ammonium salts in an emulsion liquid membrane system for which the ammonium salt functions not only as the emulsifying agent but also as the carrier for heavy metal halide complexes (Figure 5).

A 0.01 M solution of dimethyldioctadecylammonium bromide (DDAB) in toluene (2 mL) was chosen as the organic liquid membrane. The external aqueous source phase (20 mL) was a 0.50 M KCl solution with 1.0 mM Cd(II), Pb(II), and Zn(II). The internal aqueous receiving phase (2 mL) was pure water. The ultrasonication (100 W, 200 Hz for 60 s) of the membrane and receiving phases produced a W/O emulsion that was poured into the stirred source phase (600 rpm) to form a W/O/W emulsion system (14).

Results for the competitive permeation of Cd(II), Pb(II), and Zn(II) chloride complexes as a function of time are shown in Figure 6a. Selective permeation of Pb(II) was observed in agreement with the observation for the bulk liquid membrane transport system. It is noteworthy that EF for Pb(II) reached 21.5 after only 15 min in the emulsion liquid membrane system. No significant permeation of Zn(II) was observed, presumably due to low extractability of Zn(II) from the aqueous source phase by the low DDAB concentration in the membrane phase. This suggested that the membrane might also serve as a barrier to other heavy metal ion species. In agreement, transport of Cu(II), Fe(III), and Ni(II) was found to be negligible under these conditions.

The influence of the source phase chloride ion concentration upon the competitive emulsion liquid membrane transport of Cd(II), Pb(II) and Zn(II) was also examined (Figure 6b). The EF value after permeation for 10 min for Cd(II) and Pb (II) transports were enhanced as the chloride ion concentration in the source phase was increased. Consequently, selective permeation and concentration of Pb(II) was performed in the liquid membrane separation.

Polymeric Membrane Separation

Similar to the solvent extraction system mentioned above, a selective sorption of heavy metal chloride complexes has been reported for anion-exchange resins (8, 13). Thus a polymeric membrane having an anion-exchange sites (anion-exchange

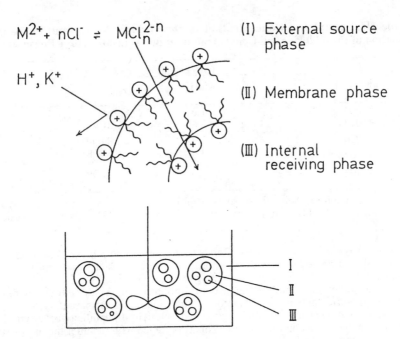

Figure 5. Emulsion Liquid Membrane and System. (Reproduced from Ref. 14. Copyright 1991 American Chemical Society.)

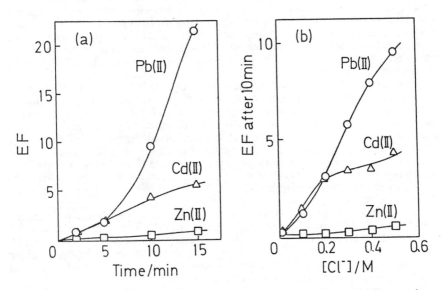

Figure 6. Enrichment Factor (EF) of Metal Ions as a Function of (a) Permeation Time and (b) Chloride Ion Concentration in the Source Phase Solution. (o) Pb(II), (Δ) Cd(II), and (□) Zn(II). (Reproduced from Ref. 14. Copyright 1991 American Chemical Society.)

membrane) is expected to show an effective separation ability for some heavy metal chloride complexes. In this separation, the metal ion is selectively sorbed (or "extracted") into the membrane phase as an anionic chloride-complex form and transfers across the membrane to the opposite-side solution. A large concentration gradient of chloride ions across the membrane causes an "uphill" transport of heavy metal chloride complexes on the basis of Donnan equilibrium as shown in equation 7.

Sorption of Heavy Metal Chloride Complexes. Figure 7 shows the sorption behavior of Cd(II) and Zn(II) on anion-exchange resin (Amberlite CG-400, Rohm & Hass Co. Ltd.) as a function of chloride ion concentration in solution (*13*). Cd(II) sorption increased significantly with an increase in the chloride ion concentration. This indicates that Cd(II) is sorbed in the form of an anionic chloride complex, such as $CdCl_n^{2-n}$. The selective sorption was noted for Cd(II) over Zn(II). As shown in Table I, the observed sorption selectivity is mainly attributed to the difference in complex formation ability of metal ion with chloride. In general, the ion-exchange resins exhibit a high sorption ability for less-hydrated ions (*23*). Thus the difference in hydrophobic nature of metal chloride complexes also affects the sorption selectivity. When the ion pair is formed between the anion-exchange site and the heavy metal chloride complexes, the following equilibrium can be derived:

$$n(RCl)R + (M^{2+})_a + 2(Cl^-)_a \overset{K_{ex}}{\rightleftharpoons} (R_n \cdot MCl_{n+2})R \tag{8}$$

Here, R^- represents the anion-exchange site in the resin, and subscript R denotes resin phase. Similar to equation 4, the relationship between log (βD) and log a_{Cl^-} is expressed by

$$\log (\beta D) = 2\log a_{Cl^-} + n\log[RCl]_R + \log K_{ex} \tag{9}$$

Figure 8 shows this relationship for sorption of Cd(II) and Zn(II). Both plots gave a straight line with a slope of 2, indicating that the reaction depicted in equation 8 indeed took place. Although the n-value was not determined due to difficulty in varying the $[RCl]_R$ value in equation 9, the n-value of 1 or 2 may be reasonable in comparison with the ion-association extraction behavior (*13*).

Polymeric Membrane Separation. On the basis of the selective interaction of heavy metal chloride complexes with the anion-exchange site in the resin, a selective transport system of heavy metal ions through an anion-exchange membrane may be feasible (*13*). Figure 9 presents the permeation principle of this system in which a 1:1 ion-pair was assumed between heavy metal chloride complex and ion-exchange site in the membrane for simplicity; n=3 in equation 8. In this permeation, the heavy metal chloride complex preferentially sorbed in the membrane phase must migrate between the fixed carriers (anion-exchange sites). If this migration occurs smoothly, a selective permeation of the heavy metal ion can be expected. As shown in equation 7, Donnan equilibrium of the system allows the heavy metal ions to be concentrated in the aqueous receiving phase.

The permeation experiments were carried out with a cylindrical glass cell shown in Figure 10. The anion-exchange membrane (Selemion AMV, Asahi Glass Co. Ltd.) was attached to the bottom of a cell (membrane area, 1.1 cm^2), and the cell, which contained 5.0 mL of pure water as a receiving-phase solution, was dipped into 97 mL of the source-phase solution containing $MgCl_2$ and 0.1 mM heavy metal ions.

Figure 11 shows a competitive permeation of Cd(II) and Zn(II) through the anion-exchange membrane as a function of chloride ion concentration in the source-phase solution. The ordinate represents the EF value after 40 h. As expected, the

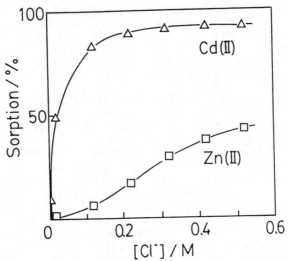

Figure 7. Effect of Chloride Ion Concentration on Metal Sorption: MgCl2 Solution (10 mL, pH 1.7) Containing 1.0 mM Cd(II) and Zn(II); Anion-exchange Resin (0.10 g); (Δ) Cd(II), (□) Zn(II). (Reproduced from Ref. 13. Copyright 1990 The Chemical Society of Japan.)

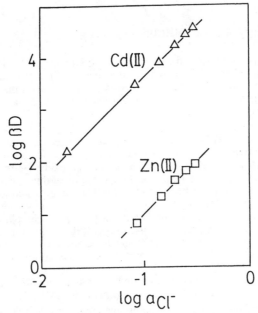

Figure 8. Relationship between Log (βD) Versus Log aCl-. (Δ) Cd(II), (□) Zn(II). (Reproduced from Ref. 13. Copyright 1990 The Chemical Society of Japan.)

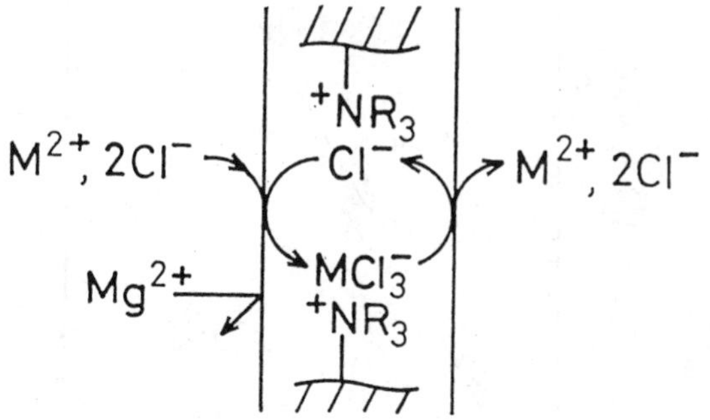

$$M^{2+}, 2Cl^- \rightarrow \quad Cl^- \leftarrow \quad \rightarrow M^{2+}, 2Cl^-$$

$$Mg^{2+} \quad MCl_3^-$$

Figure 9. Transport Mechanism for Metal Chloride Complexes through the Polymeric Anion-exchange Membrane. (Reproduced from Ref. 13. Copyright 1990 The Chemical Society of Japan.)

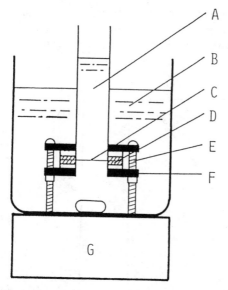

Figure 10. Cylindrical Dialysis Cell: (A) Receiving Phase Solution, (B) Source Phase Solution, (C) Membrane, (D) Silicone Packing, (E) Screw, (F) Clamp, (G) Magnetic Stirrer. (Reproduced from Ref. 13. Copyright 1990 The Chemical Society of Japan.)

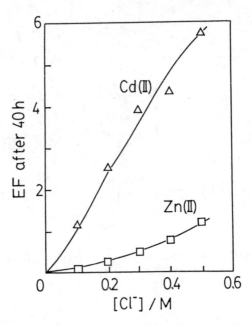

Figure 11. Enrichment Factor (EF) of Metal Ions as a Function of Chloride Ion Concentration in the Source Phase Solution. (Δ) Cd(II), (□) Zn(II). (Adapted from Ref. 13. Copyright 1990 The Chemical Society of Japan.)

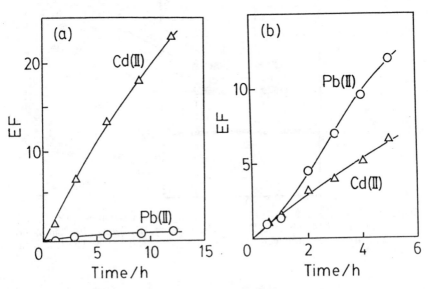

Figure 12. Enrichment Factor (EF) of Metal Ions as a Function of Permeation Time. Membrane Area: (a) 0.8 cm^2, (b) 15.0 cm^2; (o) Pb(II), (Δ) Cd(II).

permeation of Cd(II) increased significantly with an increase in the chloride concentration, and Cd(II) was selectively enriched in the receiving solution. The EF value after 40 h reached 5.8 for Cd(II) and 1.2 for Zn(II) when chloride ion concentration in the source-phase solution was 0.50 M. The low EF value of Zn(II) is attributed to the low sorptivity of Zn(II) chloride complex in the membrane phase. A similar low permeability was obtained for Pb(II) (EF value was 1.2 after 40 h), and no permeation was noted for Cu(II), Co(II), Fe(III), and Ni(II) (*24*). Although the observed permeation flux was low as compared with the liquid membrane separation, this is a first example that the polymeric membrane exhibited a selective carrier mediated transport for heavy metal ion separation, which selectivity was comparable to the liquid membrane separation.

Polymeric Plasticizer Membrane Separation

To enhance the permeation efficiency and selectivity of the polymeric membrane separation, a novel polymeric plasticizer membrane which is composed of cellulose triacetate (CTA) as a membrane support, *o*-nitrophenyl octyl ether (NPOE) as a membrane plasticizer, and trioctylmethylammonium chloride (TOMAC) as an anion-exchange carrier has been developed. Compared with the poly (vinyl chloride) plasticizer membrane which is widely used for ion-selective electrode, the CTA membrane can contain a larger amount of plasticizer due to a high affinity between CTA and NPOE (*25*). Thus the plasticizer (NPOE) solubilized in the membrane acts effectively as an organic medium for the carrier mediated membrane separation (*26-29*).

Preparation and Property of Polymeric Plasticizer Membrane. In 10 mL of chloroform, 0.20 g of CTA, 0.40 g of NPOE, and 0.40 of TOMAC were dissolved. Of this solution, 3 mL was poured into a glass culture dish (9.0 cm in diameter). After one day at room temperature, cold water was added and the membrane was peeled away from the dish. This procedure provided a stable, transparent film of 50 μm thickness. The obtained membrane was easily handled and stored under the dried condition.

The membrane transport experiments were first conducted in a cylindrical dialysis cell (membrane area, 0.8 cm^2) shown in Figure 10. The source phase solution was a 250 mL of 0.25 M MgCl$_2$ solution which contained 0.10 mM Pb(II) and Cd(II). The receiving phase was 5.0 mL of pure water. Plots of the EF value versus time for competitive transport of Pb(II) and Cd(II) are presented in Figure 12a. Selective permeation of Cd(II) over Pb(II) was observed, which is consistent with the selectivity obtained for the polymeric membrane separation. The EF for both heavy metals increased with time and reached values of 13.3 and 0.7 for Cd(II) and Pb(II), respectively after 6 h. As shown in Figure 11, the polymeric membrane separation required the permeation time of 40 h to achieve the EF value of 5.8 for Cd(II), whereas such EF value was recorded only after 2 h for the polymeric plasticizer membrane separation. Thus it is evident that the polymeric plasticizer membrane revealed a superior transport efficiency and selectivity to the conventional polymeric anion-exchange membrane.

For dialysis with a large membrane area (15.0 cm^2), three compartment cell poccessing the same cell construction shown in Figure 3 was employed. The source phase was a 250 mL of 0.25 M MgCl$_2$ solution which contained 0.10 mM Pb(II) and Cd(II). The receiving phase (10 mL) was pure water. The source phase and receiving phase solutions were circulated from reservoirs through the outer two compartments and the central compartment, respectively. Figure 12b shows the results for the competitive transport of Pb(II) and Cd(II) with time. In this dialysis, a selective permeation was noted for Pb (II) over Cd (II), which is opposite selectivity recorded in Figure 12a. Thus the EF values reached 12.0 for Pb(II) and 6.4 for Cd(II) after 5 h. Apparently the membrane area of the system affected the permeation selectivity for Cd(II) and Pb(II).

When the small membrane area $(0.8 \ cm^2)$ was used for the dialysis, concentration of Cd(II) in the membrane became very high $(4.7 \times 10^{-6} \ mol/cm^2$-membrane after 6 h). In this condition, the retention of Cd(II) in the membrane may be no longer strong, which permitted an effective Cd(II) transport through the membrane. For dialysis with the large membrane area $(15.0 \ cm^2)$, however, the Cd(II) concentration in the membrane was very low $(0.5 \times 10^{-6} \ mol/cm^2$-membrane after 5 h) . Under such condition, a strong retention of Cd(II) in the membrane occurred and this prohibited the release of Cd(II) chloride complex from the membrane phase into the aqueous receiving phase as observed for the liquid membrane separation. Whereas the permeation of Pb(II) simply increased as an increase of the membrane area due to the low extractability of Pb(II) chloride complex into the membrane. Consequently, the permeation selectivity was changed from Cd(II) to Pb(II) for dialysis with the large membrane area. It must be noted that the retention behavior of Cd(II) is affected not only by the membrane area but also by the concentration of Cd(II) in the source phase, the anion-exchanger concentration in the membrane phase, and the complex formation in the receiving phase. This type of effect was confirmed for all membrane separations adopted in this chapter (*30*).

Effect of Complexing Agents in the Receiving Phase. For practical heavy metal ion separation, a large membrane area must be utilized. To improve the Cd(II) permeability in such a system, therefore, the Cd(II) retention in the membrane phase must be reduced. Ethylenediaminetetraacetic acid (EDTA) is known as an effective complexing agent for heavy metal ions in aqueous solution. In fact, the 1:1 complex formation constants (log K) of EDTA are 16.5 for Cd(II) and 18.0 for Pb(II), respectively (*5*). Thus the addition of EDTA in the receiving phase must be effective to reduce the Cd(II) retention in the membrane phase. The EF values of Cd(II) and Pb(II) after 5 h as a function of EDTA (disodium salt) concentration in the receiving phase is shown in Figure 13 (experimental conditions are the same except for the addition of EDTA to those in Figure 12b). As expected, the permeation of Cd(II) was effectively enhanced at above the EDTA concentration of 3 mM. It is interesting to note, however, that the permeation of Pb(II) was monotonously suppressed with increasing EDTA concentration. In general, EDTA ion is highly hydrophilic and hard to extract into organic media. Thus anion-anion antiport mechanism by a coupled transport of Cl$^-$ or $MCl_m^{(m-2)-}$ with EDTA anion should be negligible in the present system. The back-extraction of the heavy metal chloride complexes from chloroform solution containing TOMAC into aqueous stripping phase revealed that the back-extraction efficiency for Pb(II) decreased significantly by the addition of EDTA in the aqueous phase. Thus a membrane extractable anionic EDTA-Pb(II) complex may be formed at membrane-receiving phase interface and interfere the Pb(II) permeation through the membrane. Consequently, a selective Cd(II) permeation was attained by the addition of EDTA in the receiving phase. This is a first instance that the polymeric plasticizer membrane exhibited a sperior separation efficiency and selectivity to the liquid membrane for Cd(II) separation (*31*).

These results demonstrate that the permeation selectivity and efficiency of heavy metal chloride complexes were controllable by both the extraction step at membrane-source phase interface and the stripping step at membrane-receiving interface. Thus the present membrane separation is quite effective for separation of Pb(II) and Cd(II) from other heavy metal ions in solutions .

Conclusions

This chapter deals with a heavy metal ion separation through three different types of functional membranes. Liquid membranes containing tetraalkylammonium ions as mobile carriers are effective for separation of toxic heavy metal ions as their metal

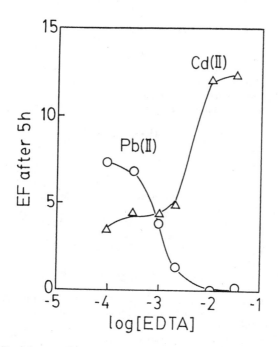

Figure 13. Enrichment Factor (EF) of Metal Ions as a Function of EDTA Concentration in the Receiving Phase Solution. (o) Pb(II), (△) Cd(II).

chloride complexes. Good selectivity for the concentration of Pb(II) was obtained in both bulk liquid membrane and highly efficient emulsion liquid membrane systems. This system can be applied to the polymeric anion-exchange membrane separation in which the anion-exchange sites act as fixed carriers for transport of heavy metal chloride complexes. The permeation selectivity of heavy metal ions is induced by a selective uptake of heavy metal chloride complexes into the anion-exchange sites in the membrane and a selective release of the metal complexes from the membrane into the receiving phase. Thus an extractive permeation similar to the liquid membrane separation was observed in the polymeric membrane separation. The CTA polymeric plasticizer membrane which contained NPOE as a membrane plasticizer and TOMAC as an anion-exchange carrier exhibited a superior function to the liquid membrane and the polymeric membrane. Although the permeation selectivity of Cd(II) and Pb(II) were influenced by the experimental conditions due to a difference in the heavy metal ion retention in the membrane, high permeation selectivity and efficiency for separation of Cd(II) and Pb(II) over other heavy metal ions were successfully performed for the polymeric plasticizer membrane separation.

Acknowledgments

The author wish to thank Professors M. Igawa (Kanagawa University) and R. A. Bartsch (Texas Tech University) for providing valuable comments. This work was supported by a Grant-in-Aid for Scientific Research from the Ministry of Education, Science and Culture of Japan, the Salt Science Research Foundation, and Nissan Science Foundation, which are gratefully acknowledged.

References

1. Warshawsky, A. In *Synthesis and Separation Using Functional Polymers*, Sherrington, D.C.; Hodge, P., Eds.; Wiley: New York, 1988; pp 325-286.
2. Kesting, R.E. *Synthetic Polymeric Membranes*; A Structural Perspective Second Edition; Wiley: New York, 1985; pp 311-328.
3. Araki, T.; Tsukube, H. *Liquid Membranes: Chemical Applications*; CRC Press: Boca Raton, FL, 1990.
4. Hayashita, T.; Takagi, M. *Talanta*, **1985**, 32, 399-405.
5. Sillen, L.G.; Martell, A.E. *Stability Constants of Metal-Ion Complexes*; The Chemical Society: London, 1964.
6. Sekine, T.; Hasegawa, Y. *Solvent Extraction Chemistry*; Marcel Dekker: New York, 1977.
7. Hogfeldt, E. In *Ion-Exchange*; Marinsky, J.A., Ed.; Marcel Dekker : New York, 1966; vol 1, pp 139-171.
8. Hayashita, T.; Noguchi, H.; Oka, H.; Igawa, M. *J. Appl. Polym. Sci.* **1990**, 39, 561-569.
9. Ide, S.; Ohki, A.; Takagi, M. *Anal. Sci.* **1985**, 1. 349-354.
10. Masana, A.; Valiente, M. *Solv. Extr. Ion Exch.* **1987**, 5, 667-685.
11. Du Preez, J.G.H.; Schanknecht, S.B.; Schillington, D.P. *Solv. Extr. Ion Exch.* **1987**, 5, 789-809.
12. Hayashita, T.; Oka, H.; Noguchi, H.; Igawa, M. *Anal. Sci.* **1989**, 5, 101-103.
13. Hayashita, T.; Kurosawa, T.; Ohya, S.; Komatsu, T.; Igawa, M. *Bull. Chem. Soc. Jpn.* **1990**, 63, 576-580.
14. Hayashita, T.; Bartsch, R.A.; Kurosawa, T.; Igawa, M. *Anal. Chem.* **1991**, 1023-1027.
15. Gordon, J.E. *The Organic Chemistry of Electrolyte Solutions*; Wiley: New York, 1975.
16. Donnan, F.G. *Z. Electrochem.* **1911**, 17. 572-581.
17. Li, N.N. U.S. Patent 3,410,794, 1968.
18. Li, N.N.; Shrier, A.L. *Recent Dev. Sep. Sci.* **1972**, 1, 163-174.
19. Frankenfeld, J.W.; Li, N.N. *Recent Dev. Sep. Sci.* **1978**, 4, 285-292.
20. Ho, W.S.; Hatton, T.A.; Lightfoot, E.N.; Li, N.N. *AIChE J.* **1982**, 28, 662-670.
21. Thien, M.P.; Hatton, T.A. *Sep. Sci. Technol.* **1988**, 23, 819-853.
22. Nakashio, F.; Goto, M.; Matsumoto, M.; Irie, J.; Kondo, K. *J. Membr. Sci.* **1988**, 38, 249-260.
23. Helfferich, F. *Ion Exchange*; McGraw-Hill: New York, 1962.
24. Hayashita, T., Saga University, unpublished results.
25. Hayashita, T.; Kumazawa, M.; Lee, J.C.; Bartsch, R.A. *Chem Lett.* **1995**.
26. Hayashita, T.; Fujimoto, T.; Morita, Y.; Bartsch, R.A. *Chem. Lett.* **1994**, 2385-2388.
27. Sugiura, M.; Kikkawa, M.; Urita, S. *Sepn. Sci. Techn.* **1987**, 22, 2263-2268.
28. Sugiura, M.; Kikkawa, M.; Urita, S. *J. Membr. Sci.* **1989**, 42, 47-55.
29. Sugiura, M. *Sepn. Sci. Techn.* **1993**, 28, 1453-1463.
30. Hayashita, T., Saga University, unpublished results.
31. Hayashita, T.; Kumazawa, M.; Yamamoto, M. *Chem. Lett.*, in press.

Chapter 22

Use of Emulsions, Microemulsions, and Hollow Fiber Contactors as Liquid Membranes

John M. Wiencek[1], Shih-Yao Hu[2], and Bhavani Raghuraman[2]

[1]Department of Chemical & Biochemical Engineering, University of Iowa, Iowa City, IA 52242
[2]Department of Chemical & Biochemical Engineering, Rutgers University, Piscataway, NJ 08855

Liquid membranes as a generic concept have primarily involved the use of U-tubes, porous solid films impregnated with a liquid carrier, or emulsified systems employed in a stirred contactor. Although such systems can display high selectivities and in some cases reasonable flux, the stability of the liquid membrane to rupture (i.e. leakage) and unwanted water transport (i.e. swell) have limited their commercial application. Our lab has focused on developing improved emulsion liquid membranes. In particular, we have investigated the possibility of employing microemulsions as liquid membranes to separate metals (especially mercury) from contaminated water. The lack of a clear advantage in such microemulsion systems suggested that a hybrid technique would provide the added stability without a loss in flux. Our current efforts utilizing emulsion liquid membranes within a hollow fiber contactor as a means to minimize swell and leakage will be discussed briefly.

This paper reviews the use of emulsions and microemulsions as liquid membranes with special emphasis placed on the separation of mercury, as $Hg(NO_3)_2$, from water using oleic acid as the extractant. Although emulsion (either macro- or micro-) liquid membranes offer advantages in terms of fast rates of separation, new modes of creating a stabilized liquid membrane utilizing hollow fiber contactors offer comparable flux in a more stable format. The paper will start with a review of the basic types of liquid membranes as currently used in research. The discussion will then focus on the author's experience with emulsified liquid membrane systems. The last section of the paper will discuss the obvious next step in liquid membrane technology, the use of emulsion liquid membranes in hollow fiber contactors.

The utility of liquid membranes has been discussed in several reviews and handbooks (1-2). Emulsion liquid membranes remain an actively researched separation technique for the removal of trace contaminants from aqueous streams. Liquid membranes are attractive because they combine extraction and stripping into a single step; thus, equilibrium constraints of partitioning are overcome by changing the chemical

0097–6156/96/0642–0319$15.00/0

nature of the substance being separated (3). When liquid soluble ion exchangers or chelators are added to the liquid membrane, a variety of metals can be effectively separated from water such as: zinc and chromium ions using Aliquat 336 as a carrier (4); silver, gold and palladium ions using macrocyclic crown ethers as a carrier (5); copper utilizing LIX reagents (6-8); and mercury using dibutylbenzoylthiourea (9) or oleic acid (10-11).

Basic Types of Liquid Membranes

U-tubes. U-tubes are a very simple implementation of the liquid membrane concept. Two miscible fluids, usually aqueous phases, are placed in separate containers. One container is the feed solution and the other is the receiving solution. These two containers are connected via an inverted or normal U-shaped tube containing an immiscible solvent (e.g. hexane or chloroform). An inverted U-tube is used if the solvent is less dense than the receiving and feed solutions (e.g. hexane vs water); whereas, a normal U-tube is used if the solvent is more dense than the receiving and feed solutions (e.g. chloroform vs water). Mixing bars and sampling drawoff ports are usually placed in the feed and receiving phase containers. The U-tube is a useful device for preliminary experimental tests on new extractants because it allows for the experimental measurement of the solute concentration in all three phases (i.e. the feed phase, the membrane phase, and the receiving phase). U-tubes are not a practical contacting device if the separation is to be utilized on a large scale. The limited surface area for mass transfer at the feed phase/solvent and the solvent/receiving phase as well as the macroscopic diffusion distances in the solvent phase within the U-tube lead to very slow rates of separation.

Supported Liquid Membranes and Hollow Fiber-Contained Liquid Membranes. Other configurations reported for simultaneous extraction and stripping include supported liquid membrane (SLM) and hollow fiber-contained liquid membrane (HFCLM). The SLM configuration has the organic phase liquid (containing the metal complexing agent) immobilized in the pores of a microporous membrane with aqueous feed phase flowing on one side and the aqueous receiving phase on the other side. A major disadvantage of SLMs is their instability mainly due to (a) loss of extractant and/or organic solvent into the flowing aqueous phase because of solubility and (b) short-circuiting of the two aqueous phases if the pressure differential across the membrane exceeds the capillary forces which hold the organic liquid in the pores. The operation will have to be interrupted to replenish lost extractant (12). In addition, SLMs are known to form emulsions at the membrane interface leading to contamination of the feed solution with extractant. The HFCLM configuration has two sets of microporous hollow fiber membranes, one carrying the feed phase and the other the strip phase (13,14) The organic liquid is contained between these two sets of fibers by maintaining the aqueous phase at a pressure higher than the organic phase but lower than its breakthrough value. As with the system described later in this paper, HFCLM offers long term stability since the membrane liquid is connected to a reservoir and is continuously replenished to make up for any loss by solubility. The major disadvantage here is the difficulty in mixing the two sets of fibers to achieve a low contained liquid membrane thickness. Presently reported thicknesses are large (approximately 500 microns) and could yield low fluxes if diffusion through the membrane is rate-limiting.

Emulsion Liquid Membranes. Extraction and stripping of metals are traditionally performed in separate operations. By use of emulsion liquid membranes (ELMs), the two steps can be accomplished in a single step. ELMs, first invented by Li (3), are made by forming a surfactant-stabilized emulsion between two immiscible phases. A water-in-oil emulsion, consisting of an oil phase with a metal extractant and an aqueous stripping reagent as an internal phase, is dispersed into an aqueous waste stream containing metal ions. Copper(II) extraction using ELMs in a stirred contactor (SC) is shown in Figure 1. Combining extraction and stripping removes equilibrium limitations inherent in conventional solvent extraction and metal concentrations in feed solutions are reduced to very low levels. After extraction, the emulsion can be demulsified to recover the enriched stripping phase. Demulsification (emulsion breaking) by application of high voltage electric fields has proven to be most successful (15). Extraction using ELMs have been reported for copper, zinc, mercury, cobalt, chromium, and nickel (4,5,7,9,11,16,17).

The main problem associated with ELM extraction is the integrity of the emulsion. Swelling and leakage of the ELM can occur after prolonged contact with the feed phase (18). Swelling occurs when the feed stream gets into the internal phase by either osmotic pressure or physical breakage and subsequent reformation of the membranes. The water content in the emulsion can thus increase from 10-20 wt. % to 30-50 wt. %. Swelling reduces the stripping reagent concentration in the internal phase which in turn lowers its stripping efficiency. Dilution of the solute that is to be concentrated in the internal phase also results in a less efficient separation. Leakage of the internal-phase contents into the feed stream because of membrane rupture not only releases extracted metals back into feed stream but further contaminates the feed stream with stripping reagents. Leakage can be minimized by making a more stable emulsion with a higher concentration of surfactant, but this makes the downstream demulsification and product recovery steps more difficult as well as increases swell. Lower shear rates would also minimize leakage, but mass-transfer resistance could then become very significant.

Experiences with Emulsions and Microemulsions: A Case Study with Mercury Separation

A number of processes exist for the treatment of aqueous streams containing heavy metals such as precipitation (19), ion exchange (20) and electrochemical methods. Unfortunately, these processes have serious disadvantages: both ion exchange and chemical precipitation result in the formation of a solid waste from which the heavy metal cannot be easily recovered. Consequently, although the aqueous stream has been reclaimed, the precipitated sludge or adsorbent must be landfilled and with it, the concomitant problems of leaching into ground water, ion-exchange with nearby soil and possible microbial conversion to a more toxic form of the metal (*e.g.* conversion of a mercury salt to the very toxic methyl-mercury) continue.

Solvent extraction is an alternative technology to remove heavy metals from wastewater. Extractants in the organic phase complex the metal ions present in the waste stream. The metals can be recovered by stripping the loaded organic phase with a different aqueous stream containing a stripping agent. The result is a more concentrated solution from which metal ions can be recovered by electrowinning or electroplating techniques (21). This technique has been widely used by the mining industry. Metal extractants are now commercially available for a variety of separation requirements. By choosing the appropriate extractant and extraction conditions, high selectivity can be achieved. In addition, extractants can usually be regenerated without losing activity, thus improving the economics of the separation.

Much effort has been expended in our labs over the last few years investigating the use of emulsion liquid membranes to carry out such wastewater treatment schemes with a special focus on the removal of mercury ions from water. Both coarse or macroemulsions as well as microemulsions were studied and compared. The advantage of emulsion liquid membrane extraction is the large surface area available for mass transfer which results in fast separations. Because the volume ratio of the feed to internal receiving phase is high, the separated metal is concentrated by factors as high as

322 CHEMICAL SEPARATIONS WITH LIQUID MEMBRANES

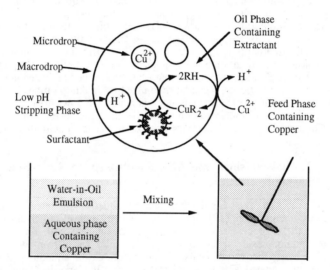

Figure 1 A schematic representation of copper ion extraction with an emulsion liquid membrane. Copper(II) is transported to the emulsion/feed phase interface and reacts with the complexing agent (RH) to form a soluble copper complex (CuR_2). This complex diffuses to the interior of the emulsion droplet until it encounters a droplet of the internal phase where the metal ion is exchanged for a hydrogen ion. The net effect is a unidirectional mass transport of the cation from the original feed to the receiving phase with counter-transport of hydrogen ions. Mercury exhibits a comparable mechanism for transport in these systems.

100. As mentioned earlier, the main disadvantage of coarse emulsion liquid membranes is the inherent instability which makes them sensitive to the mixing rate, the volume ratio of the feed phase to the membrane phase, and time. Because of this instability, leakage of internal phase into the feed phase can occur. In addition, water can be transported into the internal phase, diluting the concentrate. This phenomenon is called swell. Both leakage and swell result in reduced separation efficiency. Our studies on mercury were directed at comparing microemulsions to coarse emulsions to ascertain whether microemulsions could display anticipated benefits including reduced leakage and increased rates of separation.

Extraction Chemistry. The binding of mercury as $Hg(NO_3)_2$ to oleic acid (HR) was thoroughly characterized as a function of pH and oleic acid concentration (10). In the absence of any surfactant the binding followed a simple acid/base titration equilibrium as described by:

$$Hg^{2+} + 2\ (HR)_2\ <\!-\!>\ Hg(R \bullet HR)_2 + 2H^+ \qquad Keq,1 = 0.449$$

Thus, the reaction is essentially shifted completely to the right at low hydrogen ion concentrations (pH>2) and extraction occurs. High hydrogen ion concentrations (pH<1), result in efficient stripping as the equilibrium is shifted to the left. It is well known that dimers of oleic acid form in nonpolar organic liquids as shown in the above expression (22). The surfactant added to stabilize the emulsion liquid membrane system (as well as microemulsion system) tends to promote further aggregation of the mercury/oleic acid complexes which shifts the above equilibrium further to the right. This shift results in an enhancement in overall extraction due to the presence of the surfactant and can be described by the additional equilibrium reaction:

$$Hg(R \bullet HR)_2 \xrightarrow{\text{oxygen}} (Hg(R \bullet HR)_2)_2 \qquad Keq,2 = 1600$$

where:

$$Keq,2 = \frac{[(Hg(R \bullet HR)_2)_2]}{[Hg(R \bullet HR)_2]^2[\text{oxygen}]}$$

The term 'oxygen' refers to the concentration of molecular oxygen introduced into the organic phase by the addition of the various nonionic surfactants tested. The enhancement in extraction due to the addition of surfactant highlights the need for realistic and accurate thermodynamic data when analyzing the separation of metals in liquid membrane systems.

Perceived Advantages of Microemulsions over Coarse Emulsions. A microemulsion is an optically transparent, thermodynamically stable emulsion which can display a wide range of water volume - from completely water-continuous to completely oil-continuous in structure. A microemulsion forms spontaneously when the components of oil, water and surfactant (as well as cosurfactants in many cases) are brought into contact. While the mechanism of liquid membrane separation with a microemulsion is identical to the macroemulsion case, the physical nature of the emulsion is quite different. The internal droplets in a macroemulsion are typically 1 μm in diameter whereas the microemulsion internal droplets are usually in the range of 0.005 to 0.1 μm. Thus, the internal phase has a much larger area per unit volume in the microemulsion system which ultimately translates into faster rates of stripping into the receiving phase. The interfacial tension between the microemulsion and the feed aqueous solution is typically less than 1 dyne/cm versus approximately 10 dyne/cm for macroemulsions. Thus, smaller emulsion globules or macrodrops (see Figure 1) are formed during contact of the microemulsion with the feed phase which results in faster rates of separation than seen for the macroemulsion liquid membrane systems. The thermodynamic stability of microemulsions may lead to more stable (less leaky) formulations than seen for coarse emulsions. In addition, thermodynamic variables such as temperature may be utilized to demulsify the system, allowing ease of recovery of the concentrated solute (mercury in this case). Several of these advantages were evident in acetic acid and copper extractions (23,24), but the mercury study was the first study to carefully evaluate both macroemulsions and microemulsions side-by-side for efficacy in separation as a liquid membrane.

Efficacy of Microemulsion-based Liquid Membranes. The overall efficacy of microemulsion-based extraction of mercury from water was reported earlier (25). The results are briefly summarized here. The expected benefits of the liquid membrane process was evident in comparison to normal extraction in that residual mercury in the feed water was typically an order of magnitude lower for the microemulsion liquid membrane systems. These results highlighted the particular advantage of simultaneous extraction and stripping in circumventing normal equilibrium constraints. In addition, the microemulsion liquid membrane did display faster kinetics than the macroemulsion system (typically an order of magnitude faster). However, no particular advantage with respect to leakage was realized and, in fact, leakage was quite minimal in both the microemulsion and macroemulsion systems.

Recovery of the concentrated mercury required demulsification of the liquid membranes. The coarse emulsions could be easily demulsified by electrostatic techniques; whereas, the microemulsions required the addition of a cosolvent (18). The use of a cosolvent is very undesirable because it requires an additional separation step to remove the cosolvent (usually distillation). This particular aspect of the overall process would favor the use of macroemulsions over microemulsions.

The organic phase from the demulsified microemulsion was evaluated for its potential reuse. This organic phase was treated by distillation to remove residual butanol (used in the chemical demulsification). The organic material was then contacted with fresh stripping reagent (concentrated sulfuric acid) and allowed to form a new microemulsion. This microemulsion was then utilized again to separate mercury from the feed phase. The final feed phase mercury concentration was still much lower than the straight extraction system; thus, the internal receiving phase was still stripping and concentrating the mercury. The system was not as efficient as the fresh microemulsion system however. The lowered efficiency was directly related to the use of a cosolvent in the demulsification step. Although most of this cosolvent was distilled out of the solvent before its reuse, the small residual amount of butanol greatly increased the leakage in this system which ultimately resulted in reduced separation efficiency.

Summary of Significant Shortcomings. The emulsion liquid membranes studied each possess advantages when compared against one another. In specific, the microemulsion systems displayed separation kinetics which were typically an order of magnitude faster than the macroemulsion counterparts. However, this increased rate of separation was at the expense of product recovery. The concentrated mercury was easily recovered by electrostatic demulsification for the macroemulsion system, but required the addition of butanol to the microemulsion before mercury recovery was achieved. The requirement of chemical demulsification is a major disadvantage of the microemulsion-based liquid membrane system.

In our studies, swell was the most significant problem. For example, typical experiments for the microemulsion separation of mercury would start with an internal phase volume fraction of 0.087 yet it would end with an internal phase volume fraction of 0.535. Thus, concentration factors (final concentration in receiving phase/original concentration in feed phase) for the mercury which were expected to be on the order of 25 times were usually only on the order of 4 times. Swelling is caused by the hydration of the surfactant in the low ionic strength environment of the feed phase, followed by diffusion to the internal receiving phase and dehydration in the presence of the concentrated acid solution (i.e. osmotic driving force for water transport). Swell can only be eliminated if the means of water transport is eliminated. In this case, the surfactant must be eliminated. However, the dispersion is stabilized from leakage by the presence of the surfactant. Our attempts have since focused on minimizing the amount of surfactant in the emulsion and providing leakage stability by other means, most notably a highly porous, hydrophobic, polymeric membrane support.

A Hybrid Technique: Emulsion Liquid Membranes Utilizing Hollow Fiber Contactors

Dispersion-free solvent extraction using microporous hollow fiber contactors (HFCs) has been shown to be very effective (26-29). Hollow fiber contactors consist of microporous hollow fibers arranged in a shell-and-tube configuration. The fiber material can be either hydrophilic or hydrophobic. When a hydrophobic fiber is used, higher pressure is applied on the aqueous phase. The aqueous phase will not penetrate or wet the hydrophobic membrane. The organic phase is present in the pores. However, it cannot flow out because the aqueous phase is at a higher pressure. Thus, an aqueous-organic interface is established at the pore openings. The advantage of HFC lies in its ability to offer a very high surface area/volume ratio without dispersion or mixing of the two phases. The use of hollow fiber contactors eliminates the need for a settling stage and allows for direct scale-up due to modular design (29).

An ELM can be used to simultaneously extract and strip the metal from the wastewater by a liquid-liquid dispersion-free contacting in a HFC. This combines the advantages of emulsion liquid membrane separation (simultaneous extraction and stripping) and dispersion-free solvent extraction. The absence of high shear rates, which is encountered in stirred contactors, minimizes leakage of the internal stripping phase. The internal droplets are generally larger than the typical pore size of the membranes and hence do not come in contact with aqueous feed stream. This reduces the tendency for swelling. Figure 2 is a schematic representation of ELM extraction in a hydrophobic HFC.

Studies in our laboratory have shown the merit of ELM extraction in HFC (30). The extraction fluxes are comparable to a traditional stirred contactor (SC) ELM system (overall rates are slower due to smaller area on our lab-scale membrane module) but the separation efficiency is higher due to low swelling and leakage. Typical swelling in SC is 80-100% whereas it is typically 0-20% in HFC. There is no detectable leakage of the internal phase into the feed in the HFC. Figure 3 compares extraction of copper in the two contactors when the emulsion is formulated with 3 wt.% surfactant. The stirred tank run fails, displaying leakage as high as 40% while the leakage in the HFC is only 0.02%. The stirred tank displays a very fast initial rate of separation, but residual copper never falls below 1 ppm due to leakage. The hollow-fiber contactor did not display leakage and was able to attain ppb levels of residual copper. The emulsion stability is not a factor in the HFC and copper is removed to a level an order of magnitude lower than in a SC.

The advantage of using a HFC becomes more apparent when using less stable emulsions. In fact, it may be possible to extrapolate this concept to zero surfactant concentration where the 'emulsion' is a physical mixture of the oil phase and the stripping phase (without surfactant). Figure 4 shows that lowering the surfactant

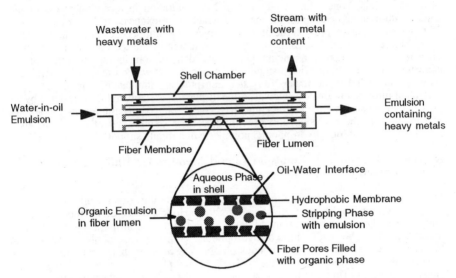

Figure 2 In a HFC, physical mixing of aqueous feed stream and emulsion is avoided. A stable interface can be maintained in a hydrophobic module by applying higher pressure at aqueous side. Metal ions are extracted from feed stream into organic phase, then stripped into internal droplets of the emulsion.

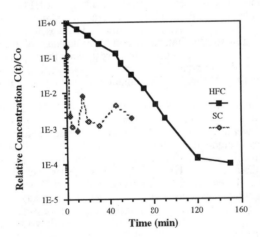

Figure 3 Comparing ELM extraction of copper in a SC and in a HFC. Aqueous phase was 1000 ppm copper solution at pH 5.36. Emulsion contained 3 wt.% ECA5025 as surfactant, 5 wt.% LIX 84 as extractant and 18 wt.% 6N H_2SO_4 as stripping agent. Solvent used was n-tetradecane. SC was operated at 400 rpm. In the HFC run, aqueous phase flowed through shell side at 40 ml/min and emulsion was stationary in the HFC tube. The ratio of aqueous phase to emulsion was 10 to 1.

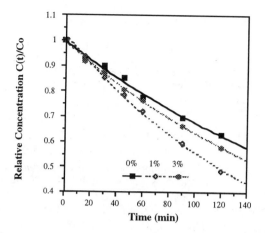

Figure 4 Effect of surfactant concentration of ELM on copper extraction in a HFC. The two phases used had the same composition as in Figure 3 except surfactant concentration was varied. Aqueous phase flowed through shell side at 70 ml/min. Organic phase flowed through tube side at 20 ml/min. . The ratio of aqueous phase to emulsion was 5 to 1.

concentration of an ELM does not reduce its extraction efficiency in a HFC. This suggests the possibility of eliminating the downstream demulsification process.

Summary

A side-by-side comparison of coarse and micro- emulsions for use as liquid membranes indicates that each possesses unique advantages. The microemulsion displays faster rates of separation yet more difficult demulsification than the coarse or macro- emulsion system. Both systems suffer from swell. A new method of contacting emulsion liquid membranes with the feed solution minimizes swell while maintaining high separation flux. The key advantage of the HFC contactor lies in its ability to stabilize the liquid membrane from leakage. Thus, surfactant concentration can be minimized and swell essentially eliminated. The system is much like a supported liquid membrane but will not produce short circuits due to solvent loss since the solvent is continuously supplied on the emulsion side of the membrane. Our lab is currently characterizing such systems.

Acknowledgments

This work was funded by the Hazardous Substance Management Research Center of New Jersey, Grant PHYS32 (A NSF Industry/University Cooperative Research Center and an Advanced Technology Center of NJ Commission on Science and Technology) and the New Jersey DEP.

References

1. Noble, R. D.; Way, J. D. *Liquid Membranes: Theory and Applications*, *ACS Symp. Ser. 347*, American Chemical Society: Washington, DC, 1987, Chapters 1 and 8.
2. Ho, W. S.; Sirkar, K. K. *Membrane Handbook*, Van Nostrand Reinhold: New York, NY, 1992, pp 595-724.

3. Li, N. N. "Separation Hydrocarbon with Liquid Membrane," U.S. Patent 3,410,794 (1968).
4. Fuller, E. J.; Li, N. N. *J. Membr. Sci.,* **1984**, *22*, 251.
5. Izatt, R.; Clark, G.; Christensen, J. *Sep. Sci. Techol.* **1987**, *22*, 691.
6. Kondo, K.; Kita, K. *J. Chem. Eng. Jn.* **1981**, *12*, 20.
7. Volkel, W.; Halwachs, W.; Schugerl, K. *J. Membr. Sci.* **1980**, *6*, 19.
8. Achwal, S.; Edrees, S. *Arabian J. Sci. Eng.* **1988**, *13*, 331.
9. Weiss, S.; Grigoriev, V. *J. Membr. Sci.* **1982**, *12*, 119.
10. Larson, K. A.; Wiencek, J. M. *I&EC Research* **1992**, *31*, 2714.
11. Boyadzhiev, L.; Bezenshek, E. *J. Membr. Sci.* **1983**, *14*, 13.
12. Danesi, P.R.; Yinger, R.; Rickert, P.G. *J. Membr. Sci.,* **1987**, *31*, 117.
13. Majumdar, S.; Sirkar, K. K.; Sengupta, A. *Membrane Handbook*, W. S. W. Ho and K. K. Sirkar, eds.; Van Nostrand Reinhold: New York, NY, 1992, Chapter 42.
14. Sengupta, A.; Basu, R.; Sirkar, K. K. *AIChE J.* **1988**, *34*, 1698.
15. Draxler, J.; Furst, W.; Marr, R. *J. Membr. Sci.* **1988**, *38*, 281.
16. Gu, Z. M.; Wasan, D.; Li, N. N. *J. Membr. Sci.* **1986**, *26*, 129.
17. Kitagawa, T.; Nishikawa, Y.; Frankenfeld, J.; Li, N.N. *Environ. Sci. and Technol.,* **1977**, *11*, 602.
18. Larson, K. A.; Raghuraman, B.; Wiencek, J. M. *J. Membr. Sci.* **1994**, *91*, 231.
19. Peters, R.; Ku, Y. *Separation of Heavy Metals and other Trace Contaminats, AIChE Symposium Series 81* American Institute of Chemical Engineers: New York, NY, 1985, p. 9.
20. Peters, R.; Ku, Y.; Bhattacharyya, D. *Separation of Heavy Metals and other Trace Contaminats, AIChE Symposium Series 81* American Institute of Chemical Engineers: New York, NY, 1985, p. 165.
21. Moyer, H.R. *Copper Sulphate via Solvent Extraction and Crystalization* Henkel Corporation: Arizona, 1979.
22. Ritcey, G. M.; Ashbrook, A. W. *Solvent Extraction, Principles and Applications to Process Metallurgy, Part I* Elsevier: New York, NY, 1984, p. 26.
23. Wiencek, J. M.; Qutubuddin, S. *Sep. Sci. Techol.* **1992**, *27*, 1211.
24. Wiencek, J. M.; Qutubuddin, S. *Sep. Sci. Techol.* **1992**, *27*, 1407.
25. Larson, K. A.; Wiencek, J. M. *Environmental Progress* **1994**, *13*, 253.
26. Alexander, P. R.; Callahan, R. W. *J. Membr. Sci.* **1987**, *35*, 57.
27. Dahuron, L.; Cussler, E. L. *AIChE J.* **1988**, *34*, 130.
28. Prasad, R.; Sirkar, K. K. *AIChE J.* **1988**, *34*, 177.
29. Prasad, R.; Sirkar, K. K. *J. Membr. Sci.* **1990**, *50*, 153.
30. Raghuraman, B.; Wiencek, J.M. *AIChE J.* **1993**, *39*, 1885.

Chapter 23

Field Testing of a Liquid-Emulsion Membrane System for Copper Recovery from Mine Solutions

Dave N. Nilsen, Gary L. Hundley, Gloria J. Galvan, and John B. Wright

Albany Research Center, U.S. Department of Energy, 1450 Queen Avenue Southwest, Albany, OR 97321-2198

The Bureau of Mines recognized the need for cost-effective technology for the recovery of metals from dilute soltuions. In this regard, liquid-emulsion membranes (LEMs) were investigated in a small, pilot plant-scale, continuous circuit for the recovery of copper from a variety of mine solutions. The mobile circuit was field tested at a copper mine with solutions containing from 120 ppm up to 1.4 g/L Cu. The solutions were representative of mine waste waters and various low-grade leach solutions. The LEM system used simple equipment designs for all units. The formulation of the emulsion membranes were optimized so that the emulsions had good stability during extraction, but were easily broken in an electrical coalescer using mild conditions. Typical results from the tests were >90 percent copper recovery while maintaining membrane swelling in the range of 4 to 8 percent. Although all of the mine solutions contained high impurity levels and suspended solids, pure copper cathodes were produced from each solution. Cost evaluations indicated the potential for the cost-effective recovery of copper from the solutions treated.

The LEM technique was first proposed by Li *(1)* in the 1960s as a possible industrial technique for the separation of substances from aqueous solutions. Initially, Li's work concentrated on the separation of hydrocarbons and, later, on the removal of dissolved constituents (phenols, phosphoric acid, sodium nitrate, and ammonia) from aqueous solutions. During the 1970s, research on this technique was extended to the removal of copper from acidic leach solutions and the extraction of uranium from wet-process phosphoric acid *(2-3)*. Research in Japan involved the extraction of NH_3, Cr^{+6}, Hg^{2+}, Cd^{2+}, and Cu^{2+} from industrial waste waters *(4)*. Draxler, Furst, Protsch, and Marr investigated the extraction of zinc from a waste water at a rayon plant in Austria *(5-6)*. However,

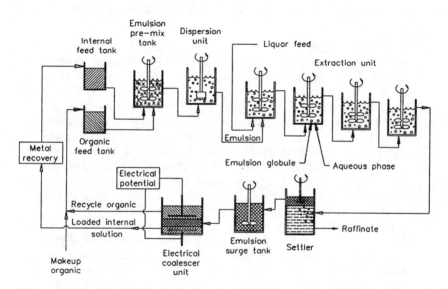

Figure 1. Simplified Flow Sheet for the Pilot Plant-scale Continuous LEM Circuit.

with the exception of the rayon plant, industrial acceptance of the LEM technique has been lacking. For one thing, little has been published related to the economics of the application of the LEM technique to dilute solution treatment. In order to supply needed information, the U.S. Bureau of Mines (USBM) is investigating the use of unsupported LEMs for the selective removal and recovery of metals from industrial-process and waste waters. At present, there is little cost-effective technology available for the selective extraction and recovery of metals from dilute solutions that contain mixtures of metals. As a part of this research, the USBM investigated the extraction and recovery of copper with LEMs from various mine solutions in field tests using a mobile pilot plant. Cost evaluations were developed for copper recovery based on the field test results.

Description of the LEM System

Unsupported LEMs were made by forming an emulsion from two immiscible phases and then dispersing that solution into a third phase (a feed solution). The emulsions used in the field tests and in other USBM research were a water-in-oil type. In a typical arrangement, the internal solution (also called the stripping solution) is emulsified into the organic phase forming an emulsion of extremely fine droplets of the internal solution dispersed in the organic phase. The principal component of the organic phase is kerosene. A conventional solvent extraction (SX) reagent (an extractant) is added to the organic phase to facilitate the selective transport of metal ions through the organic membrane. The organic phase must not be miscible with either the feed solution or the internal solution if the resultant emulsion is to remain stable. The organic phase also contains a surfactant that is used to stabilize the emulsion. The emulsion then is dispersed into the feed solution, forming globules of the emulsion. Metal is extracted at the outside surface of the globule by the SX extractant, transported across the organic membrane as a metal-organic complex, and then stripped and stored in the internal solution. Typically, a strong acid solution is used as the internal solution. Thus, extraction and stripping operations are combined in a single process in the LEM technique.

Circuit Arrangement and Equipment Designs

A simplified flow sheet for the LEM continuous circuit that was used in the field tests is shown in Figure 1. The technique used four main unit operations (emulsion generation, copper extraction, emulsion breaking, and metal recovery).

Emulsion Generation. In the generation step, recycled organic phase and internal solution were combined in a dispersion unit that generated a water-in-oil type of emulsion. The flow-through dispersion unit was similar in design to some units that are used by industry for the generation of emulsions. The residence time in this unit was only about 12 sec. The generator consisted of a slotted stator and an internal rotor. This unit was usually operated at a speed (about 5,000 rpm) that resulted in a tip speed of about 730 cm/sec, to produce internal droplets in the organic phase with an average diameter of about 5 μm.

Copper Extraction. The emulsion and feed solutions were combined in the extraction unit. Depending on the solution being processed, the flow rate of the feed solution ranged from 0.4 to 0.7 gallons per minute. A 4-stage, co-current flow unit with a superficial residence time of 20 minutes was used. Moderate agitation was used in this unit to form small globules of the emulsion and to disperse them into the feed solution. The extraction stages were fitted with a double set per shaft of relatively large (but slow turning) 3-bladed propellers, which were operated at a tip speed of about 50 cm/sec (about 100 rpm). The objective was to generate globules with an average diameter between about 0.5 to 1 millimeter. The "loaded" emulsion then was separated from the raffinate in a cylindrical settler.

Emulsion Breaking. The "loaded" emulsion then was sent to an electrical coalescer where the emulsion was broken into the two original phases that were combined in the dispersion unit. The coalescer was a rectangular vessel, similar in design to a SX settler. However, it was fitted with two horizontal electrodes. The bottom electrode was an uninsulated stainless steel (SS) screen that was in contact with the separated aqueous internal solution. This electrode was operated at ground potential. The upper electrode was a SS wire that was formed into a rectangular shape and was covered with PVC tubing to insulate the wire. A 60-hertz alternating current potential (generally 5,000 to 8,000 volts) was applied across the electrodes. Essentially no current flowed between the electrodes because the "hot" electrode was insulated. The retention time in the coalescer was about 12 minutes. In this research, an attempt was made to optimize the emulsion generation step and the formulation of the emulsion so that relatively mild conditions could be used for breaking of the emulsions, such as the use of a 60-hertz field. The recovered organic phase from the coalescer still contained the extractant and the surfactant and it was directly recycled to the emulsion generation unit.

Metal Recovery. The recovered internal solution from the coalescer contained the extracted copper and was sent to a metal recovery unit before being recycled to the dispersion unit. The mobile LEM pilot plant used electrowinning (EW) as the method for copper recovery. Since copper EW is a well known technology, no attempt was made to optimize the EW parameters or cell design during the tests. Depending on the solution being treated in the pilot plant, the EW unit was operated at current densities ranging from 5 to 20 amp/ft^2 and at a temperature of 40 °C. Cathodes were produced during all portions of the tests. The EW unit was used to remove the extracted copper and to restore the acid level of the strip solution (internal solution) used in the emulsion membrane.

Field Tests

Two separate field tests were conducted at a copper mine in Arizona. The objectives of these tests were to evaluate the technical and economical feasibility of selectively recovering copper from a variety of dilute mine solutions using the LEM technique. The first field test of the LEM mobile system was conducted in September of 1993, and the second in May-June of 1994.

Description and Conditions for Tests. During the first test, three different solutions were processed. The first part of this test used pregnant leach solution (PLS) from a heap leaching operation. That solution contained 1.43 g/L Cu. This solution was fed to the LEM system through an in-line filter for the first 91 hours of testing. Then the LEM system processed unfiltered PLS for another 24 hours. The next two tests used PLS that was diluted with mine water to 520 ppm and 320 ppm, respectively. The mine operator prepared all feed solutions that were used in these field tests. The latter two solutions were filtered prior to processing because they had a very heavy solids loading (iron was precipitating from the solutions). Even after filtering, these solutions still contained visible solids. Samples taken from the 520-ppm Cu solution indicated that the suspended solids level of the unfiltered solution was about 230 ppm, and after filtration, the feed to the system still contained about 120 ppm solids. A sample of raffinate from the settler taken during this portion of the test indicated very high solids loading in the settler (about 500 ppm). Either solids were accumulating in the settler or precipitation was continuing to take place in the circuit. Samples of the filtered, 320-ppm Cu feed solution indicated that the suspended solids in this solution was about 60 ppm. It is important to note that the solids present in the solutions did not interfere with the operation of the system during the time span of this testing.

During the second test, two different solutions were processed. The first solution contained 120 ppm Cu and the second solution 640 ppm Cu. In producing these solutions, the mine operator again combined PLS and mine water to reach concentrations that were of interest to them. These solutions also contained relatively high impurity levels (e.g., iron) and suspended solids (estimated to be approximately 100 to 200 ppm). As in the first field test, some precipitation of iron from the feed solutions occurred during the tests in the feed tanks and in the LEM equipment. This was particularly noticeable during operation with the 120 ppm Cu solution.

A summary of operating conditions is shown in table I. The organics used in the field tests contained the following components: M5640 extractant (an aldoxime obtained from Zeneca, Inc), Paranox 100 surfactant (a nonionic, polyamine compound obtained from Exxon Inc), and a mixture of Isopar M and V kerosenes (aliphatic diluents obtained from Exxon, Inc). Water-in-oil type emulsions were generated between the organic phases and the indicated internal solutions. Optimum average internal droplet size was about 5 μm in diameter.

Table I. Parameters and Conditions Used for Field Tests 1 and 2

Organic composition:	5.0 to 7.5 wt pct Acorga M5640
	1.0 wt pct Paranox 100 surfactant
	91.5 to 94 wt pct 50-50 Isopar
	M&V organic viscosity: 7.1 cp
Lean internal composition:	8.5 to 20 g/L Cu
	160 to 165 g/L H_2SO_4
Overall organic/internal ratio:	4.35 to 4.44
Average internal droplet size:	5 μm
Overall retention time in extraction stages:	20
Total elapsed run time:	196 hr (test 1)
	270 hr (test 2)
Total aqueous feed processed:	5,600 gal (test 1)
	10,950 gal (test 2)

In optimizing the parameters in the LEM technique, primarily four critical areas were used to judge LEM performance: (1) metal extraction, (2) membrane swelling, (3) membrane leakage, and (4) membrane breakage in the coalescer. Membrane swelling refers to the increase in volume of the internal phase in the emulsion membrane during the extraction process and membrane leakage refers to the rupture and transfer of the internal phase from the emulsion membrane into the aqueous feed solution. It was found (7-8) that the following parameters influenced, to some degree, the four performance areas: extractant type and concentration, surfactant type and concentration, viscosity of resultant membrane, concentration of stripping agent in the internal phase, and the surface areas of the globules (inside and outside). Some specific parameters applied to just one performance area. For example, the difference in ionic strength between the feed solution and the internal solution greatly affected membrane swelling, and the applied voltage in the coalescer influenced the breakage of the emulsion. The desirability for a stable emulsion for use in the extraction unit, but one that could easily be broken in the electrical coalescer, had opposing requirements. By necessity, there were many compromises in the selection of the optimum parameters. The USBM research focused on maximizing metal extraction while minimizing membrane swelling and leakage. Performance goals for the LEM mobile pilot plant were a copper extraction ≥90%, membrane swelling of ≤10%, and membrane leakage of <1%.

Results. Principal results from the field tests are grouped under Field Test 1 or 2, and are shown in Table II.

Table II. Summary of Field Test Results

Field test 1

PLS feed (1400 ppm):	Total process time	115 hr
	Volume processed	2800 gal
	Copper extraction	98.0 pct
	Membrane swell	5.3 pct
	Organic in raffinate	4.6 ppm
Waste water (520ppm):	Total process time	48 hr
	Volume processed	1600 gal
	Copper extraction	95.7 pct
	Membrane swell	11.1 pct
	Organic in raffinate	11.9 ppm
Waste water (320ppm):	Total process time	33 hr
	Volume processed	1200 gal
	Copper extraction	91.6 pct
	Membrane swell	11.4 pct
	Organic in raffinate	5.7 ppm

Field test 2

Waste water (320ppm):	Total process time	134 hr
	Volume processed	5550 gal
	Copper extraction	91.0 pct
	Membrane swell	8.0 pct
	Organic in raffinate	<3.0 ppm
Waste water (320ppm):	Total process time	146 hr
	Volume processed	5400 gal
	Copper extraction	91.3 pct
	Membrane swell	4.1 pct
	Organic in raffinate	<3.0 ppm

Note: (1) Membrane leakage for all field test campaigns was <1 pct.
(2) Current efficiency ranged from 93 to 95 pct for all tests.

Field Test 1. The copper extractions from all three solutions were very good: 98.0, 95.7, and 91.6 percent, respectively. Membrane swell was only 5.3 percent for the first solution and 11.1 and 11.4 percent for the final two solutions. The higher swell in the last two solutions was related to a design flaw (too much flow resistance between the extraction units and the settler at high flow rates) that resulted in the need to operate the extraction mixers at higher than desired speeds. (This problem was corrected before the second field test.) All three of the solutions that were fed to the LEM system contained more iron than copper. For example, the Cu/Fe ratios were: 0.81 for the PLS, 0.29 for the 520 ppm solution, and 0.19 for the 320 ppm solution. The same copper internal solution (electrolyte) was recycled many times and used throughout the test. The iron level in the internal solution built up to just under 2 g/L Fe and stabilized at that point. The pilot plant was designed so that the water imbalance caused by membrane swelling would result in the automatic discharge of a compensating bleed stream. During the test, the bleed stream ranged from about 5.3 to 11.4 percent of the flow rate of the internal solution. Based upon a relatively limited amount of data, the relative transfer ratios of Cu to Fe into the internal solution ranged from about 350:1 in the processing of the PLS to about 100:1 in the processing of the dilute solutions. It should be noted that since iron was precipitating in the LEM system while the dilute solutions were being processed, some entrained iron precipitates were likely carried to the copper electrolyte. This would have "artificially" lowered the relative Cu/Fe transfer rates in the dilute solution treatment. Even though the LEM system survived the processing of solutions containing iron precipitates, it would be advisable to process "stable" solutions where the precipitation of metals in the LEM system would be minimized.

One of the advantages of the LEM technique over SX is that generally, significantly less organic is lost in the raffinate with the LEM technique. This feature can be especially important from the cost standpoint in the processing of dilute solutions. A limited number of raffinate samples were taken during the field test to get an order-of-magnitude indication of the organic losses in the processing of the three solutions. The results indicated an organic content in the raffinate that ranged from about 5 to 12 ppm (total) organic. These data indicate that the loss of organic in the raffinates would be expected to be quite low with the LEM technique. By comparison, literature *(9)* indicates that typical raffinates from commercial SX units can contain from about 50 to 210 ppm total organics.

Depending on the solution being processed, the EW unit was operated with current densities from about 10 to 20 amp/ft^2 at a temperature of 40° C. The concentration of strong electrolyte was approximately 26 g/L Cu and 125 g/L H_2SO_4, and the concentration of the lean electrolyte was approximately 20 g/L Cu and 160 g/L H_2SO_4. The second and third portions of this first field test were short in duration. Therefore, these EW data were combined to obtain an overall current efficiency for the portion of the test that dealt with these solutions. Typical copper cathode analyses are given in Table III.

Field Test 2. During the second field test, waste waters containing 120 ppm Cu and 640 ppm Cu were processed. The copper extractions from both parts of this test were good, especially considering that a major focus for this test was to minimize membrane swelling. Some sacrifices in copper extractions were made in order to maintain low swelling. Even so, the overall copper extraction averaged 85 percent for the first part of the test (with 120 ppm water) and 89 percent for the second part of the test (with 640 ppm water). These averages include start-up periods when the system was far from optimum operation. The range in copper extraction for the first part of the test was between 77 to 91 percent. The range in copper extraction for the second part of the test was between 82 to 92 percent.

Table III. Cathode Analyses From Field Test
(320 and 520 ppm produced cathodes)

Metal	Concentration, pct
Al	0.00035
Co	0.0013
Fe	0.0019
Ni	0.0016
Si	<0.003
Zn	0.003
Total	<0.0111
Cu (by difference)	99.99

In both test series, the copper extraction was >90 percent when the equipment was operating at or near optimum conditions. Overall, membrane swelling averaged 9.9 percent for the first part and 4.6 percent for the second part. After adjustments had brought the LEM system to its optimum operation, copper extraction and membrane swelling for the two parts of the test were: 91.0 percent copper extraction, 8.0 percent swelling; and 91.3 percent copper extraction, 4.1 percent swelling, respectively. Although there was 5 times more iron than copper in the 120-ppm waste water, the relative transfer ratios of Cu to Fe into the internal solution was about 60:1 and was about 100:1 in the processing of the 640 ppm water. Essentially no other impurities from the feeds were transferred to the internal solutions. At essentially steady-state conditions, while processing the 120 ppm water, the lean internal solution contained about 10 g/L Cu and just 1 g/L Fe.

Another very successful aspect of this field test was the very small loss of organic phase from the emulsion membrane into the treated solution discharged from the LEM system. For all parts of the test, the level of organic in the effluent were <3 ppm (total organics). For much of the test duration, the level was <1 ppm. This is very significant from environmental and economic standpoints (especially when treating the more dilute solutions).

The EW unit was operated with several different sets of parameters to provide data for the selection of optimum parameters for the recovery of copper from dilute solutions (e.g., 120 ppm Cu). Current densities between 5 to 14 amp/ft^2, lean electrolytes containing between 8.5 and 20 g/L Cu (160 g/L H_2SO_4), and strong electrolytes containing between 10 and 33 g/L Cu (140 g/L H_2SO_4) were investigated at a temperature of 40° C. High quality copper cathodes (99.9+ percent Cu) were produced from both parts of this test.

Economic Evaluations

The evaluation method used to estimate the capital costs for the proposed plants is defined as a study-type estimate and is explained in IC 9147 *(10)*, which also defines the evaluation methodology. The predicted accuracy of such an estimate is within 30 percent of the actual plant cost. The estimated capital

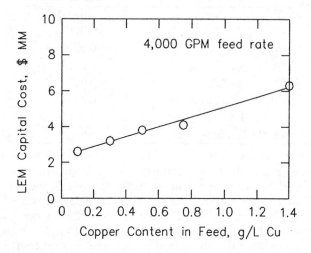

Figure 2. Capital Costs for a 4,000 GPM LEM Plant at Different Feed Concentrations of Copper.

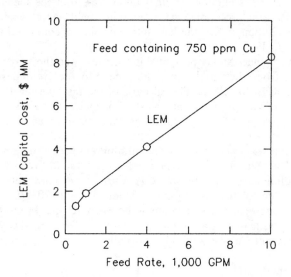

Figure 3. Capital Costs of the LEM Plant at Different Feed Flow Rates.

costs were based on third quarter, 1994 costs (Marshall and Swift [M&S] index), and where possible, costs were based on quotations from equipment manufacturers. The evaluation is based on plants operating 350 days per year, which is equivalent to 24 hours per day operation with an allowance for both scheduled and unscheduled downtime. Capital and operating costs were developed for industrial plants based on the results of the two field tests. Sensitivity studies were performed for different flow rates, different levels of copper in the feed, varying metal extractions and membrane swelling, and varying losses of organic in the raffinate. A total of 42 separate cases were considered.

Capital Costs.   Based upon the results from the field tests, cost evaluations were developed that are related to the potential use of the LEM technique for removing and recovering copper from waste waters and dilute process solutions. Figure 2 shows the effect of the copper level in the feed on the capital cost of a 4,000 GPM plant. The predicted cost of the plant would decrease as the level of copper in the feed decreased. As the copper concentration in the feed decreases, the ratio between emulsion flow rate and feed flow rate also decreases. This has the effect of resulting in lower capital costs. The cost of the LEM plant is greatly influenced by the amount of emulsion used.

Figure 3 shows the effect of feed rate on plant capital cost. In this scenario, the feed water contained 750 ppm Cu. As expected, the higher the feed rate, the more expensive the plant. However, the larger plants cost much less per gallon of treated water than do the smaller plants.

Operating Costs.   Figure 4 shows the operating cost of the LEM technique (lower curve) versus the copper content of the waste water for plants treating a flow of 4,000 GPM. The next upper curve represents the cost of LEM plus EW to recover cathode copper. (Based on recent information from several copper companies that are using SX and EW, an up-to-date cost on a well designed and operated EW circuit is 5 to 8 cents/lb.) In this figure, 7 cents/lb was used for the cost of EW. The horizontal line at the top of the figure represents the selling price of cathode copper ($ 1.42/lb on March 16, 1995). Inspection of the figure shows that even for the 100 ppm Cu waste water feed, the operating cost is estimated to be about half of the selling price of copper cathodes. It should be noted that at this point, no effort has been made to do any optimizations based on lowering the capital and/or operating costs. Therefore, it is likely that significant cost reductions can be made.

The influence of the feed rate (i.e., plant size) on the operating cost also was investigated (Figure 5). As expected, the predictions indicate that the larger the plant, the lower are the operating costs. These scenarios were based on a feed containing 750 ppm Cu. All of the projected costs are well below the value of the produced copper product.

Summary and Conclusions

A continuous, mobile, pilot plant-scale LEM circuit was designed and field tested by the USBM to extract and recover copper from relatively dilute leach

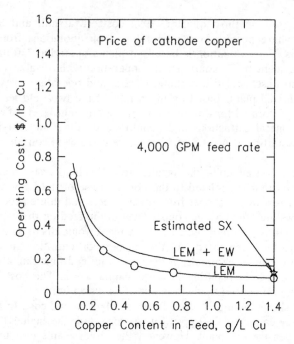

Figure 4. Operating Costs for a 4,000 GPM LEM Plant at Different Feed Concentrations of Copper.

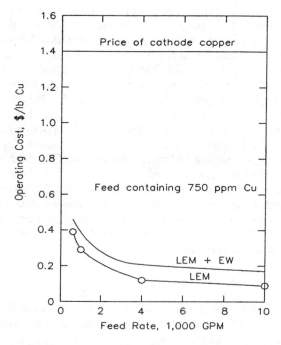

Figure 5. Operating Costs of the LEM Plant at Different Feed Flow Rates.

solutions and waste waters. The LEM system used simple equipment designs for all units and the emulsion formation was optimized so that relatively mild conditions could be used in the electrical coalescer to break the emulsion. Through the field tests, the LEM technique successfully demonstrated that copper could be selectively recovered from a range of solutions containing impurity metals and suspended solids. Good recoveries of copper were experienced in these tests (91 to 98 percent). Good membrane stabilities were demonstrated. Membrane swelling ranged from 4.1 to 11.4 percent, and membrane leakage was <1 percent in all test campaigns. Low levels of organic reported to the effluent from the LEM system (from <3.0 ppm to 11.9 ppm). This is very significant from environmental and economic standpoints (especially when treating the more dilute solutions). Pure copper cathodes (99.9+ percent Cu) were produced from all solutions tested, and high current efficiencies were experienced (93 to 95 percent). Based upon the initial economic evaluations and sensitivity studies, the economics of using LEMs for copper removal and recovery from waste waters looks attractive.

Literature Cited

1. Li, N. N. *Separating Hydrocarbons With Liquid Membranes*, U.S. Pat. 3,410,794. November 12, 1968.
2. Frankenfeld, J. W.; Cahn, R. P.; Li, N. N. *Sep. Sci. and Technol.* **1981**, *16*, 385.
3. Hayworth, H. C.; Ho, W. S.; Burns, W. A., Jr. *Sep. Sci. and Technol.* **1983**, *18*, 493.
4. Kitagawa, T.; Nishikawa,Y.; Frankenfeld, J. W.; Li, N. N. *Environ. Sci. Technol.*, **1977**, *11*, 602.
5. Draxler, J.; Furst, W.;   Marr, R. *Proceedings of the International Solvent Exchange Extraction Conference, Munich, Fed. Rep. of Germany, Sept. 11-16, 1986*; DECHEMA (Deutsche Gesellshaft Fur Chemisches Apparatewesen, Frankfurt, Fed. Rep. of Germany) 1986, Vol. I; pp 553-560.
6. Protsch, M.; Marr, R. *Proceedings of the International Solvent Extraction Conference, Denver, Colorado, U.S.A., Aug. 26-Sept. 2, 1983*; International Solvent Extraction Conference, Denver, CO, 1983; pp 66-67.
7. Nilsen, D. N.; Jong, B. W.; Stubbs, A. M. *Copper Extraction From Aqueous Solutions With Liquid Emulsion Membranes: A Preliminary Laboratory Study*; U.S. Bureau of Mines Report of Investigations 9375, 1991; 16 pp.
8. Nilsen, D. N.; Hundley, G. L. *Proceedings of the Copper 91 International Symposium, Ottawa, Canada, Aug. 18-21, 1991*; Pergamon Press, 1991, Vol. III (Hydrometallurgy and Electrometallurgy of Copper); pp 155-169.
9. Kordosky, G. A.; Dorlac, J. P. *Capital and Operating Costs for the Unit Processes of Solvent Extraction and Electrowinning as Applied to Copper Recovery*; Technical Information Paper; Henkel Corp., Tucson, AZ, undated; 15 pp.
10. Peters, F. A. *Economic Evaluation Methodology*; U.S. Bureau of Mines Information Circular 9147, 1987; 21 pp.

Chapter 24

Removal of Selenium from Contaminated Waters Using Emulsion Liquid Membranes

Kevin J. Gleason[1], Jianhan Yu[1], Annette L. Bunge[2],
and John D. Wright[1]

[1]TDA Research, Inc., 12345 West 52nd Avenue, Wheat Ridge, CO 80033
[2]Department of Chemical Engineering and Petroleum Refining,
Colorado School of Mines, Golden, CO 80401

An emulsion liquid membrane (ELM) process is presented that can rapidly reduce aqueous phase selenium concentrations from 1 mg/L to a level below 0.01 mg/L (within 15 minutes of contact time), while increasing the concentration of selenium in the internal phase by more than a factor of 40. The emulsion formulation was very stable, and swelling did not significantly dilute the selenium in the internal phase. The presence of other anions, such as sulfate, did not prevent the selenium anions from being extracted. Finally, the counter-current extraction of selenium was demonstrated. The continuous process has several advantages over the batch process, including improved rates due to the higher driving force at low concentrations, and the insensitivity of the process to leakage. The extraction rate and efficiency depend primarily on the selenium oxidation state, and subsequently on the pH of the aqueous phase.

Although selenium is an essential nutrient in minute quantities, it is highly toxic in slightly higher concentrations. The apparent Jekyll and Hyde characteristic of selenium (levels considered beneficial for humans have been shown to be deadly for waterfowl) creates problems in its regulation and management. Among the larger sources of selenium-contaminated waters, two that require immediate attention are petroleum refineries and agricultural drainage water.

The extent of the selenium problem in petroleum refineries depends on the types of crude oil processed. For example, crude oil from the San Joaquin Valley contains higher levels of selenium while other sources of crude are essentially selenium free. Because San Joaquin crude is processed in several refineries along the western coast of the US, several of these refineries have problems with selenium contamination of their process waters. Depending on the process waters (stripped sour water or final biotreated effluent) the volumetric flowrates range from 10

0097–6156/96/0642–0342$15.00/0

gallons per minute (gpm) up to 5000 gpm with selenium concentrations approaching 5 mg/L *(1)*.

Irrigated lands in the western United States are another major source of selenium contaminated waters. Because of the high mobility of the selenium oxyanions, selenium is easily leached from the soils by irrigation water *(2,3)*. The toxicity of selenium to waterfowl presents difficulties in disposing of the water artificially drained from poorly-drained clay soils, which have high natural concentrations of salts and selenium. Within the Westlands Water District of central California alone (encompassing some 600,000 acres of highly productive agricultural land), there is a current need for the disposal of about 2,000 acre-feet per year or about 2 million gallons per day (2 MM gpd) of agricultural drainage water with selenium concentrations reaching 0.5 mg/L *(4)*. The ultimate need in the district for drainage disposal is projected at about 50,000 acre-feet per year (or 50 MM gpd) in order to keep the land in production *(4)*.

Because of its known toxicity, the selenium discharge limits as set by the Clean Water Act are low (0.260 mg/L maximum, 0.035 mg/L continuous for selenium) *(5)*. The maximum allowable level in drinking water is even lower (0.01 mg/L). Yet, in amending the water quality control plan for the San Joaquin Basin to include water quality objectives for selenium, the California State Water Resources Control Board *(6)* adopted an even lower limit of 0.005 mg/L on selenium concentration in the San Joaquin River. The low limit is based on ecological impacts to fish and waterfowl *(7)*.

Emulsion liquid membranes (ELMs) represent a separation technique capable of achieving the desired level of contaminant removal. ELMs, invented by Li *(8,9)*, have been proposed as solutions to a number of waste disposal problems *(10-12)*. In this process, the contaminant is extracted from the waste aqueous phase, through an organic phase (assisted by a complexing agent), and into an internal, aqueous-stripping phase emulsified within the organic drop. Very high removals, high concentration ratios, and high extraction rates can be economically achieved. Because ELMs combine extraction and stripping into a single unit operation, the process is able to circumvent the equilibrium limitations encountered by solvent extraction. This paper presents an ELM system to remove selenium from aqueous waste streams representative of those encountered in petroleum refinery process waters and agricultural drainage water.

Selenium Chemistry

To develop an effective, optimized ELM process for selenium removal, it is necessary to understand the forms of the selenium ions present in contaminated aqueous streams. Selenium can exist in several oxidation states: -II, 0, IV, and VI *(13)*. In naturally occurring ground or surface waters, inorganic selenium is found in the IV (selenite) and VI (selenate) states *(14)*. In petroleum refinery wastewater effluents, the form of selenium depends on the process stream. In stripped sour water the selenium exists primarily in the Se(-II) oxidation state as either organoselenium compounds or selenocyanate. In the final biotreated effluent, selenium exists primarily as selenite [Se(IV)] and selenate [Se(VI)] oxyanions *(1)*.

The transformation of Se(IV) to Se(VI) is very slow, and both forms stably coexist. By contrast, in agricultural drainage waters the selenium exists almost exclusively as selenate *(3)*. For each particular oxidation state, the dominant selenium species is determined by the pH. In this study, we will focus on the selenite and selenate forms. The dissociation reactions and equilibrium constants for selenite and selenate are shown in equations 1 and 2 below *(15)*.

Selenium(IV) Dissociation:

$$
\begin{aligned}
H_2SeO_3 &\rightleftarrows HSeO_3^- + H^+ & pK_1 &= 2.64 \\
HSeO_3^- &\rightleftarrows SeO_3^{2-} + H^+ & pK_2 &= 8.27
\end{aligned}
\tag{1}
$$

Selenium(VI) Dissociation:

$$
\begin{aligned}
H_2SeO_4 &\rightleftarrows HSeO_4^- + H^+ & pK_1 &= -3 \\
HSeO_4^- &\rightleftarrows SeO_4^{2-} + H^+ & pK_2 &= 1.66
\end{aligned}
\tag{2}
$$

The equations shown above can be solved to give the speciations of selenite [Se(IV)] (Figure 1) and selenate [Se(VI)] (Figure 2) as functions of pH.

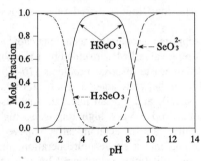

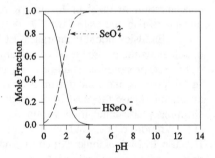

Figure 1. Se(IV) Speciation.   Figure 2. Se(VI) Speciation.

Selenous acid, H_2SeO_3, is a weak acid that dissociates into two selenite [Se(IV)] species $HSeO_3^-$ and SeO_3^{2-}. The concentration of each depends on the pH of the water. For example, at a pH of 4-7 the $HSeO_3^-$ anion will predominate, while at a pH above 10 the SeO_3^{2-} anion predominates (Figure 1). Selenic acid, H_2SeO_4, is a strong acid that forms only SeO_4^{2-} in waters with pH above 2 (Figure 2).

Unfortunately, in contaminated waters and wastewaters, innocuous anions like sulfate are inevitably present as well, usually at concentrations several orders of magnitude greater than that of the selenium anions. For example, sulfate concentrations in refinery wastewaters approach 800 mg/L *(1)*, while sulfate concentrations in agricultural drainage water are in the 3000-4000 mg/L range *(16)*. Consequently, it is difficult to remove the selenium without also having to remove far larger quantities of other, non-hazardous materials as well.

Several methods have been studied for removing selenium to meet drinking water standards including coagulation by ferric sulfate and alum, lime softening, powdered activated carbon, activated alumina, reverse osmosis, and ion exchange *(14,17,18)*. The presence of selenium species in two oxidation states significantly affects removal efficiencies. In most cases, the process can remove only one of the two species. For example, ferric sulfate coagulation removes Se(IV) but not Se(VI) *(19,20)*. In other cases, selenium removal is less efficient because the selenium in the two oxidation states compete with one another and with other anions like sulfate.

Emulsion Liquid Membrane Extraction of Selenium

This paper presents an emulsion liquid membrane (ELM) system capable of extracting both Se(IV) and Se(VI) from an aqueous waste stream containing large amounts of innocuous anions like sulfate. In an emulsion liquid membrane, small drops of an aqueous stripping phase are contained (emulsified) inside an organic hydrocarbon (oil) drop. The surfactant-stabilized oil acts as a membrane separating the wastewater (the continuous, bulk phase solution) from encapsulated droplets of aqueous solution (the internal phase) as shown in Figure 3.

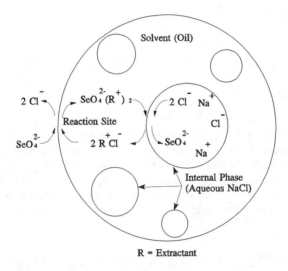

R = Extractant

Figure 3. Mechanism of ELM extraction of Se(VI).

Typically, the emulsion globules are 0.1-2 mm in diameter, and the encapsulated, aqueous droplets are 0.1-5 μm in diameter. A typical emulsion formulation for selenium extraction would consist of a high concentration of aqueous solution of NaCl dispersed in a low volatility, non-polar hydrocarbon like kerosene, and stabilized with an oil-soluble surfactant. The organic phase also contains a complexing agent for the selenium compounds. In the ELM process, a selenium anion first diffuses to the surface of the oil globule where, depending on the ionic species present (which in turn depends on the selenium oxidation state and the pH of the aqueous phase), it reacts with one or two extractant molecules and subsequently dissolves into the organic phase. The extraction reactions of the complexing agent with selenium are given below.

Se(IV) (selenite):

$$HSeO_3^- \big|_{aq} + RCl \big|_{org} \; \rightleftharpoons \; RHSeO_3 \big|_{org} + Cl^- \big|_{aq} \tag{3}$$

$$SeO_3^{2-} \big|_{aq} + 2\,RCl \big|_{org} \; \rightleftharpoons \; R_2SeO_3 \big|_{org} + 2\,Cl^- \big|_{aq} \tag{4}$$

Se(VI) (selenate):

$$SeO_4^{2-} \big|_{aq} + 2\,RCl \big|_{org} \; \rightleftharpoons \; R_2SeO_4 \big|_{org} + 2\,Cl^- \big|_{aq} \tag{5}$$

The Se-extractant complex then diffuses through the oil to the interface of one of the internal phase, sodium chloride containing, aqueous droplets, which are dispersed inside the oil globule. At this interface, the concentrated Cl⁻ ions in the interior aqueous phase strip the Se anion from the extractant, by the reactions below.

Se(IV) (selenite):

$$RHSeO_3 \big|_{org} + Cl^- \big|_{aq} \; \rightleftharpoons \; RCl \big|_{org} + HSeO_3^- \big|_{aq} \tag{6}$$

$$R_2SeO_3 \big|_{org} + 2\,Cl^- \big|_{aq} \; \rightleftharpoons \; 2\,RCl \big|_{org} + SeO_3^{2-} \big|_{aq} \tag{7}$$

Se(VI) (selenate):

$$R_2SeO_4 \big|_{org} + 2\,Cl^- \big|_{aq} \; \rightleftharpoons \; 2\,RCl \big|_{org} + SeO_4^{2-} \big|_{aq} \tag{8}$$

The stripped extractant is now free to diffuse back to the exterior oil-water interface to extract another selenium anion and the extracted selenite or selenate ion is now trapped in the interior aqueous phase. The net result of this simultaneous extraction and stripping process is an aqueous internal phase with significantly higher selenium concentrations than the original process stream. Thus, one has significantly reduced the volume of the waste stream. The overall reactions are:

Se(IV) (selenite):

$$HSeO_3^- \big|_{external} + Cl^- \big|_{internal} \; \rightleftharpoons \; HSeO_3^- \big|_{internal} + Cl^- \big|_{external} \tag{9}$$

$$SeO_3^{2-} \big|_{external} + 2\,Cl^- \big|_{internal} \; \rightleftharpoons \; SeO_3^{2-} \big|_{internal} + 2\,Cl^- \big|_{external} \tag{10}$$

Se(VI) (selenate):

$$SeO_4^{2-} \big|_{external} + 2\,Cl^- \big|_{internal} \; \rightleftharpoons \; SeO_4^{2-} \big|_{internal} + 2\,Cl^- \big|_{external} \tag{11}$$

If the Cl⁻ concentration in the internal phase is high, the selenium anions can be extracted from very dilute solutions and concentrated in the internal-phase. The driving force for the extraction is the concentration difference between the chloride ions in the internal-phase (high concentration) and the external phase (essentially zero).   In addition to selenium, other anions can complex with the extractant molecules.   For refinery wastewaters like the final effluent stream, sulfate is the principle anion of concern.   In those cases, sulfate will be extracted and stripped in reactions similar to the selenate reactions in equations 5, 8, and 11.

Because the ELM process simultaneously extracts and strips, the capacity of the emulsion for selenium is determined by the stripping anion (chloride) in the internal phase and not the concentration of the complexing agent (extractant).   That is, selenate, selenite, and sulfate do not compete for the extractant in the same manner as ion exchange and consequently, preference of one species over the others does not prevent extraction of the other species.   This is distinctly different from fixed bed ion exchange where the most highly preferred species will displace the less preferred species, causing it to bleed into the treated water.   In ELM, the concentration of extractant is based on kinetics and emulsion stability, not capacity.

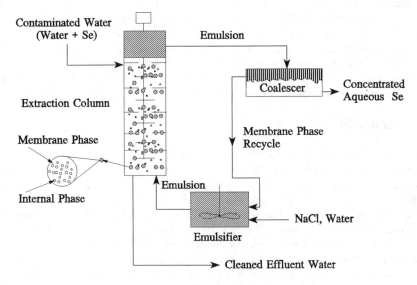

Figure 4.   ELM process for the extraction of selenium from industrial wastewaters.

The overall ELM process is shown in Figure 4.   The selenium contaminated wastewater is contacted with the emulsion in one or more stirred tanks or in a counter-current contactor extractor, as illustrated in Figure 4.   The selenium concentration in the treated wastewater is reduced to nearly zero, and the concentration in the interior phase of the emulsion increases.   The emulsion and wastewater are separated in a settler; the cleaned effluent water is ready for discharge; the emulsion phase is then taken and split in an electrostatic coalescer. After the emulsion is coalesced, a much smaller volume of greatly concentrated selenium must be treated further and disposed.   The oil phase from the coalescer is then mixed with fresh NaCl at high shear rates to reform the emulsion, and returned to the extractor. Makeup extractant and surfactant are added as necessary during the reformulation step.

The ELM process operates at room temperature and ambient pressure, and the mixing rates required in the extractors are only those required to keep the oil droplets from coalescing and stratifying. If the membrane is recycled, the only major consumable is the sodium chloride that exchanges with the selenium ions.

The ELM extraction of selenium has several advantages compared with fixed bed processes such as anion exchange or activated alumina absorption. First, the ELM process is continuous rather than semi-batch. Fixed bed extractors are susceptible to fouling and the resin degrades with use. In the ELM process, the extracting media is stirred and continuously renewed and consequently, fouling and degradation of the extracting media does not occur. The ELM process should also be more flexible in handling excursions in concentrations or flows than the fixed bed processes.

The objective of this work was to develop an emulsion liquid membrane process capable of removing selenium from contaminated waters such as petroleum refinery wastewaters or the agricultural drainage waters. The ELM process must be capable of removing both selenate and selenite from water containing a large amount of innocuous anions like sulfate. In addition, the ELM process must not be susceptible to fouling and degradation typical of fixed bed processes.

Experimental

Analytical Methods. Aqueous selenium concentrations were measured using atomic absorption spectroscopy (AA) with hydride generation. In this technique, volatile selenium hydrides are produced by sodium borohydride reduction of the acidified sample. The hydrides then are atomized in a heated quartz tube where the absorbance is measured. The sensitivity for selenium obtained by the hydride generation technique is about 0.2 nanograms/ml (based on a sample volume of 20 ml) and represents an improvement of greater than three orders of magnitude over conventional flame AA. The upper concentration limit of the AA-hydride generation technique is 0.1 mg/L. Standard selenium solutions (ranging from 0.02 to 0.10 mg/L) were used for calibration. Since the hydride generation technique measures only Se(IV), to measure Se(VI) it was necessary to carry out a pre-reduction from Se(VI) to Se(IV) prior to hydride generation analysis. In samples where both Se(IV) and Se(VI) are present, the samples were analyzed before and after the pre-reduction step and the Se(VI) was determined by difference. Sulfate concentrations were measured directly by a colorimetric technique.

Emulsion Formulation. The emulsion consists of two separate phases: the membrane (oil) phase and the internal stripping phase. The membrane phase encapsulates the aqueous internal phase, separating it from the low concentration stream of wastewater being treated (the external phase). The membrane phase is composed of a complexing agent, a surfactant, a solvent, and a co-solvent. The solvent is kerosene. The co-solvent is n-decanol. Both the complexing agent and the surfactant are proprietary materials. The extractant complexes with the anionic selenium species at the external phase interface, solubilizes them in the membrane phase, and transports the anions from the external to the internal phase. The extractant-selenium complex has a relatively low solubility in the oil phase, and tends to destabilize the emulsion. A small amount (2%) of n-decanol is used as a co-solvent to help solubilize the complex. The entire emulsion is stabilized by the addition of a surfactant. The internal phase consists of a concentrated (0.5 to 2 M) solution of NaCl. The internal phase droplets are 0.1-5 μm in diameter and are also stabilized by the surfactant. The emulsion was formed by combining 30 ml of the aqueous internal phase containing NaCl with the 90 ml of the membrane phase in a Model 23 VirTis homogenizer operated at 5000 rpm for 20 minutes.

Batch Extractions. In nearly all commercial scale operations, a continuous extraction process, either mixer-settlers or column units, would be used. However, batch extraction experiments are useful for assessing overall feasibility and for optimizing the many process variables such as emulsion formulation and volume ratios of the internal, membrane, and external phases. Consequently, the most common experiment in this study was the batch extraction. In these experiments, 500 ml of a selenium solution (1 mg/L) were prepared in the extraction vessel, either in the presence or absence of other competing anions. The prepared emulsion (50 ml) was added and the mixture was stirred at a speed of 150 rpm. In this manner the emulsion drops were uniformly dispersed in the external phase while extraction proceeded. Samples of the external aqueous phase were taken at appropriate intervals and the concentrations of Se(IV), Se(VI), and sulfate were determined.

Continuous Extractions. To investigate ELM extraction of selenium in a commercially relevant format, we constructed an agitated continuous counter-current extraction column, illustrated schematically in Figure 5. Constructed from acrylic, the column has eight equilibrium stages, separated by plates. The diameter of the column was 2.5 in. and the height of an individual section was 2.5 in. The total volume of the column was 1.35 liters and the volume of the settling section at the bottom of the column is 2.5 liters. Each stage is agitated by double impellers, supported from a 3/8 in. diameter stainless steel shaft, which is centered in a 1.4 cm diameter hole in the middle of each plate. The external phase feed and emulsion phase flow from stage to stage through the space between the shaft and the outer edge of the 1.4 cm diameter hole.

The emulsion enters the column at the lowest section, flows upward through the column driven by its buoyancy, and coalesces in the top of the column (in a stage that does not have an impeller). The feed solution containing the selenium and sulfate enters the column at the top section, flows down through the column, and is removed at the bottom. The interface between the emulsion and the feed solution is controlled by a gravity leg. The flowrates of feed solution and emulsion are checked before and after the extraction runs.

The stirring speed was controlled at 200 rpm. The size of opening in the plates between the stages was chosen based on the desired hold-up (10%) to achieve the level of concentration of selenium in the internal phase. We estimated the emulsion hold-up by measuring the time needed for the emulsion to rise from the bottom of the column to the top of the column during start-up. The effect of the drop-drop interactions (hindered flow) would be to create an actual dispersed-phase residence time that was slightly longer than our estimate.

At the beginning of the run, the water and emulsion flowrates to the column were independently set to their desired values. (These in turn set both the continuous- and dispersed-phase residence times.) After allowing time for the flow regime to reach steady state, the selenium solution was introduced into the aqueous feed. After an additional eight volumes of the continuous-phase had passed through the column and the inlet and outlet concentrations reached steady state, samples were taken from each section to measure the concentration profile.

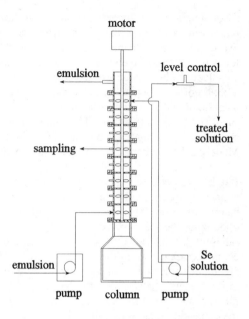

Figure 5. Schematic of the counter-current extraction system.

Results and Discussion

Equilibrium Partition Coefficients. Equilibrium solvent extraction experiments are a convenient way to screen various complexing agents for suitability as selenium extractants and to assess the potential for competition from innocuous ions like sulfate. The solvent extraction experiments allow the determination of reaction equilibrium constants and distribution or partition coefficients of the selenium species between the organic (membrane) phase and the aqueous (internal or external) phase. The equilibrium constant (K_e) and distribution coefficient (K_d) for the reaction of selenite with the complexing agent are defined in the equations given below,

$$SeO_3^{2-} + 2\ RCl \rightleftarrows R_2SeO_3 + 2\ Cl^-$$

$$K_e = [R_2SeO_3]_{org}\ [Cl^-]^2_{aq}\ /\ [SeO_3^{2-}]_{aq}\ [RCl]^2_{org} \tag{12}$$

$$K_d = [R_2SeO_3]_{org}\ /\ [SeO_3^{2-}]_{aq}$$

$$\log K_d = \log K_e + 2\ \log \{\ [RCl]_{org}\ /\ [Cl^-]_{aq}\ \}$$

where R is the complexing agent. Similar equilibrium relationships can be written for the reactions of the complexing agent with biselenite, selenate, and sulfate. In this study a $K_d > 1$ means that the anion partitions preferably into the organic phase. Since the complexing agent has a 2:1 stoichiometry with selenite (as well as selenate and sulfate) a plot of log K_d vs. log $\{[RCl]_{org} / [Cl^-]_{aq}\}$ would show a slope of 2. With a 1:1 stoichiometry, the reaction of biselenite $[HSeO_3^-]$ would show a slope of 1.

We carried out equilibrium solvent extraction experiments to evaluate the complexing reactions. The organic phase was composed of 1% extractant, 10% n-decanol, and 89% kerosene. The aqueous solutions consisted of selenite, biselenite, selenate, or sulfate anions, and varying concentrations of stripping agent (NaCl). The pH of the aqueous solutions were adjusted to assure the nearly exclusive presence of each anion. K_d was obtained by measuring the change of selenium(IV), selenium(VI), or sulfate concentrations in the aqueous solution. $[RCl]_{org}$ and $[Cl^-]_{aq}$ were calculated by mass balance. Typical results are shown in Figure 6. From Figure 6, we can anticipate the partitioning preference, or the relative amounts of anions extracted, by the complexing agent at equilibrium. Based on equilibrium measurements, the order of preference is:

$$SeO_4^{2-} > SO_4^{2-} > SeO_3^{2-} > HSeO_3^- \tag{13}$$

Among the divalent anions studied, selenate is the most preferred, followed by sulfate, and then selenite. The position of biselenite $(HSeO_3^-)$ in the partitioning preference depends on the ratio of complexing agent to stripping agent.

This implies that, while the equilibrium partitioning of each of these anions is very similar indicating that all of these compounds will be extracted to some extent, since selenate is the most preferred over the other anions, it will be extracted to a greater extent once equilibrium is reached.

Technically, the equilibrium partition coefficients are thermo-dynamic values, and as such will not always correlate with extraction rates, which are kinetic

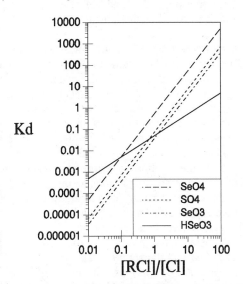

Figure 6. Relationship between K_d and $[RCl]_{org} / [Cl^-]_{aq}$ for selenate, sulfate, selenite, and biselenite.

values. However, if the partitioning kinetics for the selenium and sulfate species are not substantially different (e.g., if the rate at which partitioning reaches equilibrium for each species is instantaneous), then we can use the equilibrium partitioning results as a first approximation to the relative overall extraction rates of the various species. In this manner differences in the equilibrium partitioning should manifest themselves as slight differences in the extraction rates of each species. Therefore, based on the equilibrium partitioning data, we would anticipate that sulfate will be extracted slightly faster than selenite, but much slower than selenate.

Thus, it appears that the presence of sulfate will significantly affect the removal rate of selenium(IV), although the effect on the removal of selenium(VI) should not be as great. However, selenate, selenite, and sulfate do not compete for the extractant in the same manner as ion exchange and consequently, preference of one species over the others should not prevent extraction of the other species.

Batch ELM Extraction of Selenium. A series of batch extraction experiments were carried out to assess the feasibility of an ELM process for the removal of selenium from refinery wastewater streams. To do this we developed emulsion formulations and evaluated these formulations for the capability of reducing the aqueous phase selenium concentration below 0.05 mg/L. Next, we carried out experiments to evaluate the effect of competition between the two oxidation states, the effect of total selenium concentration of the extraction behavior, the effect of external phase pH on the transport rate, and the effect of competing anions (like sulfate) on the extraction efficiency and kinetics.

Batch ELM Extraction of Se(IV) and Se(VI). A series of batch extractions were carried out to determine the extent of extraction that we could achieve. It was also desired to determine if the extraction behavior of selenium depended on the oxidation state. In these experiments, the external aqueous phase contained either selenate [Se(VI)] or selenite [Se(IV)] (in isolation). The initial concentrations were set to 1 mg/L, typical of what one might find in refinery wastewater streams. The extractions were carried out at a pH of 10.5 to assure that both species were in the divalent state [i.e., SeO_4^{2-} for Se(VI) and SeO_3^{2-} for Se(IV)].

In the equilibrium partitioning results presented above, we observed that, at fixed concentrations of complexing agent and stripping agent (chloride), the distribution coefficient of selenate was an order of magnitude larger than the distribution coefficient of selenite.

Using the equilibrium partitioning results as a first approximation to the relative overall extraction rates of the various species, it is anticipated that selenate [Se(VI)] will be extracted faster than selenite [Se(IV)].

The results of the batch extraction experiments are presented in Figure 7, which is a plot of selenium concentration (either selenate or selenite) as a function of time during batch ELM experiments. It is clear that, in the absence of competing anions, both selenate and selenite

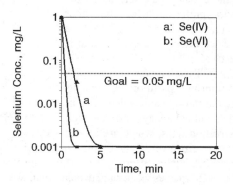

Figure 7. Selenium concentration as a function of extraction time during batch ELM experiments.

are extracted very rapidly. The emulsion liquid membrane is capable of reducing the concentration of selenium in water from 1 mg/L to less than 0.001 mg/L (our limit of detection) within 2 minutes for Se(VI) and within 5 minutes for Se(IV). Thus, our final selenium concentration is a factor of fifty less than our goal of 0.05 mg/L. The ELM process rapidly extracts either the Se(VI) or the Se(IV) oxidation state from water streams and the relative extraction behavior of the two species is consistent with the equilibrium partitioning results.

Competition Between Se(IV) and Se(VI) in the ELM Extraction of Selenium. A series of batch extractions were carried out to evaluate the competition between the two oxidation states of selenium, Se(IV) and Se(VI). In one set of batch extraction experiments we examined the extraction behavior of selenate [Se(VI)] in isolation (at 1 mg/L) and in the presence of an equal amount of selenite [Se(IV)]. In another set of experiments we studied the extraction behavior of selenite in isolation (at 1 mg/L) and in the presence of an equal amount of selenate.

We again can use the results of the equilibrium partition experiments as a first approximation of the relative extraction rates to anticipate the trends we might see. Since the K_d for selenate was an order of magnitude larger than the K_d for selenite, we might anticipate that if there was any competition between the two species, the effect would be more significant with the extraction of selenite than with the extraction of selenate.

The results of the competition experiments are presented in Figures 8 and 9. Figure 8 is a plot of selenate [Se(VI)] concentration as a function of extraction time during batch extractions. Line "b" shows the extraction behavior of selenate in isolation, and line "a" shows the extraction behavior of selenate in the presence of an equal amount of selenite. It is evident that the presence of selenite [Se(IV)] has minimal effect on the extraction behavior of selenate [Se(VI)] at the concentrations studied. We still were able to reduce the concentration to a level well below our target of 0.05 mg/L within a very short time (the first two minutes). The small difference between the two rates is probably due to the difference in total selenium concentrations, rather than to competition between the two species.

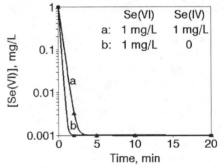

Figure 8. Effect of Se(IV) competition on the batch extraction of Se(VI).

Figure 9 is a plot of selenite [Se(IV)] concentration as a function of extraction time during batch extractions. Line "b" shows the extraction behavior of selenite in isolation, and line "a" shows the extraction behavior of selenite in the presence of an equal amount of selenate. The effect of competition between the two species appears to be more pronounced with the extraction of selenite [Se(IV)]. Again, there

354	CHEMICAL SEPARATIONS WITH LIQUID MEMBRANES

was little difference in the initial extraction rate, and we were still able to reduce the concentration below 0.05 mg/L within the first two minutes. Beyond two minutes, the extraction of selenite appears to taper off, indicating potential competition.

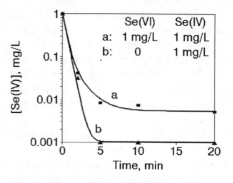

Effect of Total Selenium(IV) Concentration on Transport Rate. As suggested by the equilibrium partitioning results and as confirmed by the competition experiments, selenate

Figure 9. Effect of Se(VI) competition on the batch extraction of Se(IV).

[Se(VI)] is more easily extracted than either selenite [Se(IV)] or sulfate. There is little effect of competition on the extraction behavior of selenate. However, selenite appears to be significantly affected by competing anions. Therefore, the majority of the remaining experiments focused on the extraction behavior of selenite [Se(IV)] since it represents the worst case scenario.

It was desired to determine if the extraction behavior of Se(IV) depends on the total selenium concentration in the external phase. To investigate the effect of external phase concentration on the rate of Se(IV) extraction, we carried out batch extraction experiments at two different initial concentrations (1 and 10 mg/L). The emulsion formulation was the same as described previously. The results are given in Figure 10 (a plot of Se(IV) concentration as a function of extraction time) and show that there is little difference in the initial extraction rates between 1 and 10 mg/L selenite. Beyond two minutes, the extraction rate of the higher amount (10 mg/L) tapers off.

Recall that the distribution coefficients are sensitive to the chloride concentration in the aqueous phase. During the extraction of the higher concentrations of selenite (e.g., 10 mg/L), greater amounts of chloride ion (e.g., > 9 mg/L) are released to the external phase by the exchange with the complexing agent. Assuming that the system responds to these changes instantaneously, this decreases the distribution coefficient of Se(IV), thereby decreasing the equilibrium concentration of selenite, and ultimately the extraction rate.

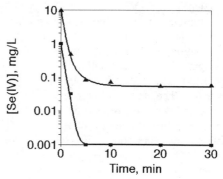

Figure 10. Effect of initial Se(IV) concentration on its batch extraction.

In spite of this, Figure 10 shows that the extraction rate of selenite at higher concentrations (e.g., 10 mg/L) is still rapid and the selenite concentration still can be reduced from 10 mg/L to near 0.1 mg/L in 5 minutes. It also suggests that a two stage process would be more effective at reducing the selenium concentration.

Effect of External Phase pH. The initial pH of the feed affects the selenium species distribution and can affect its extraction reaction stoichiometry and kinetics. As long as the pH > 2, Se(VI) will be in the fully dissociated selenate form (SeO_4^{2-}) and should be extracted very rapidly. Therefore, external phase pH should not have a significant effect on the extraction behavior of Se(VI). However, using the equilibrium partitioning results as a first approximation to relative extraction rates, one would anticipate that pH would have a dramatic effect on the extraction behavior of Se(IV). For Se(IV) to be in the fully dissociated selenite form (SeO_3^{2-}), the pH needs to be greater than 9.5. At more neutral pH, there would be significant biselenite ($HSeO_3^-$) present, and based on the measured distribution coefficients, would not be expected to be extracted as easily.

To confirm this we carried out a batch extraction with Se(IV) at pH = 8, where the selenium(IV) is evenly distributed between the monovalent $HSeO_3^-$ form and the divalent SeO_3^{2-} form. The results are shown as line "a" in Figure 11, which is a plot of selenium concentration as a function of batch extraction time. The results for the batch extractions of Se(IV) [line "b"] and Se(VI) [line "c"] at a pH = 10.5 are shown for comparison. As anticipated, the extraction rate is significantly slower than at the higher pH. It is clear that pH has a dramatic effect on the extraction rate of Se(IV).

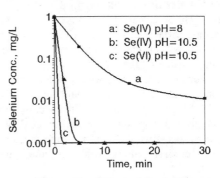

Figure 11. Effect of the initial external phase pH on the batch ELM extraction of selenium.

The results also indicate that the ELM process has a reciprocal effect on the external phase pH. As the ELM process is carried out, besides exchanging chloride ions in the internal phase and the selenium anions in the external phase, it also extracts hydroxide ions from the external phase due to the *OH* concentration difference between the external and internal phases. The result is a slight reduction in the external phase pH (from pH 10.5 to 9.3 and from pH 8 to 6) during the course of the ELM process. At the higher pH (10.5), this should not affect the extraction process since it is not changing the speciation of the selenium compounds. However, at a lower pH of 8, the reduction in pH will change the relative amounts of Se(IV) species from an equal amount of selenite and biselenite to almost exclusively biselenite. As we saw above, the extraction rate of selenite [Se(IV)] is very sensitive to pH. Therefore, as the extraction process reduces the pH, this will in turn reduce the extraction rate even further.

Effect of Sulfate on the Extraction of Selenite. In some industrial streams, the sulfate concentration can approach 1000 mg/L or more. Therefore, we investigated the effect of an extremely high concentration of sulfate on the extraction rate of selenite. The results are presented in Figure 12, which gives the simultaneous selenite and sulfate concentrations as a function of time during the batch extraction of selenite in the presence of 1000 mg/L of sulfate. We observe that the difference between the selenite and sulfate distribution coefficients is magnified during their batch ELM extraction. We also notice that the extraction of both tapers off after only a few minutes. This is primarily due to the large amount of ionic species transferred. Since the starting concentration of sulfate is 1000 times higher than the initial concentration of selenite, the amount of chloride that is transferred to the external phase is also 1000 times higher. This amount of chloride pumped into the external phase dramatically changes the distribution coefficients, and as a result changes the extraction

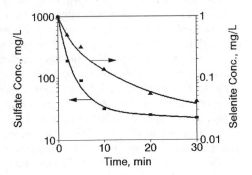

Figure 12. Effect of high concentration of sulfate (1000 mg/L) on the batch extraction of selenite.

behavior as well. Thus, when sulfate is present in large quantities with selenite in a contaminated stream, one would anticipate that the batch extraction rate of selenite would be greatly reduced. In spite of this, we were still able to reduce the concentration of selenite below our goal of 0.05 mg/L. A continuous counter-current process would be an improvement since it would maintain the driving force for extraction (i.e., the chloride concentration gradient) throughout the full extraction period.

Demonstration of a Continuous Counter-Current ELM Extraction System. Experiments were carried out to demonstrate the continuous counter-current ELM extraction of a simulated refinery wastewater stream. For the continuous extraction experiments, the external phase contained 1 mg/L Se(IV) and 1000 mg/L sulfate at an initial pH of 10.5. The emulsion consisted of membrane phase and internal phase at a ratio of $V_i/V_m = 1/3$. The membrane phase consisted of 2% complexing agent, 2% surfactant, and 2% n-decanol in kerosene. The internal phase contained 2 M NaCl as the stripping agent. The complexing agent and NaCl concentrations were chosen based on the total amount of anions (selenite and sulfate) to be extracted. The contaminated external phase feed flow rate was set to 2.7 L/hr. This was selected to set the external phase residence time through the column to 30 minutes. The ratio of external phase feed to emulsion flow rates was roughly 10/1. That is, the emulsion flow rate was 0.27 L/hr. The results of the continuous ELM extractions are shown in Figure 13, which is a plot of sulfate and selenite concentrations as a function of the external contaminated phase residence time through the column. Each data point is a measure of the concentration at each stage in the counter-current column.

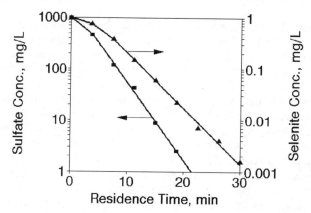

Figure 13.   Extraction of selenite and sulfate in a counter-current continuous column in the presence of 1000 mg/L of sulfate.

We had previously shown in batch extraction experiments (Figure 12) that the ELM process was capable of reducing the selenium concentration to a level below 0.05 mg/L within 30 minutes, even in the presence of an extremely large amount of sulfate (1000 mg.L). It is clear in looking at Figure 13, that the continuous counter-current ELM process was capable of reducing the selenium concentration below the 0.05 mg/L level (again, in the presence of 1000 mg/L sulfate) in half the time (15 minutes) it took the batch ELM process. In addition, whereas the batch experiments showed that the extraction rate leveled off, the continuous experiment showed a uniform extraction rate down to 0.002 mg/L concentration. With only eight stages in the continuous counter-current ELM process (equaling 30 minutes contact time), we were able to surpass the target effluent concentration of 0.05 mg/L by a factor of 1/25 in the presence of an extremely large amount of sulfate.

Another way to view the continuous counter-current data is in terms of the emulsion residence time through the column. The fresh emulsion is introduced at the bottom of the column (right side of Figure 13 or at 30 min. residence time). The external contaminated aqueous phase in contact with the fresh emulsion at the bottom of the column has a reduced selenium concentration and an elevated chloride concentration. At this point the emulsion has its highest chloride concentration. The emulsion exits at the top of the column (left side of Figure 13 or at 0 min. residence time). The external contaminated aqueous phase at the top of the column has its initial high selenium concentration and a low chloride concentration. At this point the (depleted) emulsion has its lowest chloride concentration. Recall that the chloride concentration gradient across the membrane phase drives the ELM extraction process. As we follow the <u>emulsion</u> path through the column (from right to left in Figure 13), we observe a uniform extraction rate over the majority of the column, until the top two stages of the column. At this point the internal phase has been depleted of its chloride, thus the extraction driving force is greatly reduced. This explains the apparent start-up transient in the first two stages.

Measurement of Membrane Stability. Two mechanisms, leakage and swelling, can adversely affect ELM performance. Emulsion leakage occurs when the contents of the small, highly concentrated internal phase droplets are released into the external phase. Leakage reduces the degree of removal that can be achieved by the ELM extraction system. The leakage of emulsion is related to the stability of the emulsion and is a function of the emulsion formulation and the extractor flow field.

We quantified the leakage of the emulsion by substituting LiCl for NaCl as the internal stripping agent, and then measuring the increase in continuous phase lithium concentration with time. Because lithium is totally insoluble in the membrane phase, leakage is the only way lithium can escape into the external phase. External phase lithium concentrations were measured using atomic absorption spectroscopy (AA). The results indicate that the emulsion is very stable. The emulsion leakage was 0.2% after 10 minutes, 0.35% after 20 minutes, and only 0.5% by the end of the 30 minute extraction experiment.

Emulsion swelling can also adversely affect ELM performance. Some degree of swelling is inevitable in the ELM process. Swelling is only a problem if it unacceptably reduces the driving force for the selenium extraction, unacceptably dilutes the internal-phase selenium concentration, creates handling problems, or leads to emulsion breakage.

The fractional swelling S is defined as $V_i/V_i^o - 1$, where V_i/V_i^o is a ratio of the volume of the internal phase at any given time, relative to the initial internal phase volume. The swelling was determined by measuring the change in density of the emulsion with time, and using the equation:

$$S = \left(1 + \frac{V_m}{V_i^o} \right) \frac{(\rho_e - \rho_e^o)}{(\rho_w - \rho_e)} \tag{14}$$

where V_m/V_i^o is the volumetric ratio of membrane phase to internal phase; ρ_e^o is the initial density of the emulsion, ρ_e is the density of the swelled emulsion, and ρ_w is the density of the water.

Water immigration from the external phase to the internal phase driven by the difference in osmotic pressure is the main cause of swelling. Emulsion swelling is mainly related to the concentrations of complexing agent, stripping reagent, and surfactant, as well as a function of the contact time. To evaluate the swelling behavior of the emulsion during ELM extraction of selenium, we carried out a series of batch extraction experiments with variable concentrations of surfactant (0.5 - 2%). V_i/V_m was 1/3, V_e/V_b was 1/25, and contact time was 30 minutes. The results indicate that the emulsion swelling increases linearly with increasing surfactant concentration. At 0.5% surfactant, S=0.6; at 1.0% surfactant S=0.7; and at 2% surfactant S=1.0.

Conclusions

The emulsion liquid membrane (ELM) process was shown to be an effective method for pre-concentrating dilute selenium streams, such as those found in the wastewaters from petroleum refineries.   By concentrating the selenium, the ELM process can greatly reduce the operating costs and capital costs of the coagulation/coprecipitation processes presently used to remove selenium.

We experimentally demonstrated that the ELM process could rapidly reduce the selenium concentration in an aqueous stream to well below our goal of 0.05 mg/L, while increasing the concentration of the selenium in the internal phase by more than a factor of 20.   The extraction rate and efficiency depend primarily on the selenium oxidation state and subsequently on the pH of the aqueous phase.   In particular, selenate [Se(VI)] is extracted at much higher rates than selenite [Se(IV)], and it is less susceptible to competition from sulfates.   When selenite is the major oxidation state present, extraction rates are higher and much lower final concentrations of selenium are achieved during operation at alkaline pH where the SeO_3^{2-} ion is the dominant species.   The $HSeO_3^-$ ion formed under neutral conditions is difficult to extract, and at highly acidic conditions the H_2SeO_3 is in the undissociated form.

We demonstrated that the emulsion formulation that we developed is optimal for extracting selenium.   The emulsion formulation consists of a low-cost solvent (kerosene), a low-cost proprietary surfactant, a proprietary extractant, and a higher alcohol co-solvent (n-decanol).   The internal phase reagent used to provide the driving force for extraction was NaCl, which is both non-toxic and relatively inexpensive.   The emulsion formulation proved to be very stable and swelling did not significantly dilute the selenium in the internal phase.   We also found that the extraction process was not significantly degraded by the presence of other anions such as sulfate due to the nature of the ELM process and the large capacity of the internal phase.   These competing anions do not prevent the selenium anions from being extracted.   While competing ions do slow the extraction kinetics, it is still very rapid and economical.   Finally, we demonstrated the counter-current extraction of selenium, and found that the counter-current process has several advantages over the batch process, including improved rates due to the higher driving force at low concentrations, and insensitivity of the process to leakage.

Acknowledgments

This work was supported in part by the U.S. Environmental Protection Agency under Contracts 68D30127 and 68D40069.

Literature Cited

1.   Western States Petroleum Association, personal communication, 1994.
2.   Engberg, R.A.; Sylvester, M.A.   *J. Irrigat. Drainage Engin.*, **1993**, *119*, 522-536.

3.   Fio, J.L.; Fujii, R.; Deverel, S.J. *Soil Sci. Soc. of America Journal*, **1991**, *55*, 1313-1320.

4.   Owens, L.P., personal communication, 1994.

5.   Kobylinski, E.A.; Hunter, G.L. *Chem. Engr.*, **1992**, *99*, 74-76.

6.   Marshack, J.B. *A Compilation of Water Quality Goals*, California Regional Water Quality Control Board, Central Valley Region, Sacramento, CA (1993).

7.   Ohlendorf, H.M.; Hothem, R.L.; Bunck, C.M.; Marois, K.C. *Arch. Environ. Contam. Toxicol.*, **1990**, *19*, 495-507.

8.   Li, N.N. U.S. Patent 3,410,794. (1968).

9.   Li, N.N. U.S. Patent 3,617,546. (1971)

10.   Kitagawa, T.; Nishikawa, Y.; Frankenfeld, J.W.; Li, N. *Environ. Sci. Technol.*, **1977**, *11*, 602-605.

11.   Frankenfeld, J.W.; Li, N.N. in *Recent Developments in Separation Science*, Li, N.N., Ed.; CRC Press, 1977, Vol III; pp 285-292.

12.   Raghyraman, B.; Tirmizi, N.; Wiencek, J. *Environ. Sci. Technol.*, **1994**, *28*, 1090-1098.

13.   Oppenheimer, J.A.; Eaton, A.D.; Kreft, P.H. U.S. Environmental Protection Agency, Report No. EPA/600/2-84/190 (1984).

14.   Trussel, R.R.; Trussel, A.; Kreft, P. U.S. Environmental Protection Agency, Report No. EPA/600/2-80/153 (1980).

15.   *Lange's Handbook of Chemistry*, 13th Ed, Dean, J.A. Ed; McGraw-Hill, NY, 1985.

16.   Squires, R.C.; Groves, R.G.; Johnston, W.R. *J. Irrigat. Drainage Engin.*, **1989**, *115*, 48-57.

17.   Boegel, J.V.; Clifford, D.A. U.S. Environmental Protection Agency, Report No. EPA/600/2-86/031 (1986).

18.   Maneval, J.E.; Klein, G; Sinkovic, J. U.S. Environmental Protection Agency, Report No. EPA/600/2-85/074 (1985).

19.   Sorg, T.J.; Logsdon, G.S. *J. Am. Water Works Assoc.*, **1978**, *70*, 379.

20.   Patterson, J.W. *Industrial Wastewater Treatment Technology*, 2nd Ed, Butterworth Publishers, Boston, MA, 1985, pp 395-403.

Chapter 25

Supported Liquid Membranes in Metal Separations

B. M. Misra[1] and J. S. Gill[2]

[1]Desalination Division and [2]Uranium and Rare Earths Extraction Division, Bhabha Atomic Research Centre, Bombay 400085, India

After a brief introduction to liquid membranes, studies of supported liquid membranes (SLM) and their applications in separations of various metal species relevant to nuclear research centers are described. Aspects of coupled transport in SLM and the transport model first proposed by Danesi are outlined. Choices of membrane material and solvent which improve membrane stability in a SLM system are discussed. Recent modifications of the SLM process are mentioned. Applications of SLMs in hydrometallurgy for the separation and concentration of actinides, lanthanides, and transition metals, are reviewed. A few pilot-scale studies of SLM are described which show the potential for large-scale utilization in the future.

Membrane processes, including reverse osmosis, ultrafiltration, and electrodialysis, are finding increased use in the treatment of the large volumes of low-level radioactive liquid effluents which are generated at various stages of fuel reprocessing for the nuclear fuel industry (1). These membrane processes are primarily used to achieve a significant volume reduction of the liquid wastes for subsequent processing by conventional methods. This strategy helps to reduce the equipment size for conventional plants, as well as the amount of energy and chemicals required. However, many of these membranes do not possess adequate selectivity for the separation of various radionuclides from the liquid streams. Conversely, inorganic ion exchangers and liquid membrane processes show good potential for such applications (1).

In our laboratories, research and development studies have been conducted on the separation of uranium and various lanthanides by common extractants (carriers) and of actinides by crown ether carriers using different types of liquid membrane systems. Also our studies have been directed toward determining optimal support systems for supported liquid membranes (SLM) which may offer improved flux

0097–6156/96/0642–0361$15.00/0
© 1996 American Chemical Society

BLM

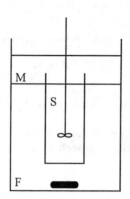

ELM

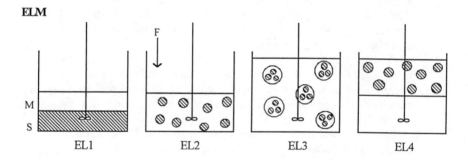

EL1   EL2   EL3   EL4

SLM

a)   b)

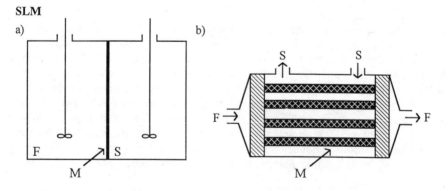

Figure 1. Different Types of Liquid Membranes: BLM = Bulk Liquid
Membrane; ELM = Emulsion Liquid Membrane; and SLM =
Supported Liquid Membrane with a) Flat Membrane and b)
Hollow Fiber Membrane Configurations. (F = feed solution; M =
membrane phase; S = stripping solution)

characteristics and stability. Some aspects of these studies of SLM systems, including recent advances in the separation of various metal ion species are presented in this chapter.

Types of Liquid Membranes.

Representations of the three general types of liquid membranes, *e.g.*, bulk liquid membranes, emulsion liquid membranes, and supported or immobilized liquid membranes, are presented in Figure 1.

Bulk Liquid Membranes (BLM). This is the simplest type of liquid membrane (*2-8*) and is utilized for fundamental studies of certain aspects of liquid membrane transport processes. In one such process, a beaker-in-a-beaker cell (Figure 1) consists of inner and outer compartments which contain the aqueous feed (F) and strip (S) solutions, respectively. The inner beaker contains the stripping solution and is surrounded by the feed solution. Both aqueous solutions contact the upper organic layer, which is the liquid membrane. Mass transfer takes place from the feed solution through the liquid membrane and into the strip solution. Bartsch *et al.* studied the transport of alkali metal cations across bulk liquid membranes in which a crown ether carboxylic acid in the organic layer served as the carrier (*2,3*).

There have been a few modifications of the BLM process in recent years. In one example (*6*), the feed and strip solutions flow down hydrophilic vertical sheets of porous viscose in an alternating sequence while the organic liquid (membrane) moves upwards between the vertical porous sheets. A modification of this system was suggested by Frasel *et al.* (*7*) wherein the viscose porous sheet was cheesecloth with Aliquat 336 as the carrier.

Emulsion Liquid Membranes (ELM). Emulsion liquid membranes or surfactant liquid membranes are essentially double emulsions, *i.e.*, water-in-oil-in-water (w/o/w) systems or oil-in-water-in-oil (o/w/o) systems. In a w/o/w ELM system, the oil phase which separates the aqueous phases of feed and strip solutions is the membrane phase. ELMs are prepared by first forming an emulsion between the two immiscible phases (EL1, Figure 1) and then dispersing the emulsion formed in a third (continuous) phase (EL2 and EL3). Thus the miscible phases are separated by an immiscible membrane phase. To form a stable emulsion, a surfactant is added. Typically, the encapsulated internal droplets in the emulsion are 1-3 μm in diameter (*9*). When the emulsion is dispersed by agitation in the external continuous phase, many small globules of the emulsion are formed. The size of these small globules depends upon the properties and concentration of the surfactant, the viscosity of the internal phase, and the intensity of the mixing (*10*). The globule size is typically in the range of 100-2000 μm. Each globule in turn contains many of the 1-3 μm internal droplets. The large number of emulsion globules provides a large mass transfer area in contact with the external continuous phase. The internal mass transfer area (typically 16 m^2/m^3) is much larger than the external mass transfer area. After transport of the desired species in the ELM process, the emulsion and external continuous phases are separated by settling (EL4) and the internal phase is then recovered by breaking the emulsion, usually with an electrostatic coalescer (*11,12*).

The ELM process has been used extensively to extract organics from aqueous feed, *e.g.*, phenol from water (*13-15*), in hydrometallurgical separations, like zinc from waste water (*16-18*) and uranium from wet process phosphoric acid (*19,20*), and in the recovery of rare earth metals (*23,24*).

The major problem of the ELM process is the emulsion stability. The emulsion must be able to withstand the shear generated by mixing during the separation process, but be easily broken to recover the internal phase. An improvement suggested for ELM (*25*) is the Emulsion Free Liquid Membrane (EFLM) in which a membrane cell is divided into two compartments by a baffle plate and the feed and strip solutions are introduced into these two compartments employing compressed air and thereby obtaining a jet for both of the streams. This module has been further modified by Gu and co-workers (*26*) in the form of the Electrostatic Pseudo-Liquid Membrane (EPLIM), which employs high voltage (~2 kV) to obtain a greater flux in metal transport.

Supported Liquid Membranes (SLM). In this type of liquid membrane, a porous polymer support membrane holds a solution of the carrier in its pores (Figure 1). The SLM is formed by impregnating the pores of a thin, porous, polymeric membrane, such as those used in ultrafiltration, with a carrier dissolved in a suitable solvent. Stability of the liquid phase membrane is provided by capillary or surface forces. The impregnated membrane acts as a common interface between the feed and strip solutions which are kept in compartments on each side of the membrane, thus aiding in selective transport of the diffusing species of interest through the membrane. This permeation can be described as simultaneous extraction and stripping operations in a single-stage process under non-equilibrium conditions.

Transport in Supported Liquid Membranes.

In the extraction of the metal ions, the carrier molecule in the membrane picks up metal ion/species from the feed solution forming a complex. This complex diffuses to the other side of the membrane where decomplexation occurs and the metal ion/species are released into the strip solution. Free carrier then diffuses back across the membrane for use in another cycle. This coupled transport through the SLM can take place by:

i) *Co-transport* in which both metal ions and counter ions are transported from the feed solution through the SLM and into the strip solution as shown in Figure 2a. If the carrier, C, is neutral, the driving force is the difference in distribution coefficients, K_d, between the feed and strip solutions. This is generally achieved by maintaining a concentration gradient of the counter ions X^-, *e.g.*, NO_3^- or Cl^-, between the feed and strip solutions. The metal ion and counter ion form a complex with the carrier C in the membrane. This complex diffuses to the other side of the membrane and the metal ion and counter ion are released into the strip solution. The chemical reaction for this coupled transport is:

$$M^{n+} + nX^- + C(\text{membrane}) \xrightarrow{\text{Extraction}} CMX_n(\text{membrane}) \tag{1}$$

$$CMX_n(\text{membrane}) \xrightarrow{\text{Stripping}} C(\text{membrane}) + M^{n+} + nX^- \tag{2}$$

The liberated carrier molecule diffuses back to the feed solution-SLM interface, picks up another metal ion and counter ion, and the process continues until the final equilibrium is attained.

ii) *Counter-transport* (Figure 2b) in which an acidic carrier, HC, loses a proton and forms a complex MC with the metal ion at the feed solution-SLM interface (Equation 3). This complex diffuses to the SLM-strip solution interface where it liberates the metal cation into the strip solution and simultaneously picks up a proton from the strip solution (Equation 4). The regenerated carrier HC diffuses back to the feed solution-SLM interface, picks up another metal ion and the process continues. The carrier molecule shuttles between the feed solution- and strip solution-SLM interfaces.

$$M^{n+} + nHC(\text{membrane}) \xrightarrow{\text{Extraction}} MC_n(\text{membrane}) + nH^+ \qquad (3)$$

$$MC_n(\text{membrane}) + nH^+ \xrightarrow{\text{Stripping}} nHC(\text{membrane}) + M^{n+} \qquad (4)$$

The driving force in a counter-transport SLM system is the pH difference between the feed and strip solutions. For efficient transport, high feed and low strip solution K_d values are needed. The K_d difference between the feed and strip solutions is maintained by a pH gradient. For divalent metal ions and an acidic carrier, the reaction at the feed solution-SLM interface is:

$$M^{2+}(\text{aq}) + 2HC(\text{membrane}) \longrightarrow MC_2(\text{membrane}) + 2H^+(\text{aq}) \qquad (5)$$

The equilibrium constant is given by the expression:

$$K = \frac{[MC_2(\text{membrane})][H^+(\text{aq})]^2}{[HC(\text{membrane})]^2[M^{2+}(\text{aq})]} \qquad (6)$$

Inverting the concentration terms in Equation 6 gives the equilibrium constant at the strip solution side of the membrane. Overall equilibrium is achieved when there is no further metal ion transport from the feed to the strip solution side of the membrane and the following relationship applies:

$$\log \frac{[M^{2+}]_{\text{strip}}}{[M^{2+}]_{\text{feed}}} = 2\,\Delta pH \qquad (7)$$

where ΔpH is the pH difference between the feed and strip solutions. If, for example, the pH values for the feed and strip solutions are 4 and 1, respectively, M^{2+} can be concentrated in the strip solution to as high as $10^6:1$ relative to its concentration in the feed solution.

Generally, the SLM technique offers many advantages over the more traditional solvent extraction for metal ion separations (*27,28*). A SLM system: i) is dispersion-free; ii) has a potential for use of a high feed to strip ratio to achieve a high

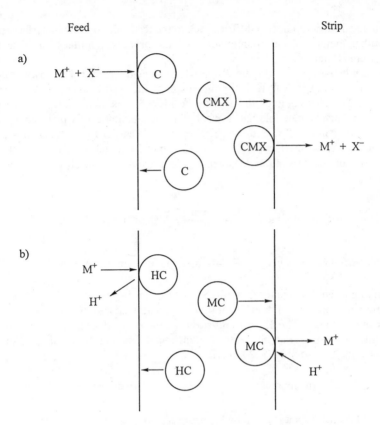

Figure 2. Transport Mechanisms: a) Co-Transport and b) Counter-Transport.

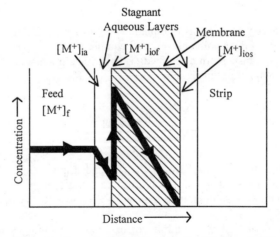

Figure 3. Concentration-Distance Profile for a SLM.

concentration of the metal ion species of interest; iii) requires a much smaller inventory of carrier (extractant); iv) provides economic use of an expensive extractant (carrier) due to low loss; v) eliminates phase separation problems; and vi) has lower capital and operation costs.

Transport Model. Danesi's treatment (29,30) of transport in a SLM has been popular and we have followed this model in our studies. A short account of this model is given here. Danesi developed a simple model in terms of a few independently measurable variables, which could describe quantitatively the permeation process for a SLM. When the distribution ratio at the membrane-strip solution interface is much lower than that at the feed solution-membrane interface, the steady state overall flux was derived by applying Fick's diffusion law. It was assumed that there were stagnant aqueous layers on both sides of the membrane (Figure 3). Since the concentration of the metal-containing species in the feed solution is much lower than that of the carrier in the membrane and the H^+ or X^- counter ions in the feed solution, three equations describing the fluxes were obtained under the assumption of linear concentration gradients.

    i) Diffusion through the aqueous boundary layer:

$$J_a = (-D_a/d_a) ([M^+]_{ia} - [M^+]_f) \tag{8}$$

where D_a is the aqueous diffusion coefficient, d_a is the thickness of the stagnant aqueous layer on the feed solution side, and $[M^+]_{ia}$ and $[M^+]_f$ are the metal ion concentrations in the stagnant aqueous layer and the feed solution, respectively.

    ii) Interfacial flux:

$$J_b = k_1[M^+]_{ia} - k_{-1}[M^+]_{iof} \tag{9}$$

where k_1 and k_{-1} are the pseudo first-order rate constants for the interfacial reaction and $[M^+]_{iof}$ is the metal ion concentration at the feed solution-membrane interface.

    iii) Membrane diffusion:

$$J_c = (-D_0/d_0) ([M^+]_{ios} - [M^+]_{iof}) \tag{10}$$

where D_0 is the membrane diffusion coefficient, d_0 is the membrane thickness, and $[M^+]_{ios}$ is the metal ion concentration at the membrane-strip solution interface. If $[M^+]_{ios}$ is considered to be zero, *i.e.*, when stripping is efficient, the membrane diffusion expression reduces to:

$$J_c = (D_0/d_0) [M^+]_{iof} \tag{11}$$

At steady state $J_a = J_b = J_c = J$ and the following expression is obtained:

$$P = (J/C) = K_d /(K_d \Delta_a + \Delta_0) \tag{12}$$

where P is defined as the permeability coefficient, $C = [M^+]_f$, $\Delta_a = d_a/D_a$ and $\Delta_0 = d_0/D_0$.

Danesi also obtained other expressions for situations in which the concentration of metal ion in the feed solution is large:

$$J = [HX]_{membrane} / n\Delta_0 \qquad (13)$$

where $[HX]_{membrane}$ is the carrier concentration in the membrane for counter-transport and n is the number of carrier molecules per metal ion in the metal-extractant complex. For a co-transport process, $[HX]_{membrane}$ in Equation 13 is replaced by $[C]_{memb}$, the concentration of the carrier in the membrane. Thus, if the concentration of metal ion in the feed solution is high compared to the extractant, H^+, or X^- concentration, the flux becomes constant for a fixed carrier concentration.

Experimentally, the permeability of a metal ion through a SLM is obtained from the following expression:

$$\ln(C/C_0) = -APT/V \qquad (14)$$

where C_0 and C are the metal ion concentrations in the feed solution at zero and t time, respectively, A is the exposed area of the membrane, V is the volume of the feed solution, and T is the temperature. In a plot of $\ln(C/C_0)$ versus A/V, P is the slope of the straight line.

Choice of Membrane Support and Solvent.

Several different membrane supports have been used to make SLMs. These include polypropylene (PP) (31-33), poly(vinylidene difluoride) (PVDF) (34), polytetrafluoroethylene (PTFE) (35), cellulose acetate (36), poly(vinyl chloride) (PVC) (37), and silicones (38,39). The requirement for a good polymeric support are high porosity, small pore size, good mechanical strength for thinness, chemical resistance, hydrophobicity, and low cost.

A SLM support can be made in three different geometries. A planar or flat geometry is useful for laboratory applications. For large-scale applications, the planar geometry is ineffective, since the ratio of surface area to the pore volume is low. On the other hand, hollow fiber and spiral-wound modules provide high surface area to volume ratios. These ratios can approach 10,000 m^2/m^3 for hollow fiber and 1,000 m^2/m^3 for spiral-wound modules.

Membrane stability is the primary problem associated with the use of SLMs. Solvent loss is most often the causative factor for membrane instability. Such solvent loss arises from evaporation and dissolution, as well as from an excessive pressure differential applied across the membrane which forces solvent (and carrier) out of the pores of the membrane. Carrier may also be lost due to irreversible side reactions within the membrane itself.

The choice of substrate plays a crucial role in the stability of the membrane. Reduction of the vapor pressure of the liquid membrane, which is particularly important in gaseous separations, may be accomplished by effective use of capillary forces (40,41). Reduction in the pore size in the membrane support can have a significant effect on the liquid vapor pressure. Thus, the vapor pressure of the liquid can be reduced, which stabilizes the liquid membrane, if the liquid is confined in pores with diameters of less than 0.2 μm. The relationship between the vapor pressure and pore size is given by the Kelvin equation:

$$r_k = (-2 \, \gamma V_m)/ RT \ln(p/p_s) \qquad (15)$$

where r_k is the pore radius, γ is the surface tension, V_m is the molar volume, p/p_s is the ratio of the vapor pressure to the vapor pressure at saturation for the liquid at temperature T, and R is the gas constant. Thus, the smaller is the pore size, the lower will be the vapor pressure p of the liquid in the pores.

Another desirable property for the SLM is good wetting of the support polymer by the liquid. The polymer material should be chosen such that the contact angle between the liquid membrane phase and the support is zero or close to zero. In this case, the liquid is said to wet the pore structure. Zisman (*42*) observed that cos θ, where θ is the contact angle, is usually proportional to the surface tension, γ, of the liquid. Linear extrapolation of the data in a plot of cos θ versus γ to zero yields a quantity called the critical surface tension, γ_c, which is characteristic of a given solid and is the highest surface tension of a liquid that will spread spontaneously on a given surface. The γ_c values of various polymer substrates are given in Table 1. The values for PP, PTFE, and silicone are 31, 18.5 and 25 dynes/cm, respectively (*42,43*). Since most commonly used SLM solvents have surface tensions in the range of 20-30 dynes/cm, a PTFE support is not a good choice, even though the chemical inertness of this polymer is attractive.

Table 1. Critical Surface Tension Values for Various Polymers (*42,43*)

Polymer	Critical Surface Tension, γ_c, (dynes/cm)
PTFE	18.5
Silicones	~25
PP	31
PVC	39
PET	43
Nylon 6,6	46

Another cause of liquid membrane instability is excessive transmembrane pressure across the membrane. This may be caused by osmotic pressure or by applied pressure. The impregnated liquid phase in a SLM can be displaced when the critical applied pressure, p_c, is reached. This shortens the lifetime of the SLM and is a very important factor in hollow fiber SLMs. For cylindrical pores, the displacement pressure is given by the Laplace-Young equation (*44*):

$$p_c = 2 \, \gamma \cos \theta / r \qquad (16)$$

Thus a maximum value of p_c is obtained when θ = 0° and no filling of the pores should occur when θ ≥ 90°. The smaller the pore size, the higher is the critical displacement pressure, p_c. Zha *et al*. (*40*) modified the Laplace-Young equation for pores of irregular size. Some p_c values for different supports are given in Table 2. A SLM must be operated within the limit of its p_c value for stability.

SLMs in Hydrometallurgy.

A number of papers have been published on the separation of lanthanides, actinides, and transition metals by SLM processes. Carriers like bis(2-ethylhexyl)phosphoric

Table 2. Critical Displacement Pressures of SLMs (40)

Membrane Material	Critical Displacement Pressure, p_c, (kPa) for				
	Kerosene	Dodecane	Heptane	Toluene	Shellsol
Celgard 2500 (PP)	313	307	215	333	356
Durapore HVHP (PVDF)	83	79	83	65	81

acid (D2EHPA) (45-47), tributyl phosphate (TBP) (48), LIX compounds (49), 2-ethylhexylphosphoric acid mono(2-ethylhexyl) ester (50-52), and crown ethers (53,54) have been employed. Also SLMs have been utilized in the separation of gases and organic solutes.

Separation of metal ions using SLMs has evoked considerable interest. Danesi studied the transport of americium (29) using CMPO as the extractant in a flat sheet SLM in a co-transport system and verified his permeability equations (Equations 12 and 13). When the americium concentration was low (10^{-6} M), the transport followed Equation 12 and when it was high (10^{-3} M), Equation 13 described the transport behavior. Their work (55) with two different thicknesses of PP membranes indicated that for a thin membrane (25 μm flat sheet), the permeation was aqueous diffusion controlled while with a thicker membrane (430 μm flat sheet), both aqueous and membrane diffusion were controlling factors.

Makamura et al. (56) studied the effect of pH and carrier concentration in the membrane on the permeability of europium in versatic acid-loaded PTFE membranes. The europium permeability was enhanced by an increase in the pH of the feed solution or with an increase in carrier concentration in the membrane, up to an maximal value after which the premability became constant. This behavior can be explained in terms of Equation 12 which reduces to Equation 17 at higher K_d values.

$$P = 1/\Delta_a \tag{17}$$

Here the permeability becomes constant and is controlled by aqueous diffusion. However when a small value of K_d makes $K_d \Delta_a$ small compared to Δ_0, Equation 12 becomes Equation 18 and the transport is membrane diffusion controlled. These

$$P = K_d/\Delta_0 \tag{18}$$

authors obtained values of 7.6×10^{-6} and 6.0×10^{-6} cm^2/s for the membrane and aqueous diffusion coefficients, respectively.

Chiarizia and Horwitz (47) investigated uranium extraction from ground water by a SLM using a phosphinic acid carrier in a PP membrane. The process was very effective in removing uranium at pH = 2. The data were fit well by the Danesi equation (Equation 12) with a D_0 value of 4.0 X 10^{-7} cm^2/s. This is an independently apparent D_0 value. By knowing the porosity and tortuosity of the membrane, the bulk diffusion coefficient can be obtained.

We have studied uranium and rare earth metal transport using TBP, D2EHPA and PC-88A extractants in silicone membrane SLMs. These silicone membranes were synthesized and characterized earlier in our laboratory (*38*). The uranyl ion transport with TBP as the carrier in kerosene gave values of the permeability, P, and the membrane diffusion coefficient, D_0, of 2.8 X 10^{-4} cm/s and 5.0 X 10^{-7} cm^2/s, respectively; whereas with D2EHPA as the carrier lower values of 1.8 X 10^{-4} cm/s and 1.0 X 10^{-7} cm^2/s were obtained. The performance of these silicone membranes was found to be comparable with commercial PP and PTFE membranes when used under similar conditions.

We also investigated Nd^{3+} transport (*57*) with PC-88A as the carrier in a silicone membrane SLM. The permeability was found to increase when the pH was increased until a saturation point was reached. This behavior was similar to that previously reported in the literature (*47,56*). A D_0 value of 3.0 X 10^{-7} cm^2/s was obtained. The data for transport of UO$_2^{2+}$, Sm^{3+}, and Nd^{3+} with TBP and D2EHPA carriers showed that the permeability increased with increasing carrier concentration, reached a maximum, and then decreased. This diminution in permeability is attributed to enhanced viscosity of the membrane liquid. Table 3 gives the flux values for UO$_2^{2+}$, Sm^{3+}, and Nd^{3+} for 1 M carrier in kerosene. It is noted that TBP as the carrier gives a higher flux in UO$_2^{2+}$ transport than does D2EHPA. In part, this may be due to the lower viscosity of a 1 M solution of TBP in kerosene.

Table 3. Transport Values for Various SLM Systems

System	Flux, J (mol/m^{2}s X 10^6)
UO$_2^{2+}$-TBP	4.8
UO$_2^{2+}$-D2EHPA	1.5
Sm^{3+}-D2EHPA	7.0
Nd^{3+}-D2EHPA	3.0
Nd^{3+}-PC-88A	5.1

Pilot-Scale Studies of SLMs.

Adoption of liquid membrane technology in process industry is increasing. However there are several limitations to such applications of liquid membranes. The primary limitations are inadequate membrane durability and low net flux. In addition, liquid membranes are single-stage processes without cascading.

The fluxes obtained with SLMs are on the order of 0.1 Kg/m^2/day . The net flux values for flat sheet modules are very small due to the low surface area. This can be improved by employing spiral-wound or hollow fiber modules. Among these, the hollow fiber modules have gained popularity recently.

A microporous hollow fiber supported liquid membrane (HFSLM) module consists of a bundle of porous hollow fibers with thin walls which can be wetted by the membrane liquid. Initial studies of hollow fibers were made with a single fiber and the aqueous feed and strip solutions were passed through the lumen and over the shell side of the hollow fiber (*29*). This was followed by a hollow fiber module containing a bundle of fibers in which organic and feed phases were passed through the shell side and lumen, respectively. The extracting organic phase wetted and filled the pores of the tube and the aqueous feed solution was maintained at a higher pressure than the organic phase to keep the reaction interface at the inside wall of the lumen. This type of module was employed by Alexander and Calhahan (59) for gold extraction. Haan et al. (60) used two such modules for extracting copper from synthetic waste water, one for extraction and the other for stripping. Sirkar and co-workers (61) used this type of module for copper and chromium separation using LIX-84 in heptane and trioctylamine in xylene, respectively.

The same group (62) employed a hollow fiber-contained liquid membrane (HFCLM) module for the removal of heavy metals, such as Cu^{2+}, Cr^{6+}, and Hg^{2+} from waste water. HFCLM combines the two-step process of extraction and stripping into one module containing two bundles of hollow fibers through which the feed and strip solutions flow. Both bundles are present in a single case through which the liquid membrane solution flows.

In nuclear plants, there is a frequent need to separate radioactive materials from the non-radioactive materials. Kathios et al. (63) employed a microporous hollow fiber module to analyze the potential of such systems for process-scale metal separations. They studied the removal of neodynium, a surrogate for americium, from 2 M nitric acid.

Recently, the separation of rare earth metals, which possess unique electronic, magnetic, and optical properties, has been investigated. There have been attempts to separate rare earth metals with hollow fiber membranes. Goto and co-workers (50) studied the separation of yttrium from holmium and erbium with D2EHPA and PC-88A, employing synergistic reagents. The same group (64) achieved separation factors for various rare earth metals with a SLM which were very similar to those reported for solvent extraction.

Pilot plant uranium and chromium recovery data for a SLM system was reported by Babcock (65). Polysulfone hollow fibers were used in the module and a tertiary amine in kerosene was the organic phase. Operational costs for the separation of uranium were $0.8/Kg for a 3.8 X 10^3 m^3/day plant. This cost estimate was lower than those for uranium recovery by solvent extraction and ion exchange.

Another pilot-scale evaluation of a SLM with hollow fibers was made by Dworzak and Naser (66). The method was applied to low-level radioactive wastes with 333 PP microporous fibers which had a surface area of 3.6 m^2.

Other pilot plant studies with SLMs were for gaseous separations including ethylene and propylene by Hughes et al. (67) and oxygen enrichment of air by Babcock (68).

With wet process phosphoric acid (WPPA), about 0.18 g/L of uranium is obtained in the leachate. For treatment of the leachate, a pilot plant ELM study (69) showed a definite advantage over conventional solvent extraction with lower solvent inventory and loss. Furthermore, the ELM separation was found to be more effective at higher temperature. This system would seem to be suitable for study in SLM systems using silicone membranes, which are stable at high temperatures.

In the separation of metals, cascading or multistage extraction is frequently employed. However a SLM is by nature a single-stage process. This drawback was overcome by Danesi and co-workers (29,70) by use of composite SLMs for the separation of metals. Here the upstaging can be done by employing two different modes of metal transport, co-transport and counter-transport. The composition of the first strip solution is such that it promotes back extraction of the metal ions by a second SLM. Two such membranes, *i.e.*, a composite SLM, can be put in series and the metal species of interest can be separated to desired levels. Thus, after the fifth stage of a Ni-Co separation system, the nickel level was reduced to $<10^{-6}$ % (*70*).

There are other applications of SLMs which have future potentials. A SLM has been used to prepare tracers from a high energy, heavy ion-irradiated gold target (*71*). These tracers have application in chemical and biological industries.

A SLM can be formed on a porous ceramic support which imbibes molten salts at ~573 °K. The resulting SLM can be used effectively for the separation of NH_3, CO_2, H_2 and He. In this application, silicone membranes of the type that we have prepared can be laminated over a ceramic support to provide new membranes with high temperature stability.

Literature Cited.

1. *Advances in the Technologies for Treatment of Low and Intermediate Level Radioactive Liquid Waste*; IAEA Technical Report Series No. 370: Vienna, 1994, pp 41-46.
2. Strzelbicki, J.; Bartsch, R.A. *J. Membr. Sci.* **1982**, *10*, 35.
3. Charewicz, W.A.; Heo, G.S.; Bartsch, R.A. *Anal. Chem.* **1982**, *54*, 2094.
4. Schlosser, S. In *Advances in Membrane Phenomena and Processes*; Mika, A.M.; Winniki, T.Z., Eds.; Wroclaw Technical University Press: Wroclaw, 1989; pp 163-183.
5. Izatt, R.M.; Lamb, J.D.; Bruening, R.L. *Sep. Sci. Technol.* **1988**, *23*, 1645.
6. Boyadzhiev, L.; Bezenshek, E.; Lararova, Z. *J. Membr. Sci.* **1984**, *21*, 137.
7. Fraser, B.G.; Pritzker, M.D.; Ligge, R.L. *Sep. Sci. Technol.* **1994**, *29*, 2079.
8. Schlosser, S.; Kossaczky, C. *J. Radioanal. Nucl. Chem.* **1986**, *101*, 115.
9. Ho, W.S.W. In *Membrane Handbook*; Ho, W.S.W.; Sirkar, K.K., Eds.; Van Nostrand Reinhold: New York, NY, 1992; pp 597-718.
10. Rautenbach, R.; Machhammer, O. *J. Membr. Sci.* **1988**, *36*, 425.
11. Kataoka, T.; Nishiki, T. *Sep. Sci. Technol.* **1990**, *25*, 171.
12. Marr, R.J.; Bart, H.J.; Draxler, J. *Chem. Eng. Process* **1990**, *27*, 59.
13. Marr, R.J.; Kopp, A. *Chem. Eng. Technol.* **1980**, *52*, 399.
14. Chang, Y.C.; Li, S.P. *Desalination* **1983**, 351.
15. Zhang, X.J.; Liu, H.H.; Lu, T.S. *Water Treatment* **1987**, *2*, 127.
16. Draxler, J.; Marr, R.J. *Chem. Eng. Process* **1986**, *20*, 319.
17. Lorbach, D.; Marr, R.J. *Chem. Eng. Process* **1987**, *21*, 83.
18. Draxler, J.; Fürst, W.; Marr, R.J. *J. Membr. Sci.* **1988**, *38*, 281.
19. Hayworth, H.C.; Ho, W.S.; Burns, W.A.; Li, N.N.. *Sep. Sci. Technol.* **1983**, *18*, 493.
20. Hayworth, H.C. *Chemtech* **1981**, 342.
21. Strzelbicki, J.; Charewicz, W. *Hydrometallurgy* **1980**, *5*. 243.
22. Gu, Z.M.; Wasan, D.T.; Li, N.N. *Surfactant Sci. Ser.* **1988**, *28*, 127.
23. Goto, M.N.; Yoshii, N.; Kondo, K.; Nakashio, F. In *Proceedings of the Symposium on Solvent Extractions;* Kyushu University, Kyushu, Japan; pp 113-118.
24. Zhang, R.H.; Xiao, L. *J. Membr. Sci.* **1990**, *5*, 249.
25. Kopp, A.; Marr, R. *Intl. Chem. Eng.* **1982**, *22*, 44.
26. Gu, Z. *J. Membr. Sci.* **1990**, *52*, 77.
27. Schulz, G. *Desalination* **1988**, *68*, 191.
28. Noble, R.D.; Koval, C.A. *Chem. Eng. Prog.* **1989**, *89*, 58.

29. Danesi, P.R. *Sep. Sci. Technol.* **1984-85**, *19*, 857.
30. Danesi, P.R.; Yinger, L.R. *J. Membr. Sci.* **1986**, *29*, 195.
31. Chiarizia, R.; Horwitz, E.P. *Solv. Extrn. Ion Exch.* **1990**, *8*, 98.
32. Babcock, W.C.; Baker, R.W.; Lachapelle, E.D.; Smith, K.L. *J. Membr. Sci.* **1980**, *7*, 71.
33. Bateman, D.R.; Way, J.J.; Larson, K. *Sep. Sci. Technol.* **1989**, *19*, 21.
34. Moreno, C.; Hrdlicka, A.; Valiente, M. *J. Membr. Sci.* **1993**, *8*, 121.
35. Longman, R.; Sifniades, S. *Hydrometallurgy* **1978**, *1*, 153.
36. Matson, S.L.; Herrick, C.S.; Ward, J.J. *I&EC Proc. Des. Dev.* **1977**, *16*, 16.
37. Smith, D.R.; Quinn, J.A. *AICHE J.* **1980**, *26*, 112.
38. Gill, J.S.; Marwah, U.R.; Misra, B.M. *J. Membr. Sci.* **1993**, *76*, 157.
39. Gill, J.S.; Marwah, U.R.; Misra, R.M. *Sep. Sci. Technol.* **1994**, *29*, 193.
40. Zha, F.F.; Fane, A.G.; Fell, C.J.D.; Shofield, R.W. *J. Membr. Sci.* **1992**, *75*, 69.
41. Deetz, D.W. In *Liquid Membranes. Theory and Applications*; ACS Symposium Series 347; Noble, R.D.; Way, J.D., Eds.; American Chemical Society; Washington, DC, 1987; pp 152-165.
42. Zisman, W.A. *Ind. Eng. Chem.* **1993**, *55*(10), 19.
43. Arkles, B. *Chemtech* **1977**, 771.
44. Adamson, A.W. *Physical Chemistry of Surfaces*; 5th Edition, Wiley: New York, NY, 1990; pp 386-392.
45. Danesi, P.R.; Horwitz, E.P.; Richert, P.G. *Sep. Sci. Technol.* **1988**, *22*, 1183.
46. Yoshizuka, K.; Sakamoto, Y.; Baba, Y.; Inoue, K.; Nakashio, F. *Ind. Eng. Chem. Res.* **1982**, *31*, 1372.
47. Chiarizia, R.; Horwitz, B.P. *Solv. Extrn. Ion Exch.* **1990**, *8*, 65.
48. Huang, T.C.; Huang, C.T. *J. Membr. Sci.* **1986**, *29*, 295.
49. Akiba, K.; Kauno, T. *Sep. Sci. Technol.* **1983**, *18*, 831.
50. Goto, M.; Kubota, F. Miyata, T.; Makashio, F. *J. Membr. Sci.* **1992**, *74*, 215.
51. Matsumoto, M.; Shimauchi, H.; Kondo, K.; Nakashio, F. *Solv. Extrn. Ion Exch.* **1987**, *5*, 301.
52. Kubota, F.; Goto, M.; Nakashio, F.; Hano, T. *Sep. Sci. Technol.* **1995**, *30*, 777.
53. Shukla, J.P.; Kumar, A.; Singh, R.K. *Radiochim. Acta* **1992**, *57*, 185.
54. Guyon, M.; Foss, J.; Guy, A.; Moutarde, T.; Chomel, R.,; Draye, M.; Lamaire, M. *Sep. Sci. Technol.* **1995**, *30*, 1961.
55. Danesi, P.R.; Chiarizira, R.; Richert, P.G. and Horwitz, E.P. *Solv. Extrn. Ion Exch.* **1985**, *3*, 111.
56. Nakamura, S.; Ohashi, S.; Akiba, K. *Sep. Sci. Technol.* **1992**, *27*, 741.
57. Gill, J.S.; Marwah, U.R.; Mukherjee, T.K. In *Abstracts of National Symposium on Organic Reagents - Synthesis and Use in Extraction Metallurgy*; ORSWEM-94, BARC, Bombay, 1994; paper no. OP-16.
58. Bartsch, R.A.; Charewicz, W.A.; Kang, S.I.; Walkowiak, W. In *Liquid Membranes. Theory and Applications*; ACS Symposium Series 347; Noble; R.D.; Way, J.D., Eds.; American Chemical Society: Washington, DC, 1987; pp 86-97.
59. Alexander, P.R.; Calhahan, R.W. *J. Membr. Sci.* **1987**, *35*, 57.
60. Haan, A.D.; Bartels, P.V.; Graauw, J. *J. Membr. Sci.* **1989**, *45*, 281.
61. Yun, C.H.; Prasad, R.; Guha, A.K.; Sirkar, K.K. *Ind. Eng. Chem. Res.* **1993**, *32*, 1186.
62. Guha, A.K.; Yun, C.H.; Basu, R.; Sirkar, K.K. *AICHE J.* **1994**, *40*, 1223.
63. Kathios, D.J.; Jarvinen, G.D.; Yarbro, S.L.; Smith, B.F. *J. Membr. Sci.* **1994**, *97*, 251.

64. Kubota, F.; Goto, M.; Nakashio, F.; Hano, T. *Sep. Sci. Technol.* **1995**, *30*, 777.
65. Babcock, W.L.; Lachapelle, E.D.; Kelly, D.J. *Abstracts of AICHE Meeting,* Denver, CO, Aug. 1983, paper no. 60E.
66. Dworzak, W.R.; Naser, A.J.; *Sep. Sci. Technol.* **1987**, *22*, 677.
67. Hughes, R.D.; Mahoney, J.A.; Steigelmann, E.F. In *Recent Developments in Separation Science*; CRC Press: Boca Raton, FL, 1986, Vol. IX; pp 173-195.
68. Babcock, W.C., *Liquid Membranes for Production of Oxygen Enriched Air*; Paper presented at the U.S. Department of Energy Membrane Technology Research and Development, Clemson University, Clemson, SC, October 1984.
69. Hayworth, H.C.; Ho, W.U.; Burns, W.A.; Li, N.N. *Sep. Sci. Technol.* **1983**, *18*, 493.
70. Danesi, P.R.; Yinger, L.R. *J. Membr. Sci.* **1986**, *27*, 339.
71. Ambe, S.; Ohkubo, Y.; Kobayashi, Y.; Iwamoto, M.; Yanokura, M.; Ambe, F. *Appl. Radiation Isotopes* **1992**, *43*, 1533.

Chapter 26

Cesium Removal from Nuclear Waste Water by Supported Liquid Membranes Containing Calix-bis-crown Compounds

Z. Asfari[1], C. Bressot[1], J. Vicens[1,3], C. Hill[2], J.-F. Dozol[2,3],
H. Rouquette[2], S. Eymard[2], V. Lamare[2], and B. Tournois[2]

[1]Ecole Européenne des Hautes Etudes des Industries Chimiques de Strasbourg, Unité de Recherche Associée 405 du Centre National de la Recherche Scientifique, 1 rue Blaise Pascal, F−67008 Strasbourg, France
[2]S.E.P./S.E.A.T.N., Centre d'Etudes de Cadarache, Commissariat à l'Energie Atomique, 13108 Saint-Paul-lez-Durance, France

Calix[4]-*bis*-crowns 1-7 are used as selective cesium-carriers in supported liquid membranes (SLMs). Application of the Danesi diffusional model allows the transport isotherms of trace level ^{137}Cs through SLMs (containing calix[4]-*bis*-crowns) to be determined as a function of the ionic concentration of the aqueous feed solutions. Compound 5 appears to be much more efficient than mixtures of crown ethers and acidic exchangers, especially in very acidic media. Decontamination factors greater than 20 are obtained in the treatment of synthetic acidic radioactive wastes. Permeability coefficient measurements are conducted for repetitive transport experiments in order to determine the SMLs stability with time. Very good results (over 50 days of stability) and high decontamination yields are observed with 1,3-calix[4]-*bis*-crowns 5 and 6.

It is well known that 137-cesium is one of the most important radionucleotides present in the nuclear wastes and its removal can reduce storage risks caused by its long half-life (2×10^6 y) (*1,2*). It has been demonstrated that crown ethers are promising metal ion extractants and carriers in liquid membranes (*3*). A new family of crown ether-type compounds is now emerging: *the calixcrowns*. These molecules combine *calixarene* and *crown ether* elements in their molecular structure. Since Alfieri *et al.(4)* reported the synthesis of the first member of a new class of macropolycyclic crown compounds with the two opposite OH groups in *p-tert*-butylcalix[4]arene bridged by a pentaethylene glycol chain, the 1,3-capping of calix[4]arenes at the lower rim has been achieved with poly(oxyethylene) chains leading to calixcrown ethers (*5*), doubly crowned calixes (*5c,6*), and a double-calix-crown (*7*). Due to the presence of a glycol chain in their framework, calixcrowns have been used as complexing agents of alkali and alkaline-earth metal cations. The selectivities of complexation were shown to

[3]Corresponding author

depend on the conformation (cone, partial cone, 1,2-alternate and 1,3-alternate) adopted by the rigidified calix[4]arene unit. For example the *partial-cone* isomer of 1,3-dimethoxy-*p-tert*-butylcalix[4]crown-5 exhibited the highest free energy for complexation of potassium cation (*5b*) and was used as a selective carrier of this cation in supported liquid membranes (*5c*). The K$^+$ selectivities of three different conformers of 1,3-diethoxy-*p-tert*-butylcalix[4]crown-5 have been measured for chemically modified field effect transistors (CHEMFETs) (*5k*) and membrane ion-selective electrodes (ISEs) (*5d*). The ionophores showed decreasing K$^+$/Na$^+$ selectivities in the order: *partial cone* > *1,3-alternate* > *cone*. The 1,3-dialkoxycalix[4]crown-6 compounds in the *1,3-alternate* conformation exhibited binding preference for cesium ion (*5e*). The X-ray crystal structure of the 1:1 complex of 1,3-dimethoxycalix[4]crown-6 with cesium picrate indicated the presence of cation/π-electron interactions (*5e*). In this contribution, we report the use of 1,3-calix[4]-*bis*-crowns 1-7 (Chart I) as selective cesium-carriers in supported liquid membranes (SLMs).

Sodium-Cesium Extractions.

Removal of cesium from medium-level radioactive wastes involves extraction of cesium from aqueous solutions which are 1 M in HNO$_3$ and 4 M in NaNO$_3$ (*8*). In a preliminary study, extraction experiments with ligands 1-7 were performed by mixing equal volumes (5 to 7 mL) of aqueous and organic solutions (calixarene : 10^{-2} M in 2-nitrophenyl hexyl ether or 2-NPHE) in sealed polypropylene tubes for one hour at room temperature (25 ± 1 °C). The aqueous solutions contained either NaNO$_3$ or CsNO$_3$ (5.0×10^{-4} M) in HNO$_3$ (1 M) to assess the selectivity toward cesium in the hypothetical presence of sodium. A measure for the selectivity was assumed to be the ratio of the distribution coefficients obtained separately for both cations :

$$\alpha_{(Cs/Na)} = \frac{D_{Cs}}{D_{Na}} \qquad \text{with}: D_M = \frac{\sum \left[\overline{M}\right]}{\sum \left[M\right]}$$

where :

$\sum \left[\overline{M}\right]$ denotes the total concentration of the metal cation (complexed and uncomplexed) in the organic phase at equilibrium,

and $\sum \left[M\right]$ denotes its total concentration in the aqueous phase at equilibrium.

$\sum \left[\overline{M}\right]$ and $\sum \left[M\right]$ were determined experimentally by analyzing aliquots (2 or 5 mL) of each phase by gamma spectrometry after centrifugation.

From the data shown in Table I, it is seen that ligands 2, 5 and 6 with six oxygen atoms in the glycol chain are much more selective for cesium over sodium than 1 and 3 with five or seven oxygen atoms in their rings, or 4 and 7, in which the polyether chains are sterically constrained by the presence of phenyl units. We have previously noted that the glycolic chains containing five oxygens in *p-tert*-butylcalix[4]-*bis*-crown-5, related to 1, are suitable for potassium and sodium cations, but are too small for the larger cesium (*6e*). In comparison, 2, 5 and 6 show 100-times higher selectivities toward cesium than do crown ethers *n*-decyl-benzo-21-crown-7 (*8*), and *tert*-butyl-benzo-21-crown-7 (*9*), which are well known for their abilities to complex and extract large alkali cations from acidic media (*8*). One explanation is that the complexes with calixarene

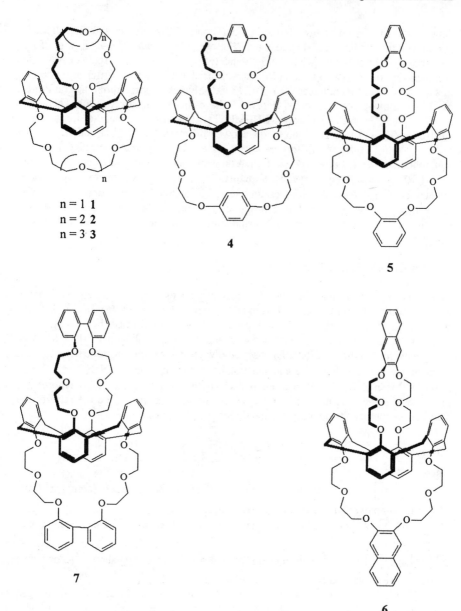

n = 1 **1**
n = 2 **2**
n = 3 **3**

4

5

7

6

1,3-calix[4]-*bis*-crown-5 **1**
1,3-calix[4]-*bis*-crown-6 **2**
1,3-calix[4]-*bis*-crown-7 **3**
1,3-calix[4]-*bis*-*p*-benzo-crown-6 **4**

1,3-calix[4]-*bis*-*o*-benzo-crown-6 **5**
1,3-calix[4]-*bis*-naphthyl-crown-6 **6**
1,3-calix[4]-*bis*-diphenyl-crown-6 **7**

Chart I : 1,3-Calix[4]-*bis*-crowns **1-7**.

Table I : Liquid-liquid Extraction Experiments : Selectivity Determination :

Aqueous feed solution: $M^+(NO_3^-)$: 5.0×10^{-4} M in HNO_3 : 1 M.
Organic solution: Extracting agent: 10^{-2} M in 2-nitrophenyl hexyl ether.

N°	Extracting agents	D_{Na}	D_{Cs}	$\alpha_{(Cs/Na)}$
1	1,3-Calix[4]-*bis*-crown-5	2×10^{-3}	0.4	--
2	1,3-Calix[4]-*bis*-crown-6	1.3×10^{-2}	19.5	1500
3	1,3-Calix[4]-*bis*-crown-7	$< 10^{-3}$	0.3	--
4	1,3-Calix[4]-*bis*-*p*-benzo-crown-6	$< 10^{-3}$	2×10^{-2}	--
5	1,3-Calix[4]-*bis*-*o*-benzo-crown-6	1.7×10^{-3}	32.5	19 000
6	1,3-Calix[4]-*bis*-naphthyl-crown-6	$< 10^{-3}$	29.5	> 29 000
7	1,3-Calix[4]-*bis*-diphenyl-crown-6	$< 10^{-3}$	7×10^{-2}	--
8	*n*-Decyl-benzo-21-crown-7	1.2×10^{-3}	0.3	250
9	*tert*-Butyl-benzo-21-crown-7	1.2×10^{-3}	0.3	250

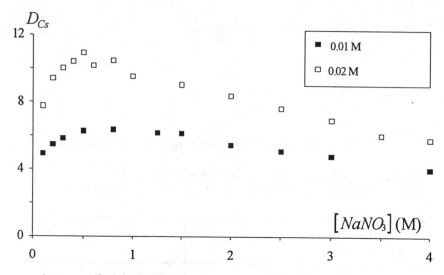

Aqueous solution : $NaNO_3$: various concentrations (pH = 2).
Organic solution : 1,3-calix[4]-*bis*-crown-6 **2** in 2-nitrophenyl octyl ether.

Figure 1. Extraction Isotherms for ^{137}Cs in Sodium Nitrate Solutions by 1,3-Calix[4]-*bis*-crown-6 **2** in 2-NPOE.

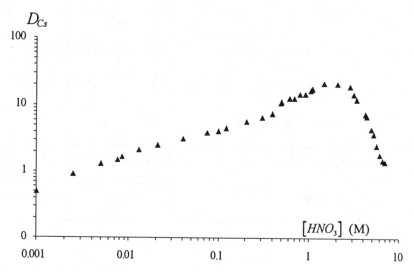

Aqueous solution : HNO_3 : various concentrations
Organic solution : 1,3-*bis*-calix[4]-crown-6 **2** (10^{-2} M) in 2-nitrophenyl octyl ether.

Figure 2. Extraction Isotherm for ^{137}Cs in Nitric Acid Solutions by 1,3-Calix[4]-*bis*-crown-6 **2** (10^{-2} M) in 2-NPOE.

derivatives are best formed because they are stabilized by the *p-bonding interactions* with the phenyl rings present in the basket-frame of **2**, **5** and **6** (*5e,9*).

Although ligands **1-7** present two potential complexation sites, the complex stoichiometry was found to be 1:1 (calixarene:cesium cation) (*10*). This behaviour may be explained by a *negative allosteric effect* which has been found in complexing systems containing two conformationally related crown ether subunits (*11*). These systems are only able to complex cations in one of their two subunits at a time because when one complex is formed, the other subunit has an unsuitable conformation to bind any species (*11*).

As shown in Figures 1 and 2 which display cesium extraction isotherms for compound **2** as a function of aqueous salt concentrations (0 < [NaNO$_3$] < 4 M, 0 < [HNO$_3$] < 7 M and cesium at trace level), back-extraction of cesium was usually allowed and favored in deionized water (where D_{Cs} is very small). This remarkable property allowed the use of **1-7** as cesium selective carriers in supported liquid membranes. Since cesium extraction was strongly enhanced in concentrated acidic media for **1-7**, as compared to mixtures of a crown **9** (*12*) and an acidic exchanger ligand, it appeared **1-7** might be used to treat medium-level radioactive wastes. The decrease of D_{Cs} for sodium nitrate concentrations greater than 0.5 M (Figure 1) or for nitric acid concentrations greater than 2 M (Figure 2) reveal both the competitive extraction of HNO$_3$ or NaNO$_3$ by the tested calixarenes and the decrease of the mean activity coefficient of trace-level cesium in a concentrated aqueous media.

Cesium Transport Through Supported Liquid Membranes (SLMs).

Selective alkali cation permeation through liquid membranes has been intensively studied to mimic natural antibiotics, to transduce chemical information into electronic signals and to treat radioactive wastes (*13*). For instance, taking advantage of the phenolic ionization of the parent calixarenes (*p-tert*-butyl- and *p-tert*-pentylcalix[n]arenes) under basic conditions, Izatt *et al.* (*14*) achieved quantitative cesium transport through bulk liquid membranes (25 % v/v CH$_2$Cl$_2$-CCl$_4$) from aqueous feed solutions of CsOH (pH > 12) to water. However, no cesium permeation was observed in the case of neutral metal nitrate feed solutions during similar experiments (*14*). Polyether-bridged calix[4]arenes were first applied to SLMs in order to study potassium/sodium permeation selectivities as compared to those of valinomycin (*13a*). We therefore decided to determine the selective ionophoric properties of calix[4]-*bis*-crowns **1-7** by measuring under similar experimental conditions (stirring rates, concentration gradients) the permeability coefficients of cesium as described in the Danesi model of trace level cation permeation through SLMs (*15*). Although the Danesi model was applied to ion-pair extraction in apolar solvents, constant permeability coefficients have also been observed in long term transport experiments (over 10 hours) using a more polar organic solvent such as 2-nitrophenyl octyl ether (2-NPOE) (*16*).

The use of **1-7** as carriers in SLMs led to coupled co-transport of cesium and nitrate ions from aqueous feed solutions of 4 M in NaNO$_3$ and 1 M in HNO$_3$ - *simulating concentrated medium-level radioactive wastes* - to deionized water (the receiving solution) because of the NO$_3^-$ concentration gradient. The decrease of ^{137}Cs radioactivity in the feed solutions was followed by regular measurements with gamma spectrometry analysis (1500 kBq.L^{-1} < ^{137}Cs initial aqueous activity < 2500 kBq.L^{-1}).

Table II : ^{137}Cs Transport Experiments
Through Flat Sheet Supported Liquid Membranes:
Permeability Determination after Six Hours of Permeation.

Aqueous feed solution: NaNO$_3$ (4 M) and HNO$_3$ (1 M).
Aqueous strip solution: deionized water.
Organic solution: Carrier : 10^{-2} M in 2-nitrophenyl octyl ether.

N°	Carriers	P_{Cs} (cm.h^{-1})
1	1,3-Calix[4]-*bis*-crown-5	9×10^{-2}
2	1,3-Calix[4]-*bis*-crown-6	1.3
3	1,3-Calix[4]-*bis*-crown-7	4×10^{-2}
4	1,3-Calix[4]-*bis*-*p*-benzo-crown-6	3×10^{-3}
5	1,3-Calix[4]-*bis*-*o*-benzo-crown-6	2.8
6	1,3-Calix[4]-*bis*-naphthyl-crown-6	2.7
7	1,3-Calix[4]-*bis*-diphenyl-crown-6	0.1
8	*n*-Decyl-benzo-21-crown-7	9×10^{-2}

Permeability coefficients P_M (cm.h^{-1}) for cesium permeation through the SLMs were graphically determined by plotting the logarithm of the ratio $\dfrac{C}{C^o}$ versus time (*15*):

$$\ln\left(\frac{C}{C^o}\right) = -\frac{\theta\,S}{V}P_M\,t \qquad (1)$$

where :

C = concentration of the cation in the feed solution at time t;
C^o= initial concentration of the cation in the feed solution;
θ = volumic porosity of the SLM (%);
S = membrane surface area (cm^2) depending on the device;
V = volume of feed and stripping solutions (cm^3) depending on the device; and
t = time (h).

The data summarized in Table II confirm the results obtained previously in sodium-cesium extraction studies (see Table I) although the organic solvent used in these experiments (2-nitrophenyl octyl ether) is a slightly more polar and more viscous diluent than 2-nitrophenyl hexyl ether. Carriers **2**, **5** and **6** with six oxygen atoms in the glycol chain are much more efficient at transporting cesium through SLMs than **1** and **3**, with five or seven oxygen atoms in their rings, or **4** and **7**, in which the polyether chains are sterically constrained.

Certain assumptions of the Danesi model (*15*) (transport controlled and limited only by diffusion) allow the permeability coefficients P_M to be evaluated by the following formula :

$$P_M \approx \frac{D_M}{\Delta_o} \qquad (2)$$

where :

D_M = distribution coefficient of the permeating cation;
Δ_o= the ratio of the organic path length (thickness of the membrane $\times$ tortuosity) to the organic diffusion coefficient: $(d_o \times \tau)/D_o$.

Equation 2 shows that P_M is proportional to D_M which directly depends on the organic carrier concentration in the SLM. That is why the cesium transport isotherms shown in Figures 3 and 4 for compound **2** reproduced the same trends as those found previously in extraction isotherms (Figures 1 and 2):

♦ An increase of cesium permeability coefficient for $[NaNO_3] < 1$ M in the feed solution (Figure 3) and for $[HNO_3] < 2$ M in the feed solution (Figure 4) as D_{Cs} increased with salt concentration in the feed solution (Figures 1 and 2);
♦ A maximum of transport kinetics due to the maximum of D_{Cs};
♦ A decrease of P_{Cs} for $[NaNO_3] > 1$ M or $[HNO_3] > 2$M, revealing three different phenomena:

   • A decrease of D_{Cs} due to sodium nitrate or nitric acid competitive extraction and transport of sodium nitrate or nitric acid;
   • A decrease of the trace level cesium mean activity coefficient with an increase in the salt concentration in the aqueous feed solution;

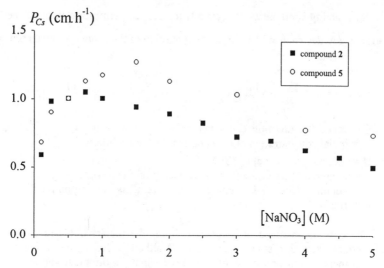

Aqueous feed solution : NaNO$_3$: various concentrations (pH$_i$ = 1.4).
Aqueous stripping solution : deionized water.
SLM : compound 2 or 5 (10^{-2} M) in 2-nitrophenyl octyl ether.

Figure 3. Transport Isotherms for ^{137}Cs from Sodium Nitrate Solutions to Deionized Water by 1,3-Calix[4]-*bis*-crown-6 2 and by 1,3-Calix[4]-*bis*-*o*-benzo-crown-6 5 (10^{-2} M) in 2-NPOE.

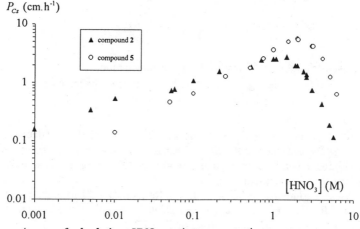

Aqueous feed solution : HNO$_3$: various concentrations.
Aqueous stripping solution : deionized water.
SLM : compound 2 or 5 (10^{-2} M) in 2-nitrophenyl octyl ether.

Figure 4 : Transport Isotherms of ^{137}Cs from Nitric Acid Solutions to Deionized Water for 1,3-Calix[4]-*bis*-crown-6 2 and for 1,3-Calix[4]-*bis*-*o*-benzo-crown-6 5 (10^{-2} M) in 2-NPOE.

•Leaching of the SLMs by the feed and the receiving solutions favored by the partitioning of the carrier, especially in the case of compound **2** which shows very poor lipophilicity.

The higher selectivity and efficiency of **5** compared to **2** can easily be noticed by comparing both transport isotherms shown in Figures 3 and 4.

Time stability of the most efficiently prepared SLMs was estimated by repeating transport experiments in which both the aqueous feed and stripping solutions were renewed every day while the SLMs remained the same. Consequently, daily partitioning of the carrier from the membrane phase to the renewed aqueous solutions caused a decrease of the carrier concentration in the SLMs, thus a decrease of D_M and, proportionally, of P_M. The evolution of the permeability coefficients *vs* the number of runs was therefore a way to evaluate leaching of the membrane by the aqueous solutions and characterize the SLMs stability with time.

As shown in Figure 5, repeated cesium transport experiments showed that **2** rapidly leached out of the SLMs ($P_{Cs} < 0.1$ cm.h^{-1} after 15 runs), because of its low partition constants between 2-nitrophenyl octyl ether (2-NPOE) and the aqueous solutions. Very good stability and efficiency were observed with the more lipophilic compounds **5** and **6**, for which the cesium permeability coefficients were 5 times higher than those of the crown ether **8** even when the concentration of the crown ether was 50 times higher in 2-nitrophenyl octyl ether (2-NPOE).

1,3-Calix[4]-*bis*-crowns **5** and **6** allowed selective removal of ^{137}Cs from sodium containing solutions. *Less than 100 mg of the ~100 g of sodium initially present in the feed solution was transported in 24 hours by compound **6**, whereas more than 95 % of the trace level ^{137}Cs was concentrated in the stripping solution.* Nitric acid transport, due to the basicity of both the organic diluent and the calixarene, could not be limited to less than 5 % (0.05 M) in 24 hours leading to concentration factors (ratio of initial waste concentration to final waste concentration) greater than 100 for a single step process.

To summarize, 1,3-calix[4]-*bis*-crowns *containing six oxygen atoms* appeared to be a promising family of carriers for the selective removal of cesium from high salinity media, such as medium-level radioactive liquid wastes, with SLMs. By choosing a highly hydrophobic organic diluent, 2-NPOE, and a very lipophilic calix[4]-*bis*-crown in the 1,3-alternate conformation suitable for cesium complexation over sodium, very selective and stable SLMs can be obtained (over a period of 50 days).

Transport of Cesium from Simulated Fission Product Solutions.

Application of calix[4]-*bis*-crowns **2**, **5** and **6** to decontaminate medium or high-level radioactive wastes, such as fission product solutions, was also investigated. Figure 6 shows the effect of the presence of six different nitrate salts (NH_4^+, Ca^{2+}, Mg^{2+}, Fe^{3+}, Al^{3+} and UO_2^{2+}) on cesium permeation from aqueous feed solutions simulating medium active concentrates (4 M in $NaNO_3$ and 1 M in HNO_3) to receiving solutions of deionized water. Except for ammonium nitrate salts, which usually behave like alkali cations and drastically hinder cesium separation in classical industrial decontamination processes (co-precipitation, ion exchange, ...), none of the other five added species (Ca^{2+}, Mg^{2+}, Fe^{3+}, Al^{3+} and UO_2^{2+}) potentially present in medium active wastes disfavored trace level cesium permeation through the SLMs.

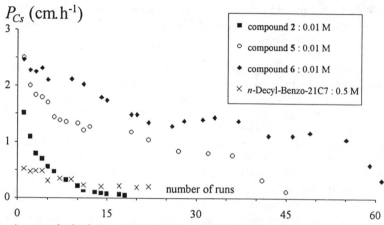

Aqueous feed solution : NaNO$_3$ (4 M) and HNO$_3$ (1 M).
Aqueous stripping solution : deionized water.
SLM : compound **2, 5** or **6** (10^{-2} M) in 2-nitrophenyl octyl ether.

Figure 5 : Repetitive Transport Experiments of ^{137}Cs from Simulated Medium-Active Liquid Wastes for 1,3-Calix[4]-*bis*-crown-6 **2**, for 1,3-Calix[4]-*bis-o*-benzo-crown-6 **5** and for 1,3-Calix[4]-*bis*-1,2-naphthyl-crown-6 **6** (10^{-2} M) in 2-NPOE.

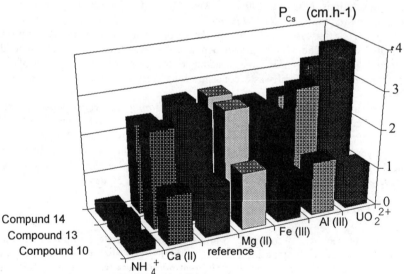

Aqueous feed solution : NaNO$_3$ (4 M), HNO$_3$ (1 M) and nitrate species (NH$_4^+$ or Ca^{2+} or Mg^{2+} or Fe^{3+} or Al^{3+} or UO$_2^{2+}$: 0.1 M).
Aqueous stripping solution : deionized water.
SLM : compound **2, 5** or **6** (10^{-2} M) in 2-nitrophenyl octyl ether.

Figure 6 : Influence of the Presence of Nitrate Salts on the Permeation of ^{137}Cs from Simulated Medium-active Liquid Wastes to Deionized Water by 1,3-Calix[4]-*bis*-crown-6 **2**, 1,3-Calix[4]-*bis-o*-benzo-crown-6 **5** and 1,3-Calix[4]-*bis*-naphthyl-crown-6 **6** (10^{-2} M) in 2-NPOE.

Table III records the transport yields of eleven main elements (out of thirty-four present in very high-level radioactive liquid wastes, such as fission products solutions) from nitric acid feed solutions (3 M in HNO_3 and 10^{-3} M in the element) to deionized water. Metal cation concentrations were determined by atomic absorption spectrometry of aliquots sampled in the receiving solutions after 24 hours of permeation. Except for rubidium, whose chemical behavior is similar to that of cesium among the other alkali cations, the high selectivity of the tested 1,3-calix[4]-*bis*-crowns toward cesium is maintained in the presence of the other 9 elements.

Conclusion.

Calix[4]-*bis*-crowns 1-7 have been used as Cs-carriers in supported liquid membranes. Application of the Danesi diffusional model allowed the transport isotherms of trace level ^{137}Cs through SLMs (containing calix[4]-*bis*-crowns) to be determined as a function of the ionic concentration of the aqueous feed solutions. Compound 5 appeared to be much more efficient than a combination of crown ether and an acidic exchanger, especially in very acidic media. This allowed a one-step treatment of synthetic acidic radioactive wastes with *decontamination factors greater than 100*. Permeability coefficient measurements were also conducted in repetitive transport experiments in order to evaluate the SML stability with time. Very good results (over 50 days of stability) and high decontamination yields were observed with 1,3-calix[4]-*bis*-crowns 5 and 6. Thus, the 1,3-calix[4]-*bis*-crowns are shown to be a new family of selective carriers for cesium removal from radioactive liquid wastes leading to 1:1 stoichiometry complexes (1,3-calix[4]-*bis*-crown : Cs).

Experimental Section

Ligands. Calix[4]-*bis*-crowns 1-7 have been prepared according to published procedures (*6b,17*).

Starting Materials for Extraction and SLMs transports. The inorganic salts used to prepare the synthetic aqueous feed solutions for extraction and transport experiments ($NaNO_3$, $CsNO_3$, ...)[137] were analytical grade products from Prolabo and Aldrich. Radioactive [22]Na and ^{137}Cs used to spike the aqueous solutions (1500 kBq.L^{-1} < initial aqueous activity < 2500 kBq.L^{-1}) were provided by the Amersham Company. The [22]Na and ^{137}Cs gamma spectra were obtained on a detection chain from Interchnique : an EGSP-2000-20R gamma spectrometer equipped with germanium detectors. The counting was always sufficiently long to insure a relative error in activity measurement of less than 5 %. The organic diluents, 2-nitrophenyl hexyl ether (2-NPHE) and 2-nitrophenyl octyl ether (2-NPOE) used to dissolve 1-7 were synthesized by Chimie Plus and used without further purification.

Transport Measurements. The same Flat-Sheet Supported Liquid Membrane (SLMs) device described by T. Stolwijk *et al.* (*18*) and shown in Figure 7 was used for the transport experiments. The volume of both aqueous solutions (feed and stripping) ranged from 45 to 55 mL depending on the glass devices manufactured by Prodilab and Verre & Science Companies. The circular membranes were about 15 to 16 cm² in area and made of polypropylene Celgard® 2500 microporous support (of 25 mm thickness

Table III : Transport Yields (after 24 hours) of Sr, Rb, Y, Zr, Ru, Rh, Cs, Ba, La, Ce and Eu Permeation from Simulated High-active Wastes to Deionized Water by 1,3-Calix[4]-*bis*-crown-6 **2**, by 1,3-calix[4]-*bis*-o-benzo-crown-6 **6** and by 1,3-calix[4]-*bis*-naphthyl-crown-6 **5** (10^{-2} M) in 2-NPOE.

Aqueous feed solution : HNO_3 : 3 M and Element : 10^{-3} M.
Aqueous stripping solution : deionized water.
SLM: compound **2**, **5** or **6** (10^{-2} M) in 2-nitrophenyl octyl ether.

		1,3- calix[4]-*bis*-crown		
		2	**5**	**6**
Element	Initial concentration in the feed solution (mg L^{-1})	Metal concentration in the receiving solution after 24 hours (mg L^{-1})		
Sr	87.6	$< 5.10^{-2}$	$< 5.10^{-2}$	$< 5.10^{-2}$
Rb	85.5	34	63	47
Y	88.9	$< 5.10^{-3}$	$< 5.10^{-3}$	$< 5.10^{-3}$
Zr	91.2	$< 5.10^{-2}$	$< 5.10^{-2}$	$< 5.10^{-2}$
Ru	101.0	$< 5.10^{-3}$	$< 5.10^{-3}$	$< 5.10^{-3}$
Rh	102.9	$< 5.10^{-3}$	$< 5.10^{-3}$	$< 5.10^{-3}$
Cs	132.9	92	125	105
Ba	137.3	$< 5.10^{-2}$	$< 5.10^{-2}$	$< 5.10^{-2}$
La	138.5	$< 5.10^{-3}$	$< 5.10^{-3}$	$< 5.10^{-3}$
Ce	140.1	$< 5.10^{-3}$	$< 5.10^{-3}$	$< 5.10^{-3}$
Eu	151.9	$< 5.10^{-3}$	$< 5.10^{-3}$	$< 5.10^{-3}$

and 45 % internal volumic porosity) soaked with a 10^{-2} M solution of 1-7 in 2-NPOE. The measurements were performed at a constant temperature of 25° C.

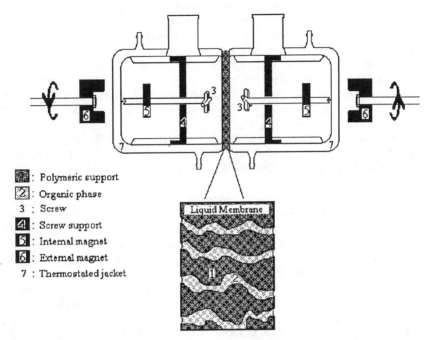

: Polymeric support
: Organic phase
3 : Screw
: Screw support
: Internal magnet
: External magnet
7 : Thermostated jacket

Liquid Membrane

Figure 7 : Flat-sheet supported liquid membrane device for transport experiments.

References

1. Cecille, L. *In Radioactive Waste Management and Disposal*, Elsevier Applied Science; New York; 1991.
2. Dozol, J. F. In *Future Industrial Prospects of Membrane Processes*, Cecille, L.; Toussaint, J. C. Eds; Elsevier Applied Science; New York; 1989.
3. Shehata, F. A. *J. Radioanal. Nucl. Chem.* **1994**, *185*, 411.
4. Alfieri, C.; Dradi, E.; Pochini, A.; Ungaro, R.; Andreetti, G. D. *J. Chem. Soc., Chem. Commun.* **1983**, 1075.
5. (a) Dijkstra, P. J.; Brunink, J. A.; Bugge, K.-E.; Reinhoudt, D. N.; Harkema, S.; Ungaro, R.; Ugozzoli, F.; Ghidini, E. *J. Am. Chem. Soc.* **1989**, *111*, 7567. (b) Ghidini, E.; Ugozzoli, F. Ungaro, R.; Harkema, S.; El-Fadl, A. A.; Reinhoudt, D. N. *J. Amer. Chem. Soc.* **1990**, *112*, 6979. (c) Nijenhuis, W. F.; Buitenhuis, E. G., De Jong, F.; Sudhölter, E. J. R.; Reinhoudt, D. N. *J. Am. Chem. Soc.* **1991**, *113*, 7963. (d) Brzozka, Z.; Lammerink, B.; Reinhoudt, D. N.; Ghidini, E.; Ungaro, R. *J. Chem. Soc Perkin Trans. 2* **1993**, 1037. (e) Ungaro, R.; Casnati, A.; Ugozzoli F.; Pochini, A. Dozol, J. F.; Hill, C.; Rouquette, H. *Angew. Chem. Int. Ed. Engl.* **1994**, *33*, 1506. (f) Cacciapaglia, R.; Casnati, A.; Mandolini, L.; Ungaro, R. *J. Chem. Soc., Chem. Commun.* **1992**, 1291. (g) Cacciapaglia, R.; Casnati, A.; Mandolini, L.; Ungaro, R. *J. Amer. Chem. Soc* **1992**, *114*, 10959. (h) Cacciapaglia, R.; Casnati, A.; Mandolini, L.; Schiavone, S.;

Ungaro, R. *J. Chem. Soc Perkin Trans.* 2 **1993**, 369. (i) Cacciapaglia, R.; Mandolini, L. *Chem. Soc. Rev.* **1993**, 221. (j) King, A. M.; Moore, C.P.; Scandanayake, K. R. A. S.; Sutherland, I. O. *J. Chem. Soc., Chem. Commun.* **1992**, 582. (k) Reinhoudt D. N.; Engbersen, J. F.; Brozozka, Z.; Van den Viekkert, H. H.; Honig, G. W. N.; Holterman, A. J.; Verker, U. H. *Anal. Chem.* **1994**, 66, 3618. (l) Kraft, D.; Arnecke, R.; Böhmer, V.; Vogt, W. *Tetrahedron* **1993**, *49*, 6019.

6. (a) Arduini, A.; Casnati, A.; Dodi, L.; Pochini, A.; Ungaro, R. *J. Chem. Soc., Chem Commun.* **1990**, 1597. (b) Asfari, Z.; Pappalardo, S.; Vicens, J. *J. Incl. Phenom.* **1992**, *14*, 189. (c) Masci, B.; Saccheo, S. *Tetrahedron* **1993**, *49*, 10739. (d) Iasi, G. D.; Masci, B. *Tetrahedron Lett.* **1993**, 6635. (e) Asfari, Z.; Harrowfield, J. M.; Sobolev, A. N.; Vicens, J. *Aust. J. Chem.* **1994**, *47*, 757.

7. Asfari, Z.; Abidi, R.; Arnaud, F.; Vicens, J. *J. Incl. Phenom.* **1992**, *13*, 163.

8. (a) I. Gerow, H.; Davis, Jr, M. W. *Sep. Sci. Technol.* **1979**, *14*, 395. (b) Gerow, H.; Smith Jr, J. E.; Davis Jr, M. W. *Sep. Sci. Technol.* **1981**, *16*, 519. (b) Blasius, E.; Nilles, K.-H. *Radio. Acta* **1984**, *35*, 173.

9. For π-cation interactions see also for example : (a) Ikeda, A.; Tsuzuki, H.; Shinkai, S. *Tetrahedron Lett.* **1994**, 8417. (b) Ikeda, A.; Shinkai, S. *J. Amer. Chem. Soc.* **1994**, *116*, 3102. (c)Ikeda, A.; Shinkai, S. *Tetrahedron Lett.* **1992**, 7385. (d) Iwamoto, K.; Shinkai, S. *J. Org. Chem.* **1992**, *57*, 7066. (e) Fujimoto, F.; Nishiyama, N.; Tsuzuki, H.; Shinkai, S. *J. Chem. Soc. Perkin Trans.* 2 **1992**, 643. (f) Araki, K.; Shimizu, H.; Shinkai, S. *Chem. Lett.* **1993**, 205.

10. Thermodynamic studies by UV and NMR from this laboratory have also shown 1:1 stoichiometry for alkali metal **1-7** complexes. These results will be published in full in due course.

11. (a) Costero, A. M.; Rodriguez, S. *Tetrahedron Lett.* **1992**, 623. (b) Costero, A. M.; Rodriguez, S. *Tetrahedron* **1992**, *48*, 6265. (c) Costero, A. M.; Pitarch, M. *J. Org. Chem.* **1994**, *59*, 2939.

12. Mc Dowell, W. J.; Case, G. N. *Anal. Chem.*, **1992**, *64*, 3013.

13. (a) W. F. Nijenhuis, A.R. van Door, A. M. Reichwein, F. de Jong, and D. N. Reinhoudt, *J. Am. Chem. Soc.*, **1991**, *113*, 3607. (b) Visser, H. C.; Reinhoudt, D. N.; De Jong, F. *Chem Soc. Rev.* **1994**, 76. (c) Verboom, W.; Rudkevich, D. M.; Reinhoudt, D. N. *Pure & Appl. Chem.* **1994**, *66*, 697. (d) Von Straten-Nijenhuis, W. F.; De Jong, F.; Reinhoudt, D. N. *Rec. Trav. Chim. Pays-Bas* **1993**, *112*, 317.(e) Brown, P. R.; Hallman, J. L.; Whaley, L. W.; Desai, D. H.; Pugia, M. J.; Bartsch, R. A. *J. Memb. Sci.* **1991**, *56*, 195.

14. (a) Izatt, R. M.; Lamb, J. D.; Hawkins, R. T.; Brown, P. R.; Izatt, S. R.; Christensen, J. J. *J. Am. Chem. Soc.***1983**, *105*, 1782. (b) Izatt, S. R.; Hawkins, R. T.; Christensen, J. J.; Izatt, R. M. *J. Am. Chem.Soc.* **1985**, *107*, 63-66.

15. Danesi, P. R. *Sep. Sci. Technol.*, **1984**, *19*, 857.

16. Hill, C. *PhD Thesis - University of Strasbourg.* **1994**.

17. Wenger, S. Asfari, Z. Vicens, J. *Tetrahedron Lett.* **1994**, *35*, 8369.

18. Stolwijk, T. B.; Sudhölter, E. J. R.; Reinhoudt, D. N. *J. Am. Chem. Soc.* **1987**, *109*, 7042.

Chapter 27

Separation of Radiotoxic Actinides from Reprocessing Wastes with Liquid Membranes

J. P. Shukla[1], A. Kumar[2], R. K. Singh[2], and R. H. Iyer[1]

[1]Radiochemistry Division, Bhabha Atomic Research Centre, Trombay, Bombay 400085, India
[2]Power Reactor Fuel Reprocessing Plant, Fuel Reprocessing Group, Bhabha Atomic Research Centre, Tarapur, P.O. Ghivali, 401504, India

Carrier-facilitated transport of actinides across bulk, supported, and emulsion liquid membranes, as well as plasticized membranes and recently developed emulsion-free liquid membranes, are reviewed. The discussion includes the effects of important experimental variables upon the solute flux for various types of liquid membranes. Applications of liquid membranes in the recovery and removal of radiotoxic actinides from the nitric acid wastes generated during reprocessing of spent fuel by the PUREX process and wastes produced by other radiochemical operations are surveyed.

Reprocessing and allied radiochemical facilities generate large volumes of liquid effluents which contain extremely low levels of biotoxic, radioactive species. Primarily, the effluents contain the lighter actinides, such as uranium, plutonium, and americium together with a host of long-lived fission product contaminants at varying levels. In view of the ALARA (As Low As Reasonably Achievable) Principle currently in effect for most nuclear installations, the potential for further removal of even trace amounts of these actinides from low alpha waste is being pursued vigorously.

Radiotoxic species are generally removed by conventional chemical techniques, such as precipitation, ion exchange, *etc*. However these methods suffer from a major drawback of ineffectiveness in the treatment of low-level waste streams.

Liquid membrane (LM) transport is a novel separation technique whose potential for application in effluent treatment in the nuclear industry is being explored (*1*). LM transport offers several advantages over other separation techniques, *e.g.*, lower capital cost and space requirements, low energy consumption, and low solvent inventory which allows the use of expensive carrier molecules. High feed/strip solution volume ratios, the possibility of utilizing feed solutions which contain suspended solids, and high separation factors are other merits of the LM transport method. A prime requirement for a successful LM system is the selection of a selective and efficient carrier for the species to be separated.

0097–6156/96/0642–0391$15.00/0
© 1996 American Chemical Society

Transport of a species through a liquid membrane is superior to solvent extraction (SX) since extraction and stripping are performed in a single unit operation. Also, liquid membrane transport is a non-equilibrium, steady state process which depends upon kinetic factors in contrast to SX which is an equilibrium process. Furthermore, even solvents with low distribution coefficients for the desired species may be utilized in LM processes. Although LM systems generally have slower rates than ion-exchange (IX) processes, the latter are particularly sensitive to the presence of suspended solids and other foulants and also must be operated in cycles.

Separations by transport across liquid membranes in various configurations, *viz.* bulk liquid membranes (BLM), emulsion liquid membranes (ELM), supported liquid membranes (SLM), and plasticized membranes (PM), are receiving ever increasing attention, particularly for the processing of dilute solutions, in the separation and recovery of metals of critical and strategic importance, and in the decontamination of low and medium-level radioactive wastes (2,3). Practical applications of such membranes in the recovery of metals from hydrometallurgical leach solutions and industrial effluents and in plutonium and americium removal from the nitric acid waste streams generated during plutonium recovery operations by the PUREX process are being explored.

Reprocessing of spent nuclear fuels produces large volumes of medium-level radioactive liquid wastes which are concentrated by distillation. The distillate is discharged into the environment and all long-lived radionuclides (^{137}Cs, ^{90}Sr, actinides, *etc.*) must be removed from the concentrate to permit its storage, after immobilization, in geological formations. The feasibility of decontaminating this concentrate by application of liquid membranes has been reported recently (4). Better management of the radioactive wastes would decrease the cost and environmental impact of reprocessing. Optimal management of the wastes calls for separation of the alpha emitters, *e.g.*, Pu, Am, Np, and Cm, and some fission products, like Tc, from the bulk of the fission and corrosion products.

This chapter summarizes the applications of LMs, particularly for the separation of radiotoxic actinides and shows their potential in waste treatment and partitioning. For decontamination of the waste solutions which are generated in nuclear facilities, one must seek a process which is both technically efficient in removing the contaminants and also economically attractive for large-scale operation.

Bulk Liquid Membranes (BLMs) and Supported Liquid Membranes (SLMs).

Synthetic waste solutions which simulate those often encountered in the PUREX process were tested in a SLM system to determine the efficiency for decontamination by removal of transuranic elements, such as americium (5). Diethyl N,N-diethylcarbamoyl methylphosphonate (DHDECMP) supported on Accurel™ or Celgard™ polypropylene (PP) hollow fibers assembled in modules showed promise in removing more than 95% of the Am and 70% of the Pu from high nitrate (6.9 M) and low acid (0.1 M) feeds into 0.25 M oxalic acid strippants. Hollow-fiber membrane supports were chosen since they offer a greater membrane area/feed volume ratio than do flat or spiral-wound membranes. For hollow fibers imbibed with tributylphosphate (TBP), plutonium and americium were transported with maximum permeabilities of about 1×10^{-3} cm/sec (6).

Separation of Pu and Am from nitrate-nitric acid feed was achieved with a SLM in which DHDECMP or *n*-octylphenyl N,N-diisobutylcarbamoyl methylphosphine oxide (CMPO) as the carriers (7). After transport of the Am into the 0.25 M oxalic acid strippant, it was recovered by sorption on an inorganic ion exchanger. Chiarizia and Danesi (8) have demonstrated the feasibility of using two SLM systems in series for the removal of lanthanides and actinides containing ^{241}Pu, ^{242}Pu, ^{237}Np (10^{-4} M) and ^{99}Tc. The first SLM (SLM-1) used the TRUEX process solvent (0.25 M CMPO + 0.75 TBP in decalin) as the supported liquid membrane and 1 M formic acid and 0.005 M hydroxylammonium formate as the strippant. For the second unit (SLM-2), 1 M Primene JM-T in decalin and 4 M sodium hydroxide were the membrane and stripping phases, respectively. With this system, the activity of the effluent was reduced to less than 100 nCi of total radionuclides/gram of solidified waste.

Highly selective transport of U(VI) (9) and Pu(IV) (10) over several fission-product contaminants has been achieved in both organic BLM and SLM systems employing TBP in dodecane as the mobile carrier. The source phase was an extremely dilute actinide nitrate solution (10^{-5} M) in 2 M nitric acid. Ascorbic acid was found to be the most efficient strippant for Pu(IV), whereas dilute ammonium carbonate was best for U(VI).

In another study, dicyclohexano-18-crown-6 (DC18C6) was shown to be an excellent carrier in liquid membranes due to its high selectivity for both U(VI) (11) and Pu (12) even in the presence of several other undesirable radiotoxic elements Table 1). Maximum transport through the macrocycle-based BLM and SLM systems (0.2 M DC18C6 in toluene as the membrane phase) was attained from a 3 M nitric acid feed solution for Pu(IV) and 7 M nitric acid for U(VI). The maximum fluxes for Pu(IV) and U(VI) were 1.6×10^{-8} and 2.9×10^{-7} mol/m^2/s, respectively (11,12). Enrichment factors of 6 or greater were readily achieved by proper manipulation of the feed solution/strip solution volume ratios (12).

Table 1. Transport of Various Products Through a Supported Liquid Membrane.(12)[a]

Fission product	Half life (years)	Activity[b] in the source phase (μCi/dm^3)	Activity in the receiving phase (μCi/dm^3)	Permeation of fission product (%)
^{106}Ru	1	4.28	0.72	16.8
^{137}Cs	28	2.63	0.44	16.5
^{125}Sb	2	0.69	0.08	11.6

[a]Conditions: Initial feed solution = 50.0 mg/dm^3 of plutonium in nitric acid; Membrane phase = 0.2 mol/dm^3 of DC18C6 in toluene; Stripping solution = 0.5 mol/dm^3 aqueous sodium carbonate; Volume of feed to stripping solution = 6.
[b]The activity of the fission product was estimated with a multichannel analyzer and a high purity germanium detector.

Subsequently, this system was applied to the recovery of uranium and plutonium from actual low-level acidic wastes generated during the plutonium purification cycle (13). The results which are presented in Table 2 reveal that maximum recoveries of U(VI) and Pu(IV) of 63 and 98 %, respectively, were obtained after transport for 6 hours from a feed solution of 4 M nitric acid. Enhancing the acidity of the feed solution to 6 M nitric acid adversely affected the separation of Pu(IV). The diminished recovery was attributed to the formation of impermeable nitrato species of the type HPu(NO$_3$)$_5$ and H$_2$Pu(NO$_3$)$_6$. On the other hand, transport of U(VI) was enhanced for higher nitric acid concentrations in the feed solution, even though the maximum recovery never exceeded 65 %. The lower U(VI) recovery was thought to be due to pore saturation of the membrane by metal species resulting from the higher concentration of uranium at the membrane interface.

Transport of Pu(IV) (14) was improved significantly by the addition of water-miscible polar liquids, such as acetonitrile or propylene carbonate, to the aqueous source phase. For example, Pu(IV) recovery was 96 % in 4 hours from a source phase of 2.7 M nitric acid containing 20 % acetonitrile, but was only 69 % during the same time period when acetonitrile was omitted from the feed solution.

Dozol *et al.* (15-17) conducted tests with genuine concentrate and two SLMs in series. For the first SLM, the membrane phase was CMPO and TBP in decalin. For the second, a solution of DC18C6 and decanol in hexylbenzene was the membrane phase. With the first SLM, a decontamination factor (DF) of 100 was achieved. The overall DF value for the two SLMs in series was greater than 400. CMPO and crown

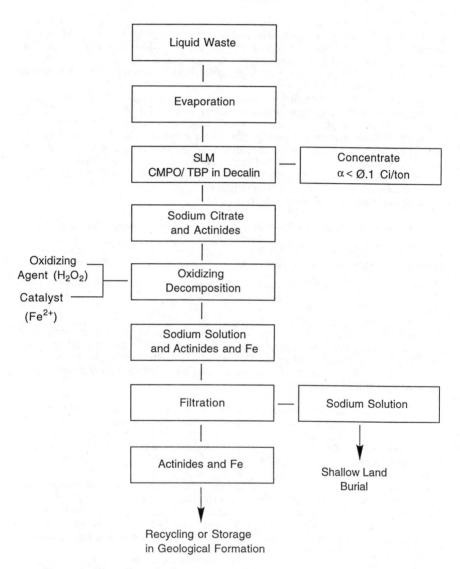

Figure 1. Flow Sheet for Treatment of Evaporation Concentrates by a SLM Technique.

Table 2. Transport of U(VI) and Pu(IV) Across a Supported Liquid Membrane from a 2AW Waste Stream as a Function of the Nitric Acid Concentration in the Feed Solution.[a]

Nitric acid concentration (M)	Transport time (h)	Permeation Coefficient P (X 10^3 m/s)		Recovery (%)	
		U	Pu	U	Pu
1	1	0.4	0.1	–	1
	6	0.3	0.04	11	8
2	1	0.4	0.7	8	6
	6	0.2	0.1	13	15
3	1	2.0	3.2	17	43
	6	1.0	0.8	57	92
4	1	4.6	4.6	36	33
	6	1.2	0.6	63	98
5	1	5.3	2.2	48	19
	6	2.2	0.4	65	73
6	1	9.3	1.3	60	22
	6	2.1	0.1	63	70

[a]Conditions: Feed solution = 2AW process effluent in nitric acid with 860.3 mg/dm^3 of U(VI), 10.4 mg/dm^3 of Pu(IV), 1.5 mCi/dm^3 of ^{137}Cs, and 0.4 mCi/dm^3 of ^{106}Ru ; Membrane phase = 0.4 M DC18C6 in toluene; Stripping solution = 1.0 M oxalic acid + 0.35 M nitric acid.

ether compounds, *e.g.*, DB18C6 and B21C7, have been shown to extract actinides and Sr-Cs, respectively, from highly saline solutions.

A flow sheet for treatment of evaporation concentrates using a SLM technique is presented in Figure 1 (*17*).

In a large-scale study related to the application of SLM technology to the removal of uranium from Hanford site groundwater (*18*), the feasibility of using large, hollow-fiber modules was demonstrated which lead to a reduction of the uranium concentration to below its threshold allowable limit of 10 ppb.

Akiba *et al.* (*19*) investigated the carrier-mediated transport of U(VI) across a SLM with trioctylphosphine oxide (TOPO) as the carrier and concluded that the uranium is extracted as $UO_2(NO_3)_2\bullet2TOPO$ and stripped into a carbonate receiving phase as $UO_2(CO_3)_3^{4-}$. Except when the carrier was very dilute, the transport rate appeared to be controlled by diffusional processes in the membrane phase, not by chemical reactions.

Recently, the potential of substituted amine carriers in the recovery of actinide from processing effluents has been explored. We have investigated transport of Pu(IV) from aqueous nitrate solutions through a SLM containing Aliquat 336 as a function of experimental variables (Shukla, J.P.; Sonawane, J.V.; Kumar, A.; singh, R.K. *Radiochim. Acta*, in press). Pu(IV) transport across the membrane reached a plateau due to carrier saturation in the membrane. Applicability of the process was tested for the recovery of actinides from oxalate-bearing wastes. Greater than 95 % of the plutonium was readily recovered from diverse raffinates in the presence of troublesome fission product contaminants. The transport of several long-lived fission products, such as ^{106}Ru, ^{137}Cs, ^{152}Eu, and ^{154}Eu which generally accompany plutonium, was assessed. No detectable amounts of these fission products were transported into the receiving phase even after 10 hours of operation. This is because these cations are not extractable by the Aliquat 336 in Solvesso 100 system.

In a related study, we investigated facilitated transport of Pu(IV) from nitric acid media across a SLM (*20*). Carriers of Amberlite LA-2, trilaurylamine (TLA), and Aliquat 336 in Solvesso 100 were utilized as representative secondary and tertiary amines and quaternary ammonium ions, respectively. Both the composition of the organic membrane solvent and type of carrier were found to influence the transport of Pu(IV). Transport across the SLM increased in the order of secondary amine <

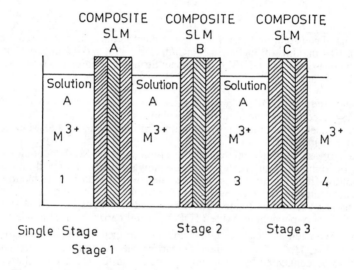

Figure 2.   Schematic Diagram of a Transport Process Which Uses
            Composite Supported Liquid Membranes in Series.

tertiary amine < quaternary ammonium ion, which is the same as that noted in liquid-liquid distribution measurements. More than 95 % recovery of Pu(IV) was achieved with the tertiary amine or quaternary ammonium ion carriers in a single run employing a feed solution which contained 5 mg/dm^3 of plutonium in 4 M nitric acid and a receiving phase of 0.1 M hydroxylamine hydrochloride in 0.3 M nitric acid. The plutonium recovery declined to 49 % with Amberlite LA-2, the secondary amine carrier. An initial Pu(IV) permeation coefficient after 2 hours of 1.1 X 10^{-3} m/s for the secondary amine carrier is about 6 times lower than that for the tertiary amine carrier TLA (6.5 X 10^{-3} m/s). For Aliquat 336, the permeation coefficient after 2 hours was 8.2 X 10^{-3} m/s which is about 7 times higher than that for the secondary amine carrier.

Amide extractants are receiving increased attention as potential complexants for tri-, tetra-, and hexavalent actinides. Their extraction capability and complete incinerability are import factors for their selection as novel extractants (*21*). These characteristics prompted us to initiate studies (J.P. Shukla, Bhabha Atomic Research Centre Report, unpublished data) on amides as carriers for 5f-block elements in SLM systems. For N-methyl N-butyl decanamide (MBDA) and N,N-di(*n*-octyl) hexanamide (DOEHA), the former proved to be the better carrier providing >90 % recovery of plutonium in a single run in the presence of common fission product contaminants, like ^{137}Cs, $^{152-154}$Eu, and ^{107}Ru. The SLM system used 4 M nitric acid as the feed solution, 0.4 M MBDA in dodecane as the membrane phase and 0.1 M hydroxylamine hydrochloride in 0.3 M nitric acid as the strippant. Under the same conditions, DOEHA as carrier afforded only a 20 % recovery of plutonium.

Danesi *et al.* investigated separations of actinides (*22-33*) with various SLM systems and carriers like the neutral bifunctional phosphorus compound CMPO (*23-31,33*), as well as acidic [di-(2-ethylhexyl)phosphoric acid = DEHPA, dinonylnaphthalenesulfonic acid = DNNSA] (*22,26-28,32,33*) and basic (Primene JM-T, a mixture of *t*-C$_{18}$H$_{37}$NH$_2$ and *t*-C$_{22}$H$_{45}$NH$_2$) carriers (*27,30*) and determined the transport mechanisms. Transport of Am(III) with CMPO as the carrier in diethylbenzene or in a mixture of decalin and diisopropylbenzene (2:1 by volume) (*22*) impregnated in microporous polypropylene (PP) flat sheet (*22-25,31*) or single hollow-fiber (*22*) Accurel™ supports (48 μm thick) was investigated. Experimental variables, such as the carrier concentration in the membrane phase and the acidity and hydrodynamic conditions (stirring rate or linear flow velocity) in the feed solution were adjusted to maximize transport. A log-log plot of the permeability coefficient versus the lithium nitrate activity in the feed solution and the carrier concentration in the membrane phase yielded straight lines in the lower portions of the plots which had the same positive slopes as did the corresponding dependencies for D$_{Am}$. Thus the transport process was controlled entirely by membrane diffusion in the lower concentration ranges.

Using the SLM technique (*32*), transport kinetics of lanthanide [Eu(III)] and actinide cations [Am(III), Cf(III)] by DEHPA dissolved in dodecane were studied. Transport of Am(III) and its mixture with Eu(III) was investigated with the composite SLM system depicted in Figure 2. A neutral carrier (CMPO) was used in the first SLM and an acidic carrier (DEHPA or DNNSA) was employed in the second SLM. The arrangement in such a composite SLM system is generally designated as "solution A (source)-SLM(A)-solution B-SLM(B)-solution A (receiving)". Aqueous solution A promotes extraction of metal cations into SLM(A) and their stripping from SLM(B), while solution B promotes stripping of metal cations from SLM(A) and their extraction into SLM(B). Thus, the single stage character of the SLM separation may be overcome by repeating the composite SLM arrangement a number of times (*26*). Series of composite SLMs were employed to perform separations of Eu(III) from Am(III) illustrating the principle of multistage separation of metal cations (*27*).

The TRUEX process solvent (0.25 M CMPO + 0.75 M TBP dissolved in decalin) is effective for transport of actinides from synthetic acidic liquid nuclear wastes (*28*). The process has been demonstrated on a laboratory scale (*22,30,31*) for actinides (U,

Pu, Am, Cm; ~ 5 X 10^{-4} M) in the presence of several fission product contaminants in a synthetic waste of PUREX origin. A solution of 0.5 M formic acid and 0.1 M sulfuric acid was utilized as the receiving phase (22,28). Some of the experimental parameters which characterized the process are listed in Table 3. With a series of 20 modules, 99.9 % of [241] Am was recovered from 2 X 10^4 L of feed solution in 24 hours. In subsequent experiments, a second SLM containing 1 M Primene JM-T in decalin was connected in series to the first SLM to overcome the reduction in strippant effectiveness caused by nitric acid transport into the strippant (29,30). The second 'helper' SLM extracted only nitric acid from the first stripping solution which was 1 M formic acid (plus 0.05 M hydroxylammonium formate in the case of Pu(IV) transport). The stripping solution for the second SLM was 4 M sodium hydroxide. Results of these laboratory-scale studies demonstrate that actinides can be efficiently removed from such effluents to the point that the resulting solution can be considered a non-transuranic waste (<100 nCi/g of disposed form).

Table 3. Parameters for a Laboratory-Scale SLM Process to Remove Americium and Other Actinides from an Acidic Nuclear Waste. (22).

Feed solution	Waste solution spiked with [241] Am
Membrane phase	0.2 M CMPO in decalin
Stripping solution	0.5 M in formic acid and 0.1 M in sulfuric acid
Volume of feed solution	20,000 L
Pumping rate	3.5 gallons/min
Linear flow velocity	14.5 cm/s
Permeability coefficient	1 X 10^{-3} cm/s
Space occupied (20 modules, pumps, tubing, etc.)	1.5 m^3 (ca. 54 ft^3)

Permeability coefficient values (29) for various actinides in flat sheet SLM (Celgard™ 2500) and hollow-fiber module (Accurel™) configurations are presented in Table 4. For the experiments conducted with synthetic nuclear waste solutions

Table 4. Permeability Coefficients for Different Actinides with Flat Sheet (Celgard™ 2500) and Accurel™ Hollow Fiber SLMs.[a] (29)

	Permeability coefficient, P^b, from various feeds					
	1 M Nitric acid		Synthetic acidic waste (AW)		Synthetic acidic waste + 0.18 M oxalic acid	
Actinide	Flat sheet[c]	Hollow fiber[d]	Flat sheet[c]	Hollow fiber[d]	Flat sheet[c]	Hollow fiber[d]
Am(III)	1.3	3.0	0.9	1.3	0.8	1.2
Pu(III)	1.4	---	---	---	---	---
Pu(IV)	1.7	---	1.2	2.5	1.0	2.5
Np(IV)[e,f]	1.2	---	7.5	5.8	0.6	6.8
U(VI)	1.4	---	1.1	4.2	1.1	4.0
Th	---	---	1.4	---	1.4	---

[a]Conditions: Membrane phase = 0.25 M CMPO + 0.75 M TBP in dodecane; Receiving phase = 0.5 M sodium citrate; Temperature = 25 °C.;

[b]Apparent permeability coefficients, $P^* = P \bullet \varepsilon$ with ε = porosity, were determined for the hollow-fiber SLM.

[c]X 10^{-3} cm/s.

[d]X 10^{-4} cm/s

[e]The feed and stripping solution contained reducing agents (0.01 M Fe^{2+} + 0.14 M $NH_2OH \bullet HNO_3$ + 0.14 M $N_2H_4 \bullet HNO_3$).

[f]For the hollow-fiber SLMs, the permeation coefficient was calculated from the initial slope of the ln (C/C_o) vs. time plot.

containing neptunium, reducing agents were present to reduce Np(V) to Np(IV). For neptunium transport with flat-sheet SLMs, the data points for plots of ln (C/C_o) vs. time gave straight lines, except for those obtained for nitric acid without reducing agents. The straight lines indicate that diffusion through the membrane is the rate-limiting step. The data obtained with hollow-fiber SLMs indicated that neptunium transport was a pseudo first-order rate process only during the first 2×10^3 seconds. During this time interval, transport of Np(IV) and Np(VI) through the membrane was rate controlling.

The mechanism of uranyl nitrate transport through a kerosene membrane with TBP as carrier was investigated (*34*) by measuring the permeability of the $UO_2(NO_3)_2 \bullet TBP$ complexes and evaluating the individual interfacial resistances. Likewise, Chaudry et al. (*35*) determined the influence of experimental variables, such as the effect of the nitric acid concentration in the feed solution, the TBP concentration in the kerosene membrane, the stripping agent (sodium nitrate, ammonium carbonate) concentration, and the temperature on the uranyl flux. They elucidated the mechanism of uranium transport and the stoichiometry of the transported species. A novel phenomenon, the so-called solubility mechanism was experimentally observed and theoretically explained (*36*) for uranium transport from a feed solution of 0.01 M uranium in nitric acid at pH 0.7-1.5 through a liquid membrane of 0-100 % of DEHPA in the hydrocarbon solvent Shellsol 2046 immobilized in a vertical Celgard™ 2500 film and into a stripping solution of 1 M sodium carbonate at pH 12.5. A simplified model was proposed to explain the non-monotonic behavior of the flux with respect to the carrier concentration.

In another study, Navratil et al. (*37-39*) used undiluted DHDECMP and also 80 volume % of TBP in dodecane as liquid membranes supported in hollow-fiber modules (a Celgard™ module with 10 fibers, each 11.5 cm long with 0.04 cm inner diameter, 0.0025 cm wall thickness, and 45 % porosity or an Accurel™ module with 2 fibers, each 11.5 cm long with 0.05 cm inner diameter, a 0.1 cm wall thickness, and 75 % porosity) for transport of uranium, plutonium, and americium. For membranes impregnated with DHDECMP, the feed solution was 7.0 M in nitrate (sodium nitrate + nitric acid) and the receiving phase was 0.25 M oxalic acid. For membranes impregnated with the TBP in dodecane, the receiving phase was 1.0 M sodium sulfate in 0.5 M sulfuric acid. With the DHDECMP carrier, the maximum recoveries of Am(III) and Pu(IV) of 97 and 70 %, respectively, were obtained from a feed solution of 6.9 M sodium sulfate in 0.1 M nitric acid. Contaminants, such as Fe(III) and Ca(II) were detrimental to actinide transport.

Recovery of actinides from nuclear waste solutions was attempted using acidic carbamoylmethylene phosphonate (CMP) and neutral carbamoylmethylene phosphine oxide carriers in SLM processes (*40*).

Stability of SLM Systems. Hofman et al. (*41*) have presented results of a pilot-scale SLM unit, which also addresses important aspects of membrane degradation. Instability of the SLMs was noted to be a major disadvantage and a solution to the problem was suggested for a uranyl nitrate system

The presence of ^{237}Np in nuclear wastes has been discussed in the context of the chemical aspects and problems associated with the potential use of membrane separation methods (*42*).

Akiba and Nakamura (*43*) studied the transport of Am(III) through two SLMs in series comprised of 0.1 M nitric acid/0.2 M diisodecylphosphoric acid + 10 % 1-decanol/1 M perchloric acid and 4 M sodium perchlorate/0.1 M dihexyl N,N-diethylcarbamoyl methylenephosphonate + 10 % 1-decanol/0.1 M nitric acid. A microporous polytetrafluoroethylene (PTFE) film of Fluoropore FP-045 was impregnated by a kerosene solution of the carrier. Efficient recovery (>97%) of Am(III) was achieved.

Stability aspects of SLMs were investigated in DC18C6-facilitated, diffusion-limited transport of uranyl ion across a flat-sheet SLM. Solvent effects on the cation flux and diffusion coefficients were evaluated (*44*). The results of this study indicated that the stability and transmembrane fluxes depend on the physico-chemical characteristics of the carrier-diluent combination, not on the characteristics of the diluent alone. The physico-chemical properties of some diluents tested in the study are given in Table 5. Greater membrane stability was obtained with a membrane solvent of low volatility and aqueous solubility. Among the various diluents tested, a mixture of *o*-dichlorobenzene and toluene (3:7 by volume) gave the best combination of stability, regeneration capability, and transport rates.

Table 5. **Boiling Point, Density, Viscosity, Surface Tension, and Water Solubility Data for Various Organic Solvents.** (*44*)

Solvent	Boiling point[a] (°C)	Surface tension (dyne-cm^2)	Density[b] (g/ml)	Viscosity (cP)	Water Solubility[c] (wt %)	(vol. %)
Benzene	80	27.6	0.872	0.608	0.178	----
Chloroform	62	27.1	1.492	0.580	0.815	0.55
o-Dichloro-benzene (DCB)	181	----	1.306	----	<0.026	----
1,2-Dichloro-ethane	84	32.2	1.250	0.791	0.877	0.71
Dichloro-methane	41	28.1	1.325	0.429	1.961	1.48
Toluene (T)	111	27.4	0.867	0.590	0.063	0.073
Xylene	138	30.1	0.866	0.764	0.019	----
DCB-T (3:7)	125		0.991			
DCB-T (4:6)	133		1.034			
DCB-T (5:5)	140		1.090			

[a]Reference 45.
[b]Determined at room temperature.
[c]From Reference 46, at 25 °C and 1 atm except for chloroform and dichloromethane (20 ° C) and 1,2-dichloroethane (23 ° C). The volume % value was calculated using the densities of the organic solvent and water.

To probe the factors which are responsible for membrane stability, Dozol *et al.* (*47*) studied the transport of radionuclides from reprocessing concentrate solutions. They suggest that the water solubility of the membrane solvent must be less than 12 g/dm^3 and the drop point of the SLM (with a magnitude proportional to the interfacial tension) must be higher than 1.1 bar to obtain a SLM which will remain stable for more than 200 hours. The diluent should also have a surface tension lower than the critical surface tension of the support. On the other hand, the viscosity of the membrane solvent does not seem to have an appreciable influence on membrane stability. Among the different membrane solvents tested, stable SLMs were obtained with isotridecanol or aromatic hydrocarbons with aliphatic groups having six or more carbon atoms, although a phase modifier must be added to the aromatic solvents to avoid the formation of a third phase.

Both the chemical composition and structural configurations of the polymeric support play an important role in cation permeability (*48*). Pu(IV) was deliberately chosen for our study since its alpha radiolysis might cause appreciable structural damage to the polymeric support. Descriptions of some of the hydrophobic supports that we have evaluated and their physical characteristics of porosity, pore size, thickness, and critical surface tension are listed in Table 6.

Table 6 Porosity, Pore Size, Thickness, and Critical Surface Tension for Porous Polymeric Supports Tested for SLM Applications. *(49)*

Porous polymer	Material	Porosity (%)	Pore size (μm)	Thickness (μm)	Critical surface tension (σ)
TE-35[a]	PTFE[b]	78	0.2	160	18
TE-36[a]	PTFE	84	0.45	160	18
TE-37	PTFE	91	1.0	160	18
HF-PP	PP[c]	70	0.2	180	35
BA-S-83	CN[d]	80	0.2	125	--

[a]Supplied with polyester supporting fabric.
[b]Polytetrafluoroethylene.
[c]Polypropylene.
[d]Cellulose nitrate.

Pu(IV) transport by DC18C6 in toluene as the supported liquid membrane was studied with polytetrafluoroethylene (TE-35, TE-36, TE-37), polypropylene (HF-PP) and cellulose nitrate (BA-S-83) membrane supports. The Pu(IV) recovery exceeded 85 % for single runs conducted with TE-35 and TE-36. With the hydrophilic CN film, only 22 % of the Pu(IV) was transported under the same conditions.

In general, cation transport is expected to increase for a larger pore size and greater porosity in the membrane support. Thus, a PTFE membrane with a pore size of 0.2 μm and 78 % porosity exhibited higher permeability than a related membrane with a pore size of 0.02 μm and 38 % porosity. On the other hand, membranes with large pore sizes exhibit poor stability. Thus the liquid membrane solution of DC18C6 in toluene could easily be forced out from the larger pores of TE-36 (0.45 μm) and TE-37 (1.0 μm) compared to TE-35 (0.2 μm). This behavior may be rationalized by the Young-Dupre equation *(50)*:

$$P_c = (2 \ \tau/a) \cos \theta \qquad (1)$$

where P_c represents the capillary pressure across the liquid membrane interface, τ is the surface tension, θ is the contact angle, and a is the radius of the cylindrical pore. The capillary pressure is inversely proportional to the pore diameter. The larger pore sizes in TE-36 and TE-37 lead to a decrease in P_c which allows leakage of the membrane solution from the pores of the support during operation (Table 7).

The imbibition rate, q, of a SLM can be correlated with the thickness of the solid support by the Rideal-Washburn equation *(51)*:

$$q = (a\tau/4\mu\delta) \cos \theta \qquad (2)$$

where μ denotes the viscosity of the imbibing liquid (water), δ is the thickness of the membrane, and a is the pore diameter. For the data in Table 7, the slower degradation of the HF-PP membrane can be attributed to its greater thickness (180 μm) compared to TE-35 (160 μm).

Concerning the effect of pore size on selectivity for Pu(IV) transport, TE-35 provided better decontamination of Pu from fission products; whereas TE-36 which has a larger pore size, exhibited poor selectivity. Experimental results are given in Table 8.

Decontamination of minor actinides from low or medium-level reprocessing wastes of PUREX origin by LMs cannot be accomplished unless the system possesses adequate radiation stability. Therefore, it is important to examine the radiation

Table 7. Permeation Coefficients for Plutonium Transport as a Function of the Pore Size and Material in the Porous Polymeric Support for Consecutive Runs.[a]

Number of runs	Elapsed time (h)	TE-35[b] (0.2 μm)	TE-36 (0.45 μm)	TE-37 (1.0 μm)	HF-PP[c] (0.2 μm)
		Permeability coefficient, $P \times 10^{-3}$ (m/s)			
1	1	5.7	4.8	2.8	3.1
	6	0.6	1.2	0.4	0.6
2	1	5.1	4.6	1.6	3.2
	6	0.5	1.1	0.1	0.6
3	1	4.8	3.8	0.9	3.0
	6	0.2	0.3	0.1	0.6
4	1	4.0	1.8	0.2	2.5
	6	0.3	0.2	0.1	0.5

[a]Conditions: Initial feed = 1.0 mg/dm³ of plutonium in 3 M nitric acid; Membrane phase = 0.4 M DC18C6 in toluene; Stripping phase = 0.1 M hydroxylamine hydrochloride.
[b]Critical surface tension = 18 nM/M.
[c]Critical surface tension = 35 nM/M.

Table 8. Transport of Various Fission Products Through SLMs as a Function of the Pore Size in the Polymeric Support.[a]

Fission product	Activity in the source phase (μCi/dm³)	Activity in the receiving phase (μCi/dm³)		Transport of fission product (%)	
		TE-35 (0.2 μm)	TE-36 (0.45 μm)	TE-35 (0.2 μm)	TE-36 (0.45 μm)
^{106}Ru	57.2	5.4	20.4	9.5	35.6
^{137}Cs	2513.2	133.2	502.6	5.3	20.0
^{125}Sb	23.3	1.9	7.0	8.0	30.0
^{90}Sr	11.1	1.7	3.9	15.0	35.0
^{144}Ce	46.4	2.3	13.9	5.0	30.0

[a]The fission product activity was estimated with a multichannel analyzer and a high purity germanium detector.

stability of the carrier and the porous polymer support for SLM systems. With this in mind, we investigated the effect of radiation on cation transport in a SLM (12). Details of the irradiation procedure for the liquid membrane solution and the porous polymer support are described elsewhere (12). Transport of U(VI) and Pu(IV) across a SLM comprised of un-irradiated HF-PP porous polymer support impregnated with a solution of DC18C6 in toluene which had been irradiated with absorbed doses of 0-71 Mrads, showed no appreciable effect of irradiation up to 46 Mrads for U(VI) and 10 Mrads for Pu(IV). After an irradiation dose of 71 Mrads, 58 % of the U(VI) could be recovered, but only 18 % of the Pu(IV). The initial U(VI) flux of 1.2×10^{-7} mol/m²/s in the absence of irradiation was reduced to 0.5×10^{-7} mol/m²/s after the liquid membrane solution received an absorbed dose of 71 Mrad. On the other hand, Pu(IV) transport decreased from an initial flux of 1.1×10^{-9} mol/m²/s to 0.2×10^{-9} mol/m²/s after the same absorbed dosage. Thus the plutonium-DC18C6-toluene system was found to be more sensitive to radiation effects than the uranium-DC18C6-toluene system.

No significant differences in the flux and percent recovery of U(VI) or Pu(IV) were noted after irradiation of impregnated membrane supports with doses up to 8 Mrads. However, examination by scanning electron microscopy revealed structural changes after irradiation of HF-PP (polypropylene) sheet impregnated with 0.4 M DC18C6 in toluene for 6-8 hours prior after exposure to an absorbed dose of 8 Mrads. Untreated supports became somewhat fragile at absorbed doses of 5 Mrads, possibly due to dehydration. Only very little additional information is available concerning the radiation stability of the PP and PTFE supports which have commonly been utilized in actinide separations by SLM processes.   Previous studies (*29,52*) indicated that when PP receives a dose of 2.5 Mrads, oxidation reactions possibly cause slow deterioration of mechanical properties, *e.g.* embrittlement. However, the exact dose which can be tolerated by a PP support depends on oxygen penetration, diffusion rates, and the trapping of radicals within the matrix.

Emulsion Liquid Membranes (ELMs).

Compared to the investigations described above for SLMs in processing actinides from spent fuel wastes, studies devoted to ELMs are very limited. ELM transport of U(VI) from a solution which was 0.01 M in nitric acid and 2 M in sodium chloride with an emulsion of 2.5 M phosphoric acid in 0.1 M DEHPA-dodecane is practically quantitative (Table 9) and equivalent to 66 stages of extraction with the DEHPA in dodecane (*53*).

Separation of Am(III), Cf(III), and Cm(III) in various ELM systems was investigated by Novikov *et al.* (*54-56*). Optimized conditions for recovery of these permeates involved an outer aqueous solution of diethylenetriaminepentaacetic acid (DTPA), citric acid, sodium nitrate, and nitric acid at pH 3.0, a membrane phase of DEHPA and SPAN 80 in carbon tetrachloride, and an inner aqueous solution of $K_{10}P_2W_{17}O_{81}$, sodium azide, and hydrazoic acid at pH 3.0.   When citric acid was added to the outer solution to enhance the extraction rate of the membrane and the actinide recovery, selectivity declined. A

Table 9. Uranium Recovery in Double Emulsion Liquid Membrane (ELM) Systems. (*53*)

Outer solution	Membrane solution	Inner solution	Mixing time (min)	Recovery (%)	Multiplication factor
0.01 M nitric acid	0.1 M DEHPA in dodecane, 3 % SPAN 80	2.5 M phosphoric acid	20	99.7	8.4
0.01 M nitric acid, 2M sodium chloride	0.1 M DEHPA in dodecane 3% SPAN 80	2.5 M phosphoric acid	20	99.0	----
0.01 M nitric acid, 2M sodium chloride	0.005 M DEHPA in dodecane	2.5 M phosphoric acid	40	99.0	66

high concentration of DEHPA in the membrane phase and addition of complexone to the source phase resulted in effective metal ion separations. The dynamic separation factors, the ratio of permeation coefficients through the emulsion membrane, were 5.4 and 16 for the pairs of Am(III)/Cm(III) and Am(III)/Cf(III), respectively, with 1 M DEHPA in the membrane and 0.02 M DTPA or 5×10^{-3} M $K_{10}P_2W_{17}O_{61}$ in the source phase (*55,56*).

Plasticized Membranes (PMs).

This class of membranes resemble those employed for dialysis or electrodialysis. However for the PM, transport selectivity is not based on the size of the permeating species or its charge, but on the specific solubility of complexes formed by the metal species and carrier in the membrane matrix. Effective separation of uranium from its mixture with iron and aluminum in dilute nitric acid was obtained with PVC sheets plasticized with TBP (*57*) and with dibutyl cresyl phosphate (DBCP) or dicresyl butyl phosphate (DCBP) (*57,58*) in a ratio of 1:3 (by weight). The U(VI) permeation coefficient was 1.5×10^{-4} cm/s when the D_u between the aqueous feed and a 50 μm thick gel membrane was 85 and the separation factors for iron and aluminum was between 6×10^2 and 6×10^4. The uranium complex imparted a relatively high resistance in the gel membrane with a permeation coefficient of 1.9×10^{-8} cm/s. Lifetimes of these mixed cresyl butyl membranes exceeded several weeks (56). The D_u values obtained after equilibrium was reached between the feed solution and plasticized membrane deceased in the order: $D_{TBP} > D_{DCBP} > D_{DCBP}$. On the other hand, the uranium flux decreased in the order: $J_{DBCP} > J_{DCBP} > J_{TBP}$ with values of about 10^{-9} mol/cm^2/sec. The authors suggested that this reversal in ordering was caused not by the distribution ratios between the feed and membrane phases, but by a high sensitivity of the species diffusion rate to a change in the structure of the active plasticizer.

Studies of uranium transport through gel membranes reported by other researchers (58,59) revealed that a change in the physico-chemical characteristics of the plasticizer in the polymer matrix does not lead to an unambiguous relationship between the transmembrane flux and the extraction equilibrium at the interface.

Emulsion Free Liquid Membranes (EFLMs).

Due to their high efficiency, ELMs would appear to be attractive alternatives to SLMs. However ELM systems often suffer from problems of low solubility and difficulties with de-emulsification. To overcome these problems, a new liquid membrane process, the emulsion free liquid membrane (EFLM), was developed by Kumar *et al.* (60). Application of this novel technique to hydrometallurgical processes is quite promising (61). Recent studies of separating heavy metals, such as Cr(VI) and Cu(II) from electroplating effluents, with an EFLM system indicate that the technique is very efficient and superior to other types of LMs.

Prompted by these results, we have evaluated the applicability of EFLMs for the treatment of low and medium-level radioactive wastes. An EFLM process was applied to decontamination of uranium from a simulated reprocessing waste through a TBP membrane (62). A schematic diagram of the process is shown in Figure 3. In this process both the feed and stripping solutions pass through the carrier solution in a jet of very fine droplets formed by mixing these liquids with compressed air. The EFLM retains the major advantages of LMs, such as high flux and selectivity and low consumption of reagents, while overcoming some LM shortcomings, such as low permeate diffusion rates, high swelling of ELMs, and instability of ELMs. TBP, a commercial extractant, was used as the uranium complexing carrier. The feed solution acidity was maintained at 2 M nitric acid and 1 M sodium carbonate was the most successful stripping solution. Experimental variables, such as the TBP concentration and the number of baffle plates in the reactor were also optimized. No appreciable change in the overall recovery of U(VI) was observed when the TBP concentration was varied in the range of 10-40 %. Changing the number of baffle plates from 5 to 15 increases the uranium recovery from 47 to 89 % in a single pass. The selectivity of uranium transport from fission product contaminants was also studied. With a uranium concentration of 2 g/dm^3 in 2 M nitric' acid, the recovery was above 90 % when a flow rate of nearly 1 L/h was maintained under the optimized recovery conditions.

Conclusions.

Studies performed with liquid membranes have demonstrated their feasibility for use in treating the low-level radiotoxic wastes which are generated in nuclear fuel reprocessing and allied radiochemical operations.

Since BLMs require a large quantity of carrier solution (which results in a thicker membrane) relative to the interfacial area where phase transfer takes place, permeation coefficients of actinides are too low for commercial viability. For the more efficient SLM systems, the hollow-fiber SLM configuration has greater commercial viability than flat-sheet SLMs due to its much higher membrane area.

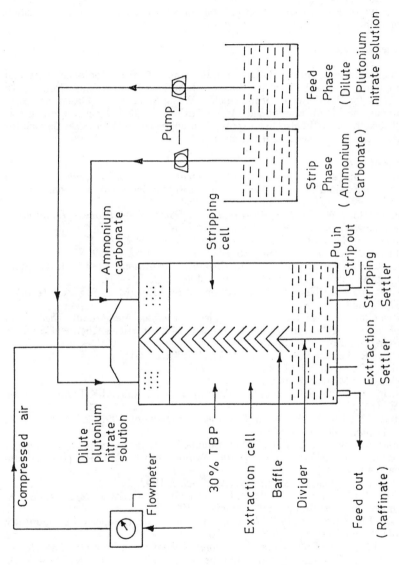

Figure 3. Schematic Diagram of an Emulsion Free Liquid Membrane (EFLM) Process.

Although very rapid separations can be accomplished with ELM systems, the process is rather complicated due to the requirement of de-emulsification to recover the encapsulated droplets. Poor stability of the emulsions is another drawback.

There is a need to evaluate LM methods in large-scale separations of minor transuranics from high-level radioactive wastes using high boiling membrane solvents and novel carriers like CMPO.

In the future, LMs are expected to play an important role in the decontamination of minor actinides from reprocessing wastes.

Acknowledgments.

The authors with to thank Dr. D.D. Sood, the Director of the Radiochemistry and Isotope Group, Shri. M.K. Rao, the director of the Fuel Reprocessing Group, Shri. M.K.T. Nair, the Past-Director of the Nuclear Waste Management Group, and Shri. D.D. Bajpai, the Plant Superindendent of PREFRE, for their keen interest in this work. The authors would also like express their appreciation to Dr. R.A. Bartsch for his assistance in the preparation of this chapter.

Literature Cited.

1. *Advances in Technologies for the Treatment of Low and Intermediate Level Radioactive Liquid Wastes*, IAEA Technical Report Series No. 370; International Atomic Energy Agency: Vienna, Austria, 1994.
2. Frankenfeld, J.W.; Li, N.N. In *Handbook of Separation Process Technology*; Wiley: New York, NY, 1987; Chapter 19, p 840-861
3. Kopunec, R.; Ngo; M..T. *J. Radioanal. Nucl. Chem*. **1994**, *183*, 181.
4. Noble, R.D.; Koval, C.A.; Pellegrio, J.J. *Chem. Eng. Progr.***1989**, *85*, 58.
5. Muscatello, A.C.; Navratil, J.D. In *Chemical Separations;* King, C.J.; Navratil, J.D. Litarvan Press: Denver, CO, 1986; Volume 2, pp 427-437.
6. Muscatello, A.C.; Navratil, J.D.; Killion, M.B.; Price, M.Y. *Sep. Sci. Technol.* **1987**, *22*, 843.
7. Muscatello, A.C.; Navratil, J.D.; Price, M.Y. In *Liquid Membranes: Theory and Applications,* ACS Symposium Series No. 347; American Chemical Society: Washington, D.C., 1987; pp 182-189.
8. Chirizia, R.; Danesi, P.R. *Sep. Sci. Technol.* **1987**, *22*, 641.
9. Shukla, J.P.; Misra, S.K. *J. Membr. Sci.* **1991**, *64*, 93.
10. Shukla, J.P.; Misra, S.K. *Ind. J. Chem.* **1992**, *31A*, 327.
11. Kumar, A.; Singh, R.K.; Shukla, J.P.; Bajpai, D.D.; Nair, M.K.T. *Ind. J.Chem.* **1992**, *31A*, 373.
12. Shukla, J.P.; Kumar, A.; Singh, R.K. *Sep. Sci. Technol.* **1992,** *27*, 447; *J. Nucl. Sci. Technol.* **1994**, *31*, 1066.
13. Kumar, A.; Singh, R.K.; Bajpai, D.D.; Shukla, J.P. *Department or Atomic Energy Symposium of Nuclear Chemistry and Radiochemistry*, IGCAR, Kalpakkam, India, 1995.
14. Shukla, J.P.; Kumar, A.; Singh, R.K. *Nucl. Sci. J.* **1993**, *30*,1.
15. Dozol, J.R.; Fymard, S.; Gambade, R.; Rosa, Gla, C.; Garza, J. *Report 13887FR*, Commission of the European Communities; 1992.
16. Dozol, J.R. *Report 13390 EN*, Commission of the European Communities; Elsevier: New York, NY, 1991; pp 95-105.
17. Dozol, J.R. *Proceedings of the International Symposium in Brussels, 1988, Report EUR 12114*, Commission of the European Communities; Elsevier: New York, NY.
18. Chiarzia, R.; Horwitz, E.P.; Rickert, P.G.; Hodgson, K.M. *Sep. Sci. Technol.* **1990**, *25*, 1571.
19. Akiba, K.; Hashimoto, H. *Anal. Sci.* **1986**, *2*, 541.
20. Shukla, J.P.; Sonawane, J.V.; Kumar, A.; Singh, R.K. *Ind. J. Chem. Technol.* **1996**, *3*, 145.

21. Nigond, L.; condamines, N.; Cordier, P.Y.; Livet, J.; Madiac, C.; Cuillerdier, C., Musicas, C.,; Hudson, M.J. *Sep. Sci. Technol.* **1995** *30*, 2075.
22. Danesi, P.R.; *Sep. Sci. Technol.* **1984-1985**, *19*, 857.
23. Danesi, P.R.; Horwitz, E.P.; Rickert, P. *J. Phys. Chem.* **1983**, *87*, 4708.
24. Danesi, P.R.; Horwitz, E.P.; Chiarizia, R.; Rickert, P. *Third Sympsium on Separation Science and Technology for Energy Applications*, Gatlinburg, TN, 1983.
25. Danesi, P.R.; Horwitz, E.P.; Rickert, P. *Proceedings of the International Solvent Extraction Conference(ISEC '83)*, Denver, CO, 1983; p 378.
26. Danesi, P.R.; Ciaetti, C. *J. Membr. Sci.* **1994**, *20*, 201.
27. Danesi, P.R.; Ciaetti, C. *J. Membr. Sci.* **1994**, *20*, 215.
28. Danesi, P.R.; Horwitz, E.P.; Rickert, P.; Chiarizia, R. *International Chemical Congress of Pacific Basin Societies*, Honolulu, HI, 1984.
29. Danesi, P.R.; Chiarizia, R.; Rickert, P.; Horwitz, E.P. *Solv. Extn. Ion Exch.* **1985**, *3*, 111.
30. Chiarizia, R.; Danesi, P.R. *Fourth Symposium on Separation Science and Technology for Energy Applications*, Knoxville, TN, 1985.
31. Danesi, P.R. *Proceedings of the International Solvent Extraction Conference (ISEC ' 86)*, Munich, Germany, 1986; Volume 1, p 527.
32. Danesi, P.R.; Reicheley-Yinger, L. *Proceedings of the International Solvent Extraction Conference (ISEC '86)*, Munich, Germany, 1986; Volume 2, p 255.
33. Danesi, P.R. *Proceedings of the International Solvent Extraction Conference (ISEC ' 90)*, Kyoto, Japan, 1990; p 264.
34. Huang, T.C.; Huang, C.T. *J. Membr. Sci.* **1986**, *29*, 295.
35. Chaudry, M.A.; Noor-ul-islam, D.M. *J. Radioanal. Nucl. Chem.* **1987**, *109*, 11.
36. Elhassadi, A.A.; Do, D.D. *Sep. Sci. Technol.* **1986**, *21*, 267.
37. Muscatello, A.C.; Navratil, J.D.; Killion, M.E. *Eighth Actinide Separtion Workshop*, BFP-3704, Boulder, CO; 1984.
38. Muscatello, A.C.; Navratil, J.D.; Killion, M.E.; Saba, M.T. *American Institute of Chemical Engineers National Meeting*, Seattle, WA, 1984; RFP 3865.
39. Navratil, J.D. *Proceedings of the International Conference on Separation of Ionic Solutes*, Smolenice, Czechoslovakia, 1985; p 55.
40. Schulz, W.W.; Navratil, J.D. *Sep. Sci. Technol.* **1985**, *19*, 927.
41. Hofman, D.L.; Craig, W.M.; Smith, J.J. *Proceeding of the International Solvent Extraction Conference (ISEC ' 90)*, Kyoto, Japan, 1990; p 265.
42. Rees, J.; Shipp, C. *Nucl. Energy* **1983**, 22, 423.
43. Akiba, K.; Nakamura, S. *Proceedings of the Symposium on Solvent Extraction*, Sendai, Japan, 1986; p 85.
44. Shukla, J.P.; Kumar, A.; Singh, R.K. *Radiochim. Acta* **1992**, *57*, 185.
45. Sekine, R.; Hasegawa, Y In *Solvent Extraction Chemistry: Fundamentals and Applications*; Marcel Dekker: New York, NY, 1973; p 928.
46. *Solubilities of Inorganic and Organic Compounds*, Volume 1, Binary Systems, Part 1; Stephen, H,; Stephen, T., Eds.; Macmillan:, New York, NY, 1963; pp 55-67.
47. Dozol, J.F.; Casas, J.; Sastre, A. *J. Membr. Sci.* **1993**, *82*, 237.
48. Kumar, A.; singh, R.K.; Shukla, J.P.; Bajpai, D.D. *Proceedings of the Conference on Recent Trends in Membrane Science and Technology*, Bombay, India, 1993; p 75
49. Zisman, W.A. *Contact Angle, Wettability and Adhesion*, Advances in Chemistry Series, Volume 43; American Chemical Society: Washington, DC, 1964, p 20.
50. Adamson, A.W. *Physical Chemistry of Surfaces*, Second Edition; Wiley: New York, NY, 1967.
51. Davis, J.T.; Rideal, E.K. *Interfacial Phenomenon*; Academic Press: New York, NY, 1961; p 419.
52. Bradley, R. *Radiation Technology Handbook*; Marcel Dekker, New York, NY, 1984.

53. Macasek, F.; Rajec, P.; Rehacek, V.; Ngoc anh, Vu.; Popovnakova, T. *J. Radioanal. Nucl. Chem. Lett.* **1985**, *96*, 529.
54. Novikov, A.P.; Bunina, T.V. Myasoedov, B.F. *Radiokhim.* **1988**, *30*, 362.
55. Novikov, A.P.; Mikheeva, M.N.; Myasoedov, B.F. *Abstracts of the International Conference on Actinides-89*, Tashkent, USSR, 1989, p 461.
56. Novikov, A.P.; Myasoedov, B.F.; Bunina, T.V.; Bukina, T.I.; Kremlyakova, N. Yu.; *Radiokhim.* **1988**, *30*, 196.
57. Block, R.; Finkelstein, A.; Kedem, D.; Vofsi, D. *Ind. Eng. Process Des. Dev.* **1967**, *6*, 231.
58. Block, R.; Kedem, O.; Vofsi, D. *Proceedings of the International Conference on Solvent Extraction Chemistry*, Gothenburg, Sweden, 1996; North Holland: Amsterdam, Holland, 1967, p 605.
59. Ketzine, L.L.; Boger, L. *Isreal Atomic Energy Commission* **1971**, *10*, 1262.
60. Jagur-Grodzinski, U.; Marian, S.; Vofsi, S. *Sep. Sci. Technol.* **1973**, *33*, 8.
61. Kumar, S.V.; Lakshmi, T.G.; Ravindram, M,; Chanda, M.; Madakavi, J.R.; Mukerjee, A.K. *National IMS Conference on Recent Trends in Membrane Science and Technology*, Bombay, India, 1993.
62. Kumar, S.V.; Lakshmi, T.G.; Ravindram, M.; Chandra, M.; Madakavi, J.R.; Mukerjee, A.K. *Ind. J. Chem. Technol.* **1995**, *2*, 21.
63. Shukla, J.P., Misra, S.K. *Eleventh National IMS Conference on Membrane Science and Technology - Perspectives and Prospects*, Jadavput University, Calcutta, India, 1994.

INDEXES

Author Index

Affiliation Index

Allied Signal Inc., 208
Bhabha Atomic Research Centre (India), 361,391
Brigham Young University, 57,130
Centre National de la Recherche Scientifique, 376
Colorado School of Mines, 1,11,270,342
Commissariat á l'Energie Atomique, 376
Ecole Centrale Paris, 103
Ecospectr (Russia), 89
Exxon Research and Engineering Company, 11,208
Kyoto Institute of Technology, 239,252
Mendeleev University of Chemical Technology, 89
N. N. Semenov Institute of Chemical Physics, 75

National Institute of Materials and Chemical Research (Japan), 167
New Jersey Institute of Technology, 115,222
Osaka City University, 142
Rutgers University, 319
Saga University, 303
TDA Research, Inc., 342
Technical University of Czestochowa, 181
Technical University of Wroclaw, 181
Technische Universität Graz, 103
Texas Tech University, 1,155
U.S. Department of Energy, 329
University of Colorado, 286
University of Iowa, 319
University of Notre Dame, 194
University of Twente, 18

Subject Index

A

Acetic acid, removal using emulsion liquid membranes, 215
Acrylic acid, use in plasma graft polymerization, 254–268
Acrylic acid grafted membranes, use in facilitated transport, 250–264
Actinide, separation from reprocessing wastes with liquid membranes, 391–406
Acyclic ligands, 167,168f
Acyclic oligoamides, 167–168,170f
Advancing front model, description, 116
Alkali metal cation facilitated transport through supported liquid membranes with fatty acids
electrogenic processes, 79–81
electroneutral transport, 76–80
experimental description, 77
kinetics, 81–85
Alkali metal perchlorate transport
lariat ether amides, 161–164
lipophilic 18-crown-6 and 15-crown-5 derivatives, 157–160

Alkene–alkane mixtures, facilitated transport in perfluorosulfonic acid membranes, 286–300
Amide compounds for highly selective transport of heavy metal ions
amino acid derivatives, 175–179
N,N'-bis(8-quinolylcarbamoyl) derivatives, 169–170,172
structure, 167
unsymmetric malonamide derivatives as heavy metal ion carriers, 171–175,176f
Amide-containing dibenzo-16-crown-5 compounds, transport of metal perchlorates, 161–165
Amine(s) as carriers, carbon dioxide facilitated transport through functional membranes prepared by plasma graft polymerization, 252–268
Amine solutions, carbon dioxide facilitated transport through supported liquid membranes, 239–250
Amino acid derivatives, use as heavy metal ion carriers, 175–179